Mario Tobias and Bing Xue (Eds.)

Sustainability in China: Bridging Global Knowledge with Local Action

MDPI

This book is a reprint of the special issue that appeared in the online open access journal *Sustainability* (ISSN 2071-1050) in 2014 and 2015 (available at: http://www.mdpi.com/journal/sustainability/special_issues/local-action).

Guest Editors
Mario Tobias
Institute for Advanced Sustainability Studies (IASS)
Berliner Street 130, D-14467 Potsdam
Germany

Bing Xue
Institute of Applied Ecology at Chinese Academy of Sciences
No.72, Wenhua Road
Shenyang, 110016
China

Editorial Office
MDPI AG
Klybeckstrasse 64
Basel, Switzerland

Publisher
Shu-Kun Lin

Managing Editor
Le Zhang

1. Edition 2015

MDPI • Basel • Beijing • Wuhan

ISBN 978-3-03842-113-9 (Hbk)
ISBN 978-3-03842-114-6 (PDF)

Table of Contents

List of Contributors

Daniel Béland: Johnson-Shoyama Graduate School of Public Policy, University of Saskatchewan, Saskatoon, SK S7N 5B8, Canada

Bu Office of Development and Reform, Yancheng Institute of Technology, Yancheng 224051, China

Bin Chen: China Energy Research Society, Beijing 100045, China; School of Environment, Beijing Normal University, Beijing 100875, China

Wenjiang Chen: School of Philosophy and Sociology, Lanzhou University, Lanzhou 730000, China

Xingpeng Chen: College of Earth and Environmental Sciences, Lanzhou University, Tianshui South Road 222 #, Lanzhou 730000, China; Research Institute for Circular Economy in Western China, Lanzhou University, Tianshui South Road 222 #, Lanzhou 730000, China

Huijuan Dong: Key Lab of Pollution Ecology and Environmental Engineering, Institute of Applied Ecology, Chinese Academy of Sciences, Shenyang 10016, China

Egrinya Eneji: Department of Soil Science, Faculty of Agriculture, University of Calabar, Calabar PMB 1115, Nigeria

Huihui Feng: College of Earth and Environmental Sciences, Lanzhou University, Lanzhou 730000, China; Institute for Applied Material Flow Management (IfaS), Trier University of Applied Science, Birkenfeld 55761, Germany

Yong Geng: Key Lab of Pollution Ecology and Environmental Engineering, Institute of Applied Ecology, Chinese Academy of Sciences, Shenyang 10016, China

Grunow Rhein-Ruhr-Institut für Sozialforschung und Politikberatung e.V., Universität Duisburg-Essen, Heinrich-Lersch-Str. 15, 47057 Duisburg, Germany

Ting Guan: School of Public Affairs, Zhejiang University, Yuhangtang Road 688, Hangzhou 310058, China

Xiaodan Guo: School of Economics and Management, North China Electric Power University, Chang Ping District, Beijing 102206, China

Xiaopeng Guo: School of Economics and Management, North China Electric Power University, Hui Long Guan, Chang Ping District, Beijing 102206, China

Peter Heck: Institute for Applied Material Flow Management, University of Applied Sciences Trier, Campusallee 9926, 55768 Neubrücke, Germany

KhuFaculty of Engineering and Physical Sciences, University of Surrey, Civil Engineering (C5), Guildford, Surrey GU2 7XH, UK

Qi Lei: School of Economics and Management, North China Electric Power University, Chang Ping District, Beijing 102206, China

Hengji Li: Scientific Information Center for Resources and Environment, Lanzhou Branch of the National Science Library, Chinese Academy of Sciences, Tianshui Middle Road 8 #, Lanzhou 730000, China

Yongjin Li: School of Philosophy and Sociology, Lanzhou University, Lanzhou 730000, China; Institute of Arid Agroecology, School of Life Sciences, Lanzhou University, Lanzhou 730000, China; Research Center for Circular Economy in Western China, Lanzhou University, Lanzhou 730000, China; Lanzhou Center of Literature and Information, Chinese Academy of Sciences, Lanzhou 730000, China

Li Liang: School of Economics and Resource Management, Beijing Normal University, Beijing 100875, China; China Energy Research Society, Beijing 100045, China

Lin School of Economics and Resource Management, Beijing Normal University, Beijing 100875, China; China Energy Research Society, Beijing 100045, China

Jie Liu: Jilin Province Environmental Monitoring Center, 2063 Tailai St, Changchun 130011, China

Jintong Liu: Key Laboratory of Agricultural Water Resources, Center for Agricultural Resources Research, Institute of Genetics and Developmental Biology, Chinese Academy of Sciences, Shijiazhuang 050022, China

Lee Liu: Geography Program, School of Environmental, Physical & Applied Sciences, University of Central Missouri, Warrensburg, MO 64093, USA

Zuoxi Liu: Key Lab of Pollution Ecology and Environmental Engineering, Institute of Applied Ecology, Chinese Academy of Sciences, Shenyang 10016, China; University of Chinese Academy of Sciences, Beijing 100049, China

David López-Carr: Department of Geography, University of California, Santa Barbara, CA 93106, USA

Chengpeng Lu: Key Lab of Pollution Ecology and Environmental Engineering, Institute of Applied Ecology, Chinese Academy of Sciences, Shenyang 10016, China; Institute of Applied Ecology, Chinese Academy of Sciences, Shenyang 110016, China

Chenyu Lu: College of Geography and Environment Science, Northwest Normal University, Lanzhou 730070, China

Shichao Luo: School of Economics and Resource Management, Beijing Normal University, Beijing 100875, China; China Energy Research Society, Beijing 100045, China

Fengjiao Ma: Key Laboratory of Agricultural Water Resources, Center for Agricultural Resources Research, Institute of Genetics and Developmental Biology, Chinese Academy of Sciences, Shijiazhuang 050022, China

Hong Miao: School of Resources and Environment, Ningxia University, Yinchuan 750000, China

Zhilin Mu: Environment Protection and Resources Conservation Committee of the National People's Congress, Beijing 100805, China

Jiaxing Pang: College of Earth and Environmental Sciences, Lanzhou University, Tianshui South Road 222 #, Lanzhou 730000, China; Research Institute for Circular Economy in Western China, Lanzhou University, Tianshui South Road 222 #, Lanzhou 730000, China

Hua-peng Qin: Key Laboratory for Urban Habitat Environmental Science and Technology, School of Environment and Energy, Peking University Shenzhen Graduate School, Shenzhen 518055, China

IX

Wanxia Ren: Key Lab of Pollution Ecology and Environmental Engineering, Institute of Applied Ecology, Chinese Academy of Sciences, Shenyang 10016, China

Qiong Su: Key Laboratory for Urban Habitat Environmental Science and Technology, School of Environment and Energy, Peking University Shenzhen Graduate School, Shenzhen 518055, China

Nv Tang: Key Laboratory for Urban Habitat Environmental Science and Technology, School of Environment and Energy, Peking University Shenzhen Graduate School, Shenzhen 518055, China

Jie Tao: School of Economics and Management, North China Electric Power University, No. 2 Bei Nong Road, Beijing 102206, China

Mario Tobias: The Potsdam Chamber of Commerce (IHK), 14467 Potsdam, Germany; Institute of Automotive Management and Industrial Production, Technische Universität Braunschweig, 38106 Braunschweig, Germany

Chunjuan Wang: College of Geography and Environment Science, Northwest Normal University, Lanzhou 730070, China

Lijian Wang: Department of Social Security, School of Public Policy and Administration, Xi'an Jiaotong University, Xi'an 710049, China; College of Geography and Environment Science, Northwest Normal University, Lanzhou 730070, China

Minpeng Xiong: School of Economics and Management, North China Electric Power University, Chang Ping District, Beijing 102206, China

Bing Xue: Key Lab of Pollution Ecology and Environmental Engineering, Institute of Applied Ecology, Chinese Academy of Sciences, Shenyang 10016, China; Institute for Advanced Sustainability Studies (IASS), Potsdam 14467, Germany

Qingyou Yan: School of Economics and Management, North China Electric Power University, No. 2 Bei Nong Road, Beijing 102206, China

Jianxing Yu: School of Public Affairs, Zhejiang University, Yuhangtang Road 688, Hangzhou 310058, China

Jiahai Yuan: School of Economics and Management, North China Electric Power University, Hui Long Guan, Chang Ping District, Beijing 102206, China

Zilong Zhang: College of Earth and Environmental Sciences, Lanzhou University, Tianshui South Road 222 #, Lanzhou 730000, China; Research Institute for Circular Economy in Western China, Lanzhou University, Tianshui South Road 222 #, Lanzhou 730000, China

Zhenguo Zhang: College of Economics and Management, Dalian Nationalities University, Dalian 130011, China

Changhong Zhao: School of Economics and Management, North China Electric Power University, Chang Ping District, Beijing 102206, China

Weili Zhu: College of Geography and Environment Science, Northwest Normal University, Lanzhou 730070, China

About the Guest Editors

Prof. Dr. Dr. **Mario Tobias** received his Diploma Degree in Biology at the Technische Universität Braunschweig in 1997, and then received his Doctoral Degree (PhD in Biology) in 2000, as well as a Diploma Degree in Business Administration in 2002 at the Technische Universität Braunschweig. In 2007, he received his second Doctoral Degree (PhD in Business Administration) at the Freie Universität Berlin. From 2000-2005, he was a business manager of Environment and Sustainability in Bundesverband der Informationswirtschaft, Telekommunikation und neuen Medien (BITKOM), and then served as an executive board member with responsibility in business area Technologies and Services of BITKOM from 2005 to 2010. From 2011 to 2014, he served as Secretary General of the Institute for Advanced Sustainability Studies (IASS) in Potsdam. Currently, since 2014, he has been the Managing Director of the Potsdam Chamber of Commerce and Industry (CCI Potsdam). Since 2015, he has been an Adjunct Professor at the Institute of Automotive Management and Industrial Production at Technische Universität Braunschweig.

Dr. **XUE Bing**, born in January 1982, received his PhD degree in Geography from Lanzhou University of China in June 2009. Currently, he is an Associate Professor and the Principal Investigator (Head) of the Research Center for Industrial Ecology & Sustainability at the Institute of Applied Ecology (IAE) of the Chinese Academy of Sciences (CAS). Since January 2014, he has also been a joint research fellow at the Institute of Advanced Sustainability Studies (IASS) Potsdam, Germany. His research interests mainly focus on the interactions of the human-environmental system and sustainable climate governance, based on the techniques of urban and environmental computing, by employing trans-disciplinary and inter-disciplinary approaches. In 2011, he received the Green Talent Award from the German Federal Ministry of Education and Science (BMBF), and was awarded as a Humboldtian in 2014 by the Alexander von Humboldt Foundation. He also serves voluntarily for various international societies and local communities; he is a reviewer for international journals, a policy consultant expert for Chinese local governments, *etc*.

Preface

China's road to sustainability has attracted global attention. Since the "Reform & Opening Up" policy, China's rapid pace of both urbanization and industrialization has made its being the second largest economy but meantime a heavy environmental price has been paid over the past few decades for addressing the economic developmental target. Today, as the biggest developing country, China needs to take more responsibilities for constructing its local ecological-civilization society as well as for addressing the global challenges such as climate change, resources scary and human beings well-fare; therefore, we need to have deeper understandings into China's way to sustainability at very different levels, both spatially and structurally, concerns ranging from generating sustainable household livelihoods to global climate change, from developing technological applications to generate institutional changes. In this spirit, this publication, "Sustainability in China: Bridging Global Knowledge with Local Action" aims to investigate the intended and spontaneous issues concerning China's road to sustainability in a combined top-down and bottom-up manner, linking international knowledge to local-based studies.

It goes without say that we are deeply grateful to all the authors who contributed to this publication. We would also like to thank the anonymous reviewers for their valuable comments. Sincerely appreciation goes to the editorial team of the journal Sustainability of MDPI- namely Editor-in-Chief Prof. Dr. Marc A. Rosen, Managing Editor Dr. Le Zhang, Assistant Editor Yaqiong Guo, Ms. Vicky Hu, Ms. Hui Liu, Ms. Shuang Zhao, Ms. Jing Li, Ms. Jie Gu and Ms. Anna Chen.

Last but not the least, we would like to thank the support from the Institute for Advanced Sustainability Studies (IASS) Potsdam, the Alexander von Humboldt Foundation, the International Postdoctoral Exchange Fellowship Program under China Postdoctoral Council (20140050), and the Natural Science Foundation of China (41471116, 41101126, 71303230).

Bing Xue and Mario Tobias
Guest Editors

Assessing the Financial Sustainability of China's Rural Pension System

Lijian Wang and Daniel Béland

Abstract: Considering the rapid growth of China's elderly rural population, establishing both an adequate and a financially sustainable rural pension system is a major challenge. Focusing on financial sustainability, this article defines this concept of financial sustainability before constructing sound actuarial models for China's rural pension system. Based on these models and statistical data, the analysis finds that the rural pension funding gap should rise from 97.80 billion Yuan in 2014 to 3062.31 billion Yuan in 2049, which represents an annual growth rate of 10.34%. This implies that, as it stands, the rural pension system in China is not financially sustainable. Finally, the article explains how this problem could be fixed through policy recommendations based on recent international experiences.

Reprinted from *Sustainability*. Cite as: Wang, L.; Béland, D. Assessing the Financial Sustainability of China's Rural Pension System. *Sustainability* **2014**, *6*, 3271-3290.

1. Introduction

Shifting demographics are creating major concerns about the long-term financial sustainability of old-age pension schemes all around the world. These concerns are particularly pressing in China. For instance, a recent study shows that, in 2012, China's urban pension fund revenues could not cover expenditures in 19 of its 32 provinces [1]. Additionally, Gao [2], Sin [3] and Ma [4] estimated the size of the pension fund shortage in urban China to be 2.824 trillion, 9.15 trillion and 18.3 trillion Yuan, respectively. As these figures suggest, the lack of the financial sustainability of China's urban pension system is a well-established reality. In this article, we ask whether China's new rural pension system suffers from similar financial shortfalls and challenges.

When dealing with this issue, we must keep in mind two crucial realities about rural pensions in China. First, rural poverty remains a key social problem in China. For example, more than 22% of the elderly rural population live in poverty [5]. This reality is consistent with the Chinese saying that people are "getting old before getting rich" [6]. Second, changes in cultural beliefs and family structures have weakened the traditional family, which is the traditional source of social and economic support for the elderly in rural areas [7]. Considering these two remarks, it is important for China to operate a sustainable rural pension system to ensure the long-term economic security of current and future cohorts of rural pensioners.

This attention to sustainability issues is particularly essential today, because, in 2011, China started implementing the New Type of Rural Social Endowment Insurance (NTRSEI) across the country. Considering the major financial flaws that recently led to the demise of the Old Rural Social Endowment Insurance (ORSEI) [8] and the current demographic challenges facing a rapidly aging China [9], sound financial foundations of the new rural pension system must be laid now to avoid future pension policy failures that could hurt the country, especially its already vulnerable elderly rural population. This is why, in the new context of NTRSEI, we need to conduct a prospective study about the financial sustainability of China's new rural pension system.

Financial sustainability is a core principle of social security [10,11]. Drawing on the available literature [12–14], we define the financial sustainability of China's rural pension system simply as a positive financial state in which fund revenues exceed fund expenditures. Existing international studies provide detailed, sophisticated analyses of pension fund revenue and expenditure featuring long-term actuarial estimate models, infinite-horizon models and generational accounting, among other techniques [15–18]. For instance, Grande established a general duty pension income and expenditure model [19]; Annika [20] and Yasar [21] constructed an actuarial model for Swedish and Turkish public pensions, respectively. Based on existing models, Barr [22] proposed that a pension credit crisis was spreading all over the world; Fedotenkov [23] and Gerrans [24] took Europe and Australia, respectively, as examples to demonstrate the existence of a global pension crisis. Regarding China's pension financing, Béland and Yu analyzed contemporary pension politics in China and observed the latest developments of China's financial pension paradigm. James [25] identified transition costs and fund devaluation as two of the most important financial challenges facing China's pension system while advising policymakers to link pensions, financial markets and state-owned enterprises (SOE) reform. Selden assessed China's pension reform from the perspective of economic development and the need to overcome an enduring urban-rural divide. Wang [26] calculated the implicit pension debt and transition costs of China's pension system using computable general equilibrium analysis. Zhou [27], Gao [28] and Zheng [29] also measured China's pension fund shortage. The results of these studies showed that a financial pension crisis has appeared in China. To meet such financial challenges, scholars offered policy suggestions. For instance, Zeng [30] advised raising the statutory retirement age; Hu [31] suggested that China should issue new regulations to improve pension investment, and Ge proposed a plan to collect social security taxes in China. Although China is currently looking at potential reform options to improve the sustainability of its pension system, including its rural component, comprehensive reforms have not yet been enacted, and concerns about the financial sustainability of rural pensions remain strong and unlikely to vanish any time soon [32,33].

The literature directly analyzing the sustainability of China's rural pension system can be divided into two main categories. First, qualitative studies focus on the status, problem and effect factors of that system. For instance, Xie discusses the sustainability of rural pensions in Yunnan Province, while stressing the fact that population aging is a key factor affecting its sustainability [34]. In another qualitative study, Kou explores the sustainable development capacity of China's rural pension system from an institutional and a financial perspective [35]. In another qualitative study, Liu stresses the role of factors, such as government policies and the management of social security accounts, in the sustainability of the rural pension system [36].

Second, quantitative studies have contributed to the scholarly discussion about pension sustainability in rural China. For example, Qian and his colleagues develop an actuarial model regarding the financial sustainability of the new rural social pension insurance fund. Based on this model and their analysis, they claim that fund is financially unsustainable [37]. As for Li, he analyzes the financial situation of the new rural pension system and its pressures on future government expenditures, arguing that the level of sustainability of China's rural pension system is very low [38]. In another quantitative study, Feng identifies the main factors, such as the contribution rate and the rate of return on investments, that have a significant impact on the sustainability of the country's rural pension system [39]. Finally, Xue finds that both central and local governments can afford their subsidies to new rural social endowment insurance, as long as the Chinese economy can achieve sustainable, stable growth [40]. Although the literature on the financial sustainability of old-age pensions in China is growing, too little attention has been paid to its ever-expanding rural pension system. In order to help fill this gap, the following article offers a brief overview of the international debate on pension sustainability, formulates an analytical framework for financial sustainability, constructs an actuarial model for China's rural pension system, measures its pension financial gap and, finally, provides an answer to our basic research question about the sustainability of that system before formulating policy recommendations.

2. The International Pension Sustainability Landscape

Because old-age pensions involve long-term financial commitments on the part of employers, workers and, especially, governments, pension reform is a key policy area in which the concept of sustainability has, in recent times, proven increasingly influential [41]. This situation is related to the rise of sustainability as a key concept that is ever present in debates about both environmental and socio-economic issues, which points to the distinction between environmental and human sustainability. Clearly, the financial soundness of old-age pension systems belongs to the realm of human sustainability, which is less studied, but every bit as crucial as environmental sustainability. In the best of worlds, citizens, experts and policymakers should care about both sides of the sustainability coin [42]. In this article, we focus on pension finance as an issue of human sustainability.

In an era of accelerated population aging, pension sustainability has become a major policy concern all around the world [43–46]. This is especially true in East Asia, Europe and, to a lesser extent, North America, where population aging is considered a crucial challenge to the financial integrity of existing pension systems [47]. Considering this, in the name of financial sustainability and in a context of genuine demographic and fiscal concerns, major pension reforms have taken place in countries as diverse as Japan, Canada, Italy and Sweden. Sweden is an especially striking case, because, in the 1990s, after long negotiations between political parties, that country's pension system was reshaped to guarantee its long-term financial sustainability through the enactment of automatic adjustment mechanisms tying changing benefit levels to demographic and economic variables. Simultaneously, the Swedish reform attempted to protect low-income workers and retirees so that the quest for long-term financial sustainability in old-age pensions was not achieved on their backs [48]. Ironically, however, the Swedish reform might not be sustainable politically, as automatic cuts in benefits are unpopular, which is putting strong pressure on elected officials to devise a system that might not be as much on "autopilot" as what the founders of the country's new pension system believed in the 1990s, when they laid its foundation [49].

A lesser known, yet equally striking, case of pension reform aimed at improving financial sustainability took place in Canada in the mid-1990s. At that time, alarming actuarial reports about the future of the Canada Pension Plan (CPP) pushed federal and provincial officials to contemplate a major reform of the earning-related pension system. Because Canada's contribution rates were much lower than those of other developed countries, such as France, Germany and Sweden, Canadian policymakers agreed that the main way to improve the long-term financial sustainability of the CPP was to gradually increase the contribution rate, something that was completed by 2003. Additionally, indirect benefit cuts, a new way to invest pension trust fund surpluses and other, more technical changes were adopted. As a result of this reform, CPP program will now be financially sustainable for a period of at least 75 years [50,51].

In contrast, the United States has yet to enact comprehensive reform to improve the long-term sustainability of its federal Social Security program. This situation is related to the strong partisan divide over whether benefit cuts or the generation of new revenues should be enacted to make Social Security sustainable for the decades to come. Ironically, however, the lack of significant pension reform in the United States over the last three decades is related to the fact that this country was one of the first to tackle pension sustainability through the adoption of changes to Social Security as early as in 1977 and 1983. Enacted during the Reagan years (1981–1989), the second wave of changes took a bipartisan form and featured a gradual increase in the retirement age from 65 to 67, set to take place between 2000 and 2027 [52]. Although U.S. Social Security is not facing a short-term financial crisis related to such policy changes, the debate about how to guarantee the long-term sustainability of that program for at least the next 75 years has been taking place in the United States since the Clinton years (1993–2001) [53].

These remarks about Sweden, Canada and the United States raise a number of issues about the quest for financial sustainability in pensions. First, as the Swedish example suggests, reforms aimed at improving the financial sustainability of pension systems have clear redistributive consequences, and policymakers need to understand how such reforms may affect low-income individuals and other segments of the population. Second, as the Canadian case shows, pension reform enacted in the name of financial sustainability does not have to be centered primarily on benefit cuts, as generating higher revenues is a legitimate way to address future pension shortfall. Finally, as the United States' 1983 reform points out, increasing the retirement age is another way to improve the long-term sustainability of public pension systems. Although these are issues that Chinese experts and policymakers should keep in mind as they move forward, the first thing to do as far as pension sustainability is concerned is to assess the scope of the potential fiscal challenges facing the country's rural pension system. This is exactly what we do in the following sections, before spelling out some of the policy implications of our quantitative analysis.

3. Analytical Framework

3.1. China's Rural Pension System

Over the past two decades, China has witnessed significant efforts to establish an effective rural pension system. The history of the rural pension system can be divided into three phases: the Old Rural Social Endowment Insurance, or ORSEI (1981–2008); the New Type of Rural Social Endowment Insurance, or NTRSEI (2009–2013); and the Social Endowment Insurance for Urban and Rural Residents, or SEIURR (from 2014) [54]. Importantly, SEIURR has yet to be implemented across the entire country. Furthermore, the core content of NTRSEI and SEIURR is identical as far as rural residents are concerned, which means NTRSEI remains as the statutory rural pension system in China.

Following *The Guidance to Carry Out the NTRSEI Pilot Project*, issued in September, 2009, non-student rural residents aged 16 and older who are ineligible for participation in another public old-age social security program can voluntarily enroll in NTRSEI. Currently, individual contributions vary from 100 to 500 Yuan a year, and participating rural residents aged 60 and over receive a basic monthly pension of 55 Yuan each. The main features of China's rural pension system are summarized in Table 1.

Table 1. Main characteristics of China's pension system.

Categories	New Type of Rural Social Endowment Insurance
Year of creation	2009
Basic approach	Fully funded and pay-as-you-go
Basic principles	Guaranteed basics, wide coverage, elasticity and sustainability
Guiding ideology	Sharing the pension burden among individuals, collectivities and the state
Participation	Voluntary
Eligible population	Non-student rural residents aged 16 and older and citizens not eligible to enroll in any other old-age pension scheme
Fund sources	Individual contributions, collective subsidies and government subsidies
Yearly individual contributions	Choice between 100, 200, 300, 400 and 500 Yuan and above
Government payment subsidies	Yearly local government subsidies for pension contributory benefits cannot be lower than 30 Yuan per person, and they are automatically added to each individual pension account
Pension structure	Basic pension and individual account pension
Eligibility conditions	Reaching 60 years old
Basic benefit	55 Yuan per month, and local governments can increase that basic amount at will
Individual account pension	The quotient of the total amount of money in the individual account and 139
Competent authority	Ministry of Human Resources and Social Security

Sources: People's Republic of China Social Insurance Law; Guidance to Carry Out the NTRSEI Pilot Project, and Guidance to Carry Out the Old-age Insurance for Urban Residents Pilot Project.

According to Table 1, NTRSEI funds should be financed through individual contributions, collective subsidies and government subsidies; an NTRSEI pension includes a basic pension and an individual account pension. In the construction process of China's NTRSEI, there exists many risks, such as the typically low education level of rural residents, the flawed nature of agency governance, the underdevelopment of information systems and the unsystematic nature of fund operations. These issues make the study of the sustainability of China's rural pension system necessary.

Conducting new research on the sustainability of that system is particularly important for a number of additional reasons. First, that system covers more individuals than any other pension scheme in China. Second, in part for that reason, policymakers are increasingly interested in the future of the rural pension system. Third, compared to the literature on China's urban pension system, there is much less scholarship on rural pension development and sustainability in China. Finally, China's rural pension system relies more on *ad hoc* governmental support than on sound actuary principles, which means that balancing fund revenues and expenditures is especially important for the Chinese government, from a fiscal standpoint.

3.2. Relationship among Variables

The first variable needed to assess the sustainability of any pension system is the estimated rural pension fund revenues. According to the Financial Management Regulation of the NTRSEI issued by the Ministry of Finance and the Ministry of Human Resources and Social Security, such revenues include individual contributions, collective subsidies, state subsidies, interests, transfers from the central or local government and other revenues. In practice, the rural pension fund revenues are only composed of individual contributions, government payment subsidies and interests [55]. The second variable at hand is rural pension fund expenditures. As started in the above-mentioned regulation, they comprise pension expenditures, transfers from the central or local government, turned-over revenues and other expenditures. Yet, actual expenditures only include the basic pension and individual account pension expenditures [55]. In view of these remarks, we obtain Figure 1.

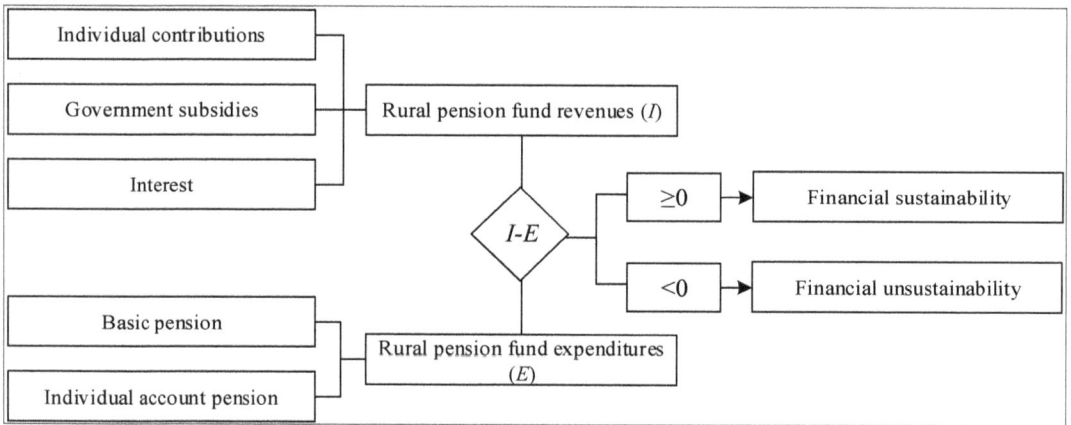

Figure 1. Connotation of financial sustainability of rural pension system in China.

3.3. Measured Indicators

In line with fundamental NTRSEI principles, the accumulated amount of pension benefits an insured person receives starting the 140th month after retirement and continuing until death has no corresponding income item, and there is no social pooling account. To conform to the social insurance actuarial model, we assume that there is a virtual social pooling account with zero income, and let I_t, I_{gt}, I_{jt}, I_{rt}, I_{st}, E_t, E_{zt}, E_{ct}, Y_t and E_{bt}, respectively, denote rural pension fund revenues, individual contributions, government subsidies, interests, the virtual social pooling account income, rural pension fund expenditures, normal spending by individual accounts, overspending by individual accounts, accumulated amounts in individual accounts and basic pension expenditures, at year t. We obtain:

$$I_t = I_{gt} + I_{jt} + I_{rt} + I_{st} \tag{1}$$

$$E_t = E_{zt} + E_{ct} + E_{bt} \tag{2}$$

$$F_t = I_t + Y_{t-1} - (E_t + Y_t) = I_{gt} + I_{jt} + I_{rt} + I_{st} + Y_{t-1} - (E_{zt} + E_{ct} + E_{bt} + Y_t) \tag{3}$$

where F_t is the rural pension funding gap at year t.

Following The Guidance to Carry Out the NTRSEI Pilot Project and Financial Management Regulation of the NTRSEI, we further have:

$$Y_t = Y_{t-1} + I_{gt} + I_{jt} + I_{rt} - E_{zt} \tag{4}$$

due to:

$$I_{st} = 0 \tag{5}$$

then:

$$F_t = -E_{ct} - E_{bt} \tag{6}$$

4. Model Specification

4.1. Basic Hypotheses

Because of the account specificity of the rural pension system, we can only assess its funding gap by calculating overspending in both the basic pension and individual accounts. Before doing so, we put forward the following hypotheses: (1) the policy framework of China's current rural pension system remains stable until 2050; (2) both central and local state subsidies are collectively referred to as state subsidies; (3) state subsidies operate on a pay-as-you-go basis; (4) to facilitate forecasting and due to the regular increase in rural per capita net income, we calculate the ratio between yearly individual contributions and yearly rural per capita net income; (5) according to the *People's Republic of China Social Insurance Law*, the parameters of the rural pension system must be adjusted on a regular basis according to the average personal income growth and inflation level; and (6) demographic forecasting methods and empirical research on China have made much progress in recent decades. Borrowing from existing research [38,39,55], we simply assume that the age distribution of the insured population is the same as the one of the total rural population and that the coverage rate of the new rural pension system is stable over time.

8

4.2. Calculating Models

4.2.1. Overspending in Individual Accounts

Let L'_t, $L_{x,t}$, a, b, O_t, M_t, $\overline{W}_t$, C_1, C_2, C_3, I'_t, I'', h, $K_{b-t,t}$, g_y, i, $Q_{b,t}$, n, E_{zct} and $l_{y,t}$, respectively, denote the number of participants at year t, the number of rural population aged x at year t, the minimum age of new participants, the retirement age, the coverage rate of the rural pension system at year t, the total amount of individual account savings at year t, the average rural per capita net income at year t, the individual contribution rate, the rate of collective subsidies, the rate of state subsidies, normal premiums at year t, supplementary premiums at year t, the number of supplementary payment years, future individual contributions of participants aged x, the average rate of increase of rural per capita net income, the interest rate, the individual account pension of participants aged b, the calculating coefficient of the individual account pension and the level of individual account expenditures at year t. Based on all this, we obtain:

$$L'_t = \sum_{x=a}^{b-1} L_{x,t} \times O_t \tag{7}$$

$$M_t = \overline{W}_t \times (C_1 + C_2 + C_3) \tag{8}$$

According to Formulas (7) and (8), we have:

$$I'_t = L'_t \times M_t = \sum_{x=a}^{b-1} L_{x,t} \times O_t \times \overline{W}_t \times (C_1 + C_2) \tag{9}$$

The Guidance to Carry Out the NTRSEI Pilot Project allows participants whose number of contribution years is lower than 15 to add supplementary contributions to their individual account, hence:

$$I''_t = \sum_{i=b-h+1}^{b-1} L_{i,t} \times O_t \times (i+h-b) \times \overline{W}_0 \times C_1 \tag{10}$$

$$I_t = I'_t + I''_t = \sum_{i=a}^{b-1} L_{i,t} \times O_t \times \overline{W}_t \times (C_1+C_2) + \sum_{i=b-h+1}^{b-1} L_{i,t} \times O_t \times (i+h-b) \times \overline{W}_0 \times C_1 \tag{11}$$

$$K_{b-t,t} = (h-t) \times \overline{W}_0 \times C_1 \times (1+i)^t + \sum_{j=1}^{t} \overline{W}_0 \times (1+g_y)^{t-j+1} \times (C_1+C_2) \times (1+i)^{t-j+1} \tag{12}$$

so that:

$$Q_{b,t} = \frac{K_{b-t,t}}{n} \times 12$$

$$= \frac{(h-t) \times \overline{W}_0 \times C_1 \times (1+i)^t + \sum_{j=1}^{t} \overline{W}_0 \times (1+g_y)^{t-j+1} \times (C_1 + C_2) \times (1+i)^{t-j+1}}{n} \times 12 \tag{13}$$

As stipulated by the *People's Republic of China Social Insurance Law*, "if individuals participating in basic pension insurance passed away due to illness or non-work-related reasons, their dependents can receive funeral subsidies and survivor's pension". Therefore, we have:

$$E_{zct} = \sum_{k=1}^{t} L_{b+k-1,t} \times O_{t-k} \times Q_{b,t-k+1} \times (1+m)^{k-1} + \sum_{k=1}^{t} (L_{b+k-2,t} - L_{b+k-1,t+1}) \times K_{b-1,t-k+1} \times \frac{n-12(k-1)}{n} \tag{14}$$

When $n/12$ is an integer and $y - b \leq n/12$,

$$E_{ct,y} = 0 \tag{15}$$

When $n/12$ is an integer and $y - b > n/12$,

$$E_{ct,y} = 12 \times \{\sum_{x=a}^{b-1}(h-t) \times \overline{W}_0 \times C_1 \times (1+i)^t + \sum_{j=1}^{t} \overline{W}_0 \times (1+g_y)^{t-j+1} \times (C_1 + C_2 + C_3) \times (1+i)^{t-j+1}\}/n \tag{16}$$

When $n/12$ is a non-integer and $y - b \leq [n/12]$,

$$E_{ct,y} = 0 \tag{17}$$

When $n/12$ is a non-integer and $y - b = [n/12] + 1$,

$$E_{ct,y} = \{1 - \frac{n}{12} + [\frac{n}{12}]\} \times 12 \times$$
$$\{\sum_{x=a}^{b-1}(h-t) \times \overline{W}_0 \times C_1 \times (1+i)^t + \sum_{j=1}^{t} \overline{W}_0 \times (1+g_y)^{t-j+1} \times (C_1 + C_2 + c_3) \times (1+i)^{t-j+1}\}/n \tag{18}$$

When $n/12$ is a non-integer and $y - b > [n/12] + 1$,

$$E_{ct,y} = 12 \times \{\sum_{x=a}^{b-1}(h-t) \times \overline{W}_0 \times C_1 \times (1+i)^t + \sum_{j=1}^{t} \overline{W}_0 \times (1+g_y)^{t-j+1} \times (C_1 + C_2 + C_3) \times (1+i)^{t-j+1}\}/n \tag{19}$$

According to Formulas (15)–(19), we obtain:

$$E_{ct} = 12 \times \sum_{y=b}^{\omega-1} O_t E_{ct,y} \times (\frac{L_{y,t} + L_{y+1,t+1}}{2}) \tag{20}$$

4.2.2. Basic Pension

Let $Q_{t,1}$ and k_1, respectively, denote the basic pension standard at year t, and the average adjustment rate of the basic pension standard. We have:

$$E_{bt} = Q_{t,1} \times L_t' = Q_{1,1} \times k_1^{t-1} \times O_{t,1} \times \sum_{x=b}^{\omega-1} \frac{L_{x,1} + L_{x+1,2}}{2} \qquad (21)$$

4.2.3. Rural Pension Funding Gap

According to Formulas (6), (20) and (21), we have:

$$F_t = -E_{ct} - E_{bt} = -12 \sum_{y=b}^{\omega-1} O_t E_{ct,y} \times (\frac{L_{y,t} + L_{y+1,t+1}}{2}) - Q_{1,1} \times k_1^{t-1} \times O_{t,1} \times \sum_{x=b}^{\omega-1} \frac{L_{x,1} + L_{x+1,2}}{2} \qquad (22)$$

5. Results

5.1. Data Collection

(1) Time frame. The medium-term plan for the pension system in China extends to 2020, and the long-term plan extends to the 100th anniversary of China's 1949 revolution [56]. Considering these time frames, we take 2014–2049 as our calculating period.

(2) Participants. Based on *The Guidance to Carry Out the NTRSEI Pilot Project*, parameter values for the minimum age for new participants, retirement age, the calculating coefficient of the individual account pension and the number of supplementary payment years are 16, 60, 139 and 15, respectively.

(3) Interest rate. The interest rate set by *The Guidance to Carry Out the NTRSEI Pilot Project* is equal to China's average yearly interest rate on deposits. Averaging the China's yearly rate from 1999 to 2010, and using it as i, we obtain $i = 0.0243$.

(4) Rural per capita net income. Using the *China Statistical Yearbook*, we can find historical data on rural per capita net income and feed these data into a grey system model [57]. We then obtain the rural per capita net income for 2014 to 2049 and the average rate of increase of per capita rural net income.

(5) Basic pension standard at initial year. Survey data [58] show that the average monthly pension currently received by rural elderly residents is 59.95 Yuan, so $Q_{t_0,1} = 59.95$.

(6) Individual contribution rate, collective subsidies rate and state subsidies rate. From our survey in six counties, we found that the average annual value of individual contributions is 192.04 Yuan. Based on that survey, we also know that, in practice, there are no actual collective subsidies. As for the yearly state subsidies, according to *The Guidance to Carry Out the NTRSEI Pilot Project*, they are worth 30 Yuan per person. Based on these, we have $C_1 = 3.73, C_2 = 0, C_3 = 0.58$.

(7) The average adjustment rate of the basic pension. Considering the recent nature of NTRSEI, we do not have good historical data about the evolution rate of the basic pension. Using the rate of increase of the per capita consumption expenditure of rural households from 2000 to 2010, displayed in the *China Statistical Yearbook* and the *People's Republic of China Social Insurance Law*, we have $m = 10.70\%$ using the simple moving average (SMA) methods.

(8) Coverage rate of the new rural pension system. From our survey, we estimate that, in the counties under study, the coverage rate of the rural pension system varied between 85% and 99%. Considering the actual situation in rural China, we estimate that $O_t = 95\%$.

(9) Rural demographic data. The basic demographic data are obtained from the *China Population Statistics Yearbook 2013*; the population forecasting methods are borrowed from a recent article [59].

5.2. Calculating the Rural Pension Funding Gap

On the basis of our data collection and Formulas (20)–(22), we have Table 2.

Table 2. The rural pension funding gap in China, 2013–2020 (in billion Yuan).

Year	Overspending in Individual Accounts	Basic Pension	Rural Pension Funding Gap	Year	Overspending in Individual Accounts	Basic Pension	Rural Pension Funding Gap
2014	0	97.800	−97.800	2032	10.661	796.636	−807.297
2015	0	111.164	−111.16	2033	11.920	880.609	−892.529
2016	0	125.108	−125.11	2034	13.256	968.415	−981.671
2017	0	141.306	−141.31	2035	14.617	1054.158	−1068.78
2018	0	157.426	−157.43	2036	16.034	1141.636	−1157.67
2019	0	171.777	−171.78	2037	17.395	1219.556	−1236.95
2020	0	188.270	−188.27	2038	18.776	1302.434	−1321.21
2021	0	203.413	−203.41	2039	19.963	1382.961	−1402.92
2022	0	229.779	−229.78	2040	21.217	1460.986	−1482.2
2023	3.324	265.868	−269.19	2041	22.439	1540.021	−1562.46
2024	3.758	301.991	−305.75	2042	23.567	1642.905	−1666.47
2025	4.362	343.570	−347.93	2043	24.691	1731.715	−1756.41
2026	4.960	390.395	−395.36	2044	26.246	1840.496	−1866.74
2027	5.651	436.897	−442.55	2045	27.516	1974.377	−2001.89
2028	6.430	497.762	−504.19	2046	29.173	2169.385	−2198.56
2029	7.247	562.057	−569.30	2047	31.228	2433.705	−2464.93
2030	8.317	638.644	−646.96	2048	34.372	2704.511	−2738.88
2031	9.381	714.883	−724.26	2049	38.881	3023.426	−3062.31

Data source: numbers in Table 2 are derived from the authors' calculations.

Table 2 indicates that (1) nationwide overspending in individual accounts would appear in 2023 and grow rapidly until 2049. Specifically, the overspending in individual accounts would rise from 3.32 billion Yuan in 2023 to 38.88 billion Yuan in 2049, thus increasing by 9.92% a year; (2) The total amount spent on the basic pension would increase from 97.80 billion Yuan in 2014 to 3023.43 billion Yuan in 2049, an average yearly increase of 10.30%. The main reason for this rapid anticipated increase in basic pension spending is that the number of rural elderly residents and the average amount of the basic pension are both rising; (3) The rural pension funding gap is the sum of overspending in individual accounts and the basic pension. The rural pension funding gap should increase from 97.80 billion Yuan in 2014 to 3062.31 billion Yuan in 2049, which represents an annual growth rate of 10.34%. Considering the above remarks, it is clear that, as it stands, China's rural pension system is not financially sustainable.

6. Analysis of the Results

China is now an aging society with more than 91 million elderly people in rural areas alone [60]. Considering the objective to fight elderly poverty and the rapid increase in the elderly population, the need for a sustainable rural pension system is more pressing than ever. As far as rural pensions are concerned, the recent expansion of NTRSEI coverage can be understood as a major success story [61–63]. Yet, the above analysis suggests that NTRSEI is anything but financially sustainable. Considering this, in order to increase the sustainability of China's new rural pension system, we urge policymakers to explore the following two reform paths:

First, the state could increase its direct fiscal support for that system, a move that would be entirely consistent with the *People's Republic of China Social Insurance Law*, which clearly stipulates that "The government shall supplement any shortage in the basic pension insurance fund." Our first suggestion to address the rural pension funding gap is therefore to transfer money from the state budget to the new rural pension system. In light of this policy advice, the question is whether the Chinese state has the ability to cover such a funding gap out of its general revenues.

According to *The Guidance to Carry Out the NTRSEI Pilot Project* and the *People's Republic of China Social Insurance Law*, the fiscal transfer from the central state to NTRSEI is the sum of the rural pension funding gap and state subsidies. Let $Q_{t,2}$, k_2, I_{jt} and E_{ht} denote the state subsidies standard at year t, the increasing rate of the state subsidies standard, the total number of state subsidies at year t and the transfer payment demand from the state budget at year t, respectively. We get:

$$Q_{t_0,2} = \overline{W}_{t0} \times C_3 \tag{23}$$

$$I_{jt} = Q_{t,2} \times L_t = Q_{1,2} \times k_2^{t-1} \times O_{t,2} \times \sum_{x=a}^{b-1} \frac{l_{x,1} + l_{x+1,2}}{2} \tag{24}$$

$$E_{ht} = |F_t| + I_{jt} = E_{ct} + E_{bt} + I_{jt}$$

$$= 12 \times \sum_{y=b}^{\omega-1} O_t \times E_{ct,y} \times (\frac{l_{y,t} + l_{y+1,t+1}}{2}) + Q_{1,1} \times k_1^{t-1} \times O_{t,1} \times \sum_{x=b}^{\omega-1} \frac{l_{x,1} + l_{x+1,2}}{2} + Q_{1,2} \times k_2^{t-1} \times O_{t,2} \times \sum_{x=a}^{b-1} \frac{l_{x,1} + l_{x+1,2}}{2} \quad (25)$$

Using the same data collection and Formulas (23)–(25), we obtain Figure 2.

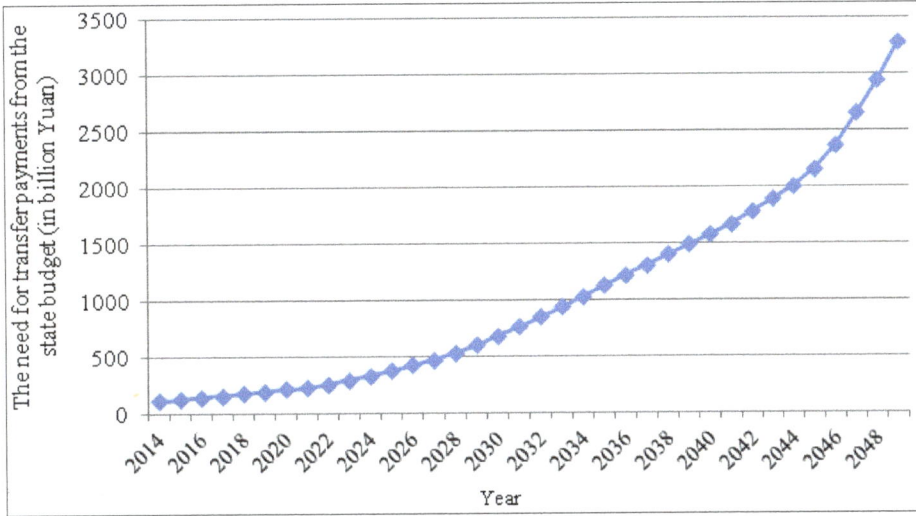

Figure 2. The need for transfer payments from the state budget for the rural pension system, 2014–2049. Data source: Numbers in Figure 2 are derived from our own calculations.

The need for transfer payments from the state budget for the rural pension system would continue to increase between 2014 and 2049, climbing from 111.89 billion Yuan in 2014 to 3280.12 billion Yuan in 2049, an annual rate of increase of 10.13%. Importantly, there should be no financial crisis in rural pensions as long as the Chinese state transfers such sums to NTRSEI.

Second, in addition to relying on state subsidies to support the rural pension system, policymakers can alter its parameters and design to mitigate factors favoring unsustainability, just as Sweden, Canada and the United States did. Concrete potential policy suggestions to do just this include the following:

(1) Creating dynamic adjustment mechanisms.

The rural pension system should meet the basic economic needs of rural residents, but the increase in spending triggered mainly by population aging should remain under control. Retirement age and account structure are two parameters policymakers may adjust to control such cost increases. For instance, the average payment period of the individual account pension could be shortened simply by increasing the statutory retirement age, which would automatically reduce overspending in the individual accounts. Regarding account structure, it would be appropriate to create a social pooling account for the basic pension. After setting up the social pooling account, the government would put the money to fund the annual basic pension into that account at the beginning of each year. Such a change would be likely to reduce the level of unsustainability of the rural pension system.

(2) Improving the investment practices.

Increasing the return rates of the rural pension fund is just another way to promote greater fund revenues. With this in mind, we advise policymakers that (i) the central government allows the National Council for Social Security Fund (NCSSF) to invest money from the rural pension fund. The NCSSF has the experience and the technological knowhow to make such investments. In fact, "The annual return of China's Social Security Fund in the past 12 years stood at 8.4%" [64]. This means that the NCSSF should be able to invest rural pension money in an effective way to increase returns; (ii) To further increase these rates of return, the State Development Bank could issue high-yield special bonds that the pension authorities could buy using money from the rural pension fund; (iii) New investment methods, such as contracted deposits and consignment loans, could also be adopted to further improve returns.

(3) Designing a national pension redistribution mechanism.

China is a large and country in which levels of economic and social development can vary greatly from region to region. This situation creates the coexistence of rural pension fund gaps in some regions of the country and fund surpluses in other regions. To address this territorial and financial issue, authorities could set up a national adjustment fund that could redistribute money across the country by using fund surpluses in some regions to address funding gaps elsewhere in the country.

The *People's Republic of China Social Insurance Law* states that "the People's government of the provinces, autonomous regions and municipalities can combine and implement the New Type of Rural Social Endowment Insurance and Social Endowment Insurance for Urban Residents according to their actual situation." On 7 February 2014, China's Premier, Li Keqiang, also chaired an executive meeting of the State Council, which decided to establish a national Social Endowment Insurance for Rural and Urban Residents. It should be noted that the combination of the rural and the urban pension systems will affect their sustainability. We know that: (i) such an integration may cause rural pension benefits to increase; (ii) due to the different financial capacity of rural and urban residents, the unified pension system has to set a minimum individual contribution standard based on rural residents' capacity; and (iii) the institutional frameworks between NTRSEI and SEIURR are basically identical. Considering all this, we find that the combination of the rural and the urban pension systems should increase pension expenditures and decrease pension revenues. This situation should further undermine the sustainability of the rural pension system. Although one could use a similar method as the one featured in this paper to calculate the sustainability of China's new integrated pension system, such a system has yet to be implemented everywhere across the country. Hence, this paper only provides a case study to evaluate the sustainability of China's pension system. Our results are relevant for the implementation of that system relevant across China.

7. Policy Implications

The financial sustainability of old-age pension schemes is an unavoidable issue in contemporary market economies [65]. As suggested at the beginning of this paper, the risk of financial unsustainability in pensions is widespread, which has forced governments in countries, such as Sweden, Canada, United States and well beyond, to enact major reforms [66].

This global push for pension sustainability already materialized in China, concerning the basic old-age insurance for enterprise employees. Because the individual accounts of China's basic old-age insurance for enterprise employees are "empty" and filled with "implicit debt" [67], experts and policymakers are rightly concerned about the financial sustainability of China's pension system. This is true because, in China, as elsewhere, avoiding the financial risk of pension unsustainability remains a major challenge. It is in this uncertain financial context that, in 2009, China established its new rural pension system, before implementing it across the country in 2012. Unfortunately, our results show that this new system is not financially sustainable. Taking into account foreign pension reforms, such as those discussed in Section 2, China should take strong and early action to tackle the current sustainability crisis in rural pensions. Drawing on such international experiences, the policy implications of our findings are as follows:

(1) From a quantitative perspective, our results help explain why China's new rural pension system is unsustainable, while assessing the approximate scope of that problem. The paper not only evaluates the size of China's rural pension funding gap, but also identifies the sources of this funding gap, which is essential information for policymakers needing to adjust the basic parameters of NTRSEI to make it more financially sustainable. This means that now is the time for China to improve social security concepts and governance while drawing on recent foreign pension reforms.

(2) Our analysis shows the discrepancy between official principles and the actual operation of China's new rural pension system. The basic, country-wide principles of NTRSEI include financial sustainability and policy compatibility with existing economic and social development levels. Yet, there is a serious budget shortfall in China's rural pension system, which is at odds with the idea of financial sustainability. Considering the contradiction between this principle and financial reality on the ground, China can optimize NTRSEI by adjusting the basic parameters of the rural pension system to achieve the financial sustainability principle. International experiences also show that the sooner China acts, the better. In particular, China can learn from Canada to take more forceful measures, such as increasing contribution rates and expanding pension fund investment now, rather than waiting for the rural pension system to face an immediate financial crisis.

(3) From a governance perspective, policymakers must realize the need for greater financial sustainability in NTRSEI. Our paper could help foster this broader understanding by quantifying the large scope of the problem facing such policymakers. It is hoped this paper helps policymakers deepen their understanding of financial sustainability issues in rural pensions, so that they can truly address them instead of putting their head in the sand and denying the existence of the problem. Financial sustainability is not only about words, but about concrete policy practices, which is why Chinese experts and policymakers should draw on international experience in domestic pension reform, including when dealing with rural pensions.

(4) More concretely, learning once again from international experience, China should make sure the value of individual account assets does not fall in the short run, while reforming the mechanism through which the interest rates for individual accounts are set. In the long run, policymakers can ask the National Social Security Fund Council to invest and operate rural social endowment insurance funds while improving the operation of the system. Simultaneously, based on the experience of pension reform elsewhere around the world, China should further extend coverage for the social pension funds, increase the age of retirement, develop its capital markets and improve the legal and policy framework surrounding the investment of social pension funds.

(5) Creating a sustainable pension system is not simply about cost containment, which may lead to growing poverty among the rural elderly. This is why we recommend the improvement of the system's sustainability in the context of adequate pension benefits. Our specific approach is about increasing governmental investment and actively using various financing methods to generate financial sustainability. Part of this sustainability project is to maintain pension benefits that are equal to the basic economic needs of Chinese rural residents. Given that premise, pushing back the retirement age, enhancing the contributions of the working-age population and increasing pension fund investments are better solutions than direct benefit cuts. Hence, our policy recommendations should not cause growing poverty among the rural elderly.

(6) Increasing the central government's subsidies is the most reliable method to improve the revenue stream available to fund rural pensions. Considering the existence of uneven local fiscal capacities and the related resistance to the integration of the pension system all around China, the central government's subsidies are the most appropriate and the fairest policy choice. In this context, the central government should also increase basic monthly pensions cross the country, especially in the poorer central and western regions. Meanwhile, the central government should provide more individual contribution subsidies. In poverty-stricken counties, border counties, regions inhabited by ethnic minorities and other areas facing particular socio-economic challenges, China's central government should even consider paying the entire subsidies available for rural pensions.

8. Conclusions

The new rural pension system is one of the most important social policy systems in China, and great progress has been achieved in implementing NTRSEI. However, experience clearly shows that financial sustainability is a key factor to consider when achieving long-term social policy objectives is concerned. Using social insurance actuarial techniques, this article shows that the rural pension funding gap should rise from 97.80 billion Yuan in 2014 to 3062.31 billion Yuan in 2049, an increase of slightly more than 10% a year. Considering this, the new rural pension system is unsustainable, at least in its current form. However, as suggested above, based on recent international experiences, there are several ways in which policymakers can make the Chinese rural pension system more sustainable over time. We hope our findings and policy recommendations help Chinese pension experts and policymakers create a more sustainable rural pension system, while using a better actuarial framework to assess its long-term sustainability.

Acknowledgments

The authors thank Rachel Hatcher for the copy editing and the reviewers for their comments and suggestions. This research was supported by the Shaanxi Social Science Fund (No. 13G046), the China Postdoctoral Science Foundation (No. 2013M532068), the Shaanxi Postdoctoral Science Foundation and the Fundamental Research Funds for the Central Universities. Daniel Béland acknowledges support from the Canada Research Chairs Program.

Author Contributions

Daniel Béland designed the research project; Lijian Wang conducted the research and analyzed the data; Lijian Wang and Daniel Béland wrote the paper together. Both authors read and approved the final draft of the article.

Conflicts of Interest

The authors declare no conflict of interest.

References

1. Zheng, B. *China Pension Report 2013*; Economic & Management Press: Beijing, China, 2013. (In Chinese)
2. Gao, J.; Ding, K. The model of fund gap in China's fundamental pension insurance and its applications. *Syst. Eng. Theory Methodol. Appl.* **2006**, *15*, 49–54. (In Chinese)
3. Sin, Y. *China: Pension Liabilities and Reform Options for Old Age Insurance*; The World Bank: Washington, DC, USA, 2005.
4. Ma, J.; Zhang, X.; Li, Z.; Xiao, M.; Xu, J.; He, D. Defusing national asset-liability risks in the medium and long term. *Caijing* **2012**, *15*, 35–46. (In Chinese)
5. Xu, J.; Xu, Y. Elderly poverty in view of life course theory. *Sociol. Stud.* **2009**, *6*, 122–144. (In Chinese)
6. Hvistendahl, M. Can China Age Gracefully? A Massive Survey Aims to Find Out. *Science* **2013**, *341*, 831–832.
7. Chou, R.J.-A. Filial piety by contract? The emergence, implementation, and implications of the "Family Support Agreement" in China. *Gerontologist* **2011**, *51*, 3–16.
8. Hurst, W.; O'Brien, K.J. China's contentious pensioners. *China Q.* **2002**, *170*, 345–360.
9. Zhang, N.J.; Guo, M.; Zheng, X. China: Awakening giant developing solutions to population aging. *Gerontologist* **2012**, *52*, 589–596.
10. Billig, A.; Ménard, J.C. Actuarial balance sheets as a tool to assess the sustainability of social security pension systems. *Int. Soc. Secur. Rev.* **2013**, *66*, 31–52.
11. Grech, A.G. Assessing the sustainability of pension reforms in Europe. *J. Int. Comp. Soc. Policy* **2013**, *29*, 143–162.

12. Chadwick, D.G.; Wilson, M.B.; Anderson, C.F. Shaping oral health care in North Carolina with East Carolina University's community service learning centers. *North Carol. Med. J.* **2014**, *75*, 36–38.

13. Spence, A. Achieving Financial Sustainability in Today's Changed World. *Philanthropist* **2011**, *24*, 97–102.

14. Ayayi, A.; Sene, M. What drives microfinance institution's financial sustainability. *J. Dev. Areas* **2010**, *44*, 303–324.

15. Boulier, J.F.; Huang, S.; Taillard, G. Optimal management under stochastic interest rates: The case of a protected defined contribution pension fund. *Insur. Math. Econ.* **2001**, *28*, 173–189.

16. Blake, D. Pension schemes as options on pension fund assets: Implications for pension fund management. *Insur. Math. Econ.* **1998**, *23*, 263–286.

17. Geyer, A.; Ziemba, W.T. The Innovest Austrian pension fund financial planning model InnoALM. *Oper. Res.* **2008**, *56*, 797–810.

18. Gollier, C. Intergenerational risk-sharing and risk-taking of a pension fund. *J. Public Econ.* **2008**, *92*, 1463–1485.

19. Grande, G.; Visco, I. A public guarantee of a minimum return to defined contribution pension scheme members. *J. Risk* **2011**, *13*, 3–43.

20. Sundén, A. The Swedish experience with pension reform. *Oxf. Rev. Econ. Policy* **2006**, *22*, 133–148.

21. Yaşar, Y. The crisis in the Turkish pension system: a post Keynesian perspective. *J. Post Keynes. Econ.* **2013**, *36*, 131–152.

22. Barr, N.; Diamond, P.; Engel, E. Reforming Pensions: Lessons from Economic Theory and Some Policy Directions. *Economía* **2010**, *11*, 1–15.

23. Fedotenkov, I.; Meijdam, L. Crisis and Pension System Design in the EU: International Spillover Effects via Factor Mobility and Trade. *Economist (Leiden)* **2013**, *161*, 175–197.

24. Gerrans, P. Retirement savings investment choices in response to the global financial crisis: Australian evidence. *Aust. J. Manag.* **2012**, *37*, 415–439.

25. James, E. How can China solve its old-age security problem? The interaction between pension, state enterprise and financial market reform. *J. Pension Econ. Finance* **2002**, *1*, 53–75.

26. Wang, Y. *Implicit Pension Debt, Transition Cost, Options and Impact of China's Pension Reform: A Computable General Equilibrium Analysis*; The World Bank: Washington, DC, USA, 2001.

27. Zhou, W. An Accurate Analysis of the Risk facing the Account Rate System of Endowment Insurance and a New Design for the Account Rate System. *J. Quantitative Tech. Econ.* **2008**, *24*, 91–97. (In Chinese)

28. Gao, J. The Actuarial Model and Its Applications for China IPD. *Econ. Math.* **2004**, *21*, 120–128. (In Chinese)

29. Zheng, B. 60 Years of Social Security in China: Achievements and Lessons. *China J. Popul. Sci.* **2009**, *5*, 2–19. (In Chinese)

30. Zeng, Y. Effects of Demographic and Retirement-Age Policies on Future Pension Deficits, with an Application to China. *Popul. Dev. Rev.* **2011**, *37*, 553–569.
31. Hu, J. An empirical approach on regulating China's pension investment. *Eur. J. Law Econ.* **2013**, *36*, 1–22.
32. Lee, J.; Midgley, J.; Zhu, Y. *Social Policy and Change in East Asia*; Lexington Books: Lanham, MD, USA, 2013.
33. Lu, J. Policy Selection in Social Security during the Financial Crisis. *China Econ.* **2010**, *7*, 9–11. (In Chinese)
34. Xie, H.; Yang, S. Sustainability Analysis of the New Rural Pension System Based on the Tendency toward Population Aging. *Acad. Explor.* **2014**, *3*, 106–110. (In Chinese)
35. Kou, T.; Wan, M. Institutional Improvement and Financial Support: The Sustainable Development ability of the Rural Pension System. *Financial Econ. Issue* **2011**, *1*, 96–100. (In Chinese)
36. Liu, J. Discussion on the Basic Points and Key Strategies of Sustainable Development in the New Rural Social Endowment Insurance System. *Soc. Sec. Stud.* **2010**, *3*, 42–46. (In Chinese)
37. Qian, Z.; Bu, Y.; Zhang, Y. Actuarial Analysis of the New Rural Social Pension Insurance System in the Context of the Aging Problem. *Economist* **2012**, *8*, 58–65. (In Chinese)
38. Li, J. On the Financial Situation of the New Rural Old-Age Insurance in the context of Urbanization and Aging: 2011–2050. *Insur. Stud.* **2012**, *5*, 111–118. (In Chinese)
39. Feng, T.; Li, M. Simulation and Forecast of the Balance in the New Type Rural Social Endowment Insurance Funds. *J. Pub. Manag.* **2010**, *7*, 100–110. (In Chinese)
40. Xue, H. Assessment of the Sustainability of the Financial Security Capacity of the New Rural Social Endowment Insurance: Policy Simulation Perspective. *China Soft Sci.* **2012**, *5*, 68–80. (In Chinese)
41. Cox, R.H.; Béland, D. Valence, Policy Ideas, and the Rise of Sustainability. *Governance* **2013**, *26*, 307–328.
42. Pfeffer, J. Building Sustainable Organizations: The Human Factor. *Acad. Manag. Perspect.* **2010**, *24*, 34–45.
43. Jacobs, A.M. *Governing for the Long Term: Democracy and the Politics of Investment*; Cambridge University Press: New York, NY, USA, 2011.
44. Cremer, H.; Pestieau, P. Reforming our pension system: Is it a demographic, financial or political problem? *Eur. Econ. Rev.* **2000**, *44*, 974–983.
45. Disney, R. Crises in public pension programmes in OECD: What are the reform options? *Econ. J.* **2000**, *110*, 1–23.
46. Blake, D.; Mayhew, L. On The Sustainability of the UK State Pension System in the Light of Population Ageing and Declining Fertility. *Econ. J.* **2006**, *116*, 286–305.
47. Béland, D.; Shinkawa, T. Public and Private Policy Change: Pension Reform in Four Countries. *Policy Stud. J.* **2007**, *35*, 349–371.
48. Marier, P. *Pension Politics: Consensus and Social Conflict in Ageing Societies*; Routledge: London, UK, 2008.

49. Weaver, R.K. Pensions on Autopilot? Sustaining Automatic Stabilizing Mechanisms in Public Pensions. In Proceedings of the Annual Meeting of the American Political Science Association, Chicago, IL, USA, 29 August–1 September 2013.

50. Little, B. *Fixing the Future: How Canada's Usually Fractious Governments Worked Together to Rescue the Canada Pension Plan*; University of Toronto Press: Toronto, ON, Canada, 2008.

51. Béland, D.; Myles, J. Varieties of Federalism, Institutional Legacies, and Social Policy: Comparing Old-Age and Unemployment Insurance Reform in Canada. *Int. J. Soc. Welf.* **2012**, *21*, S75–S87.

52. Light, P.C. *Still Artful Work: The Continuing Politics of Social Security Reform*; McGraw-Hill: New York, NY, USA, 1995.

53. Béland, D.; Waddan, A. *The Politics of Policy Change: Welfare, Medicare, and Social Security Reform in the United States*; Georgetown University Press: Washington, DC, USA, 2012.

54. Gustafson, K.; Baofeng, H. Elderly Care and the One-Child Policy: Concerns, Expectations and Preparations for Elderly Life in a Rural Chinese Township. *J. Cross Cult. Gerontol.* **2014**, *29*, 1–12.

55. Feng, T.; Yang, Z. Forecasting the New Type of Rural Social Endowment Insurance Fund in View of Land Circulation. *J. Quant. Tech. Econ.* **2013**, *30*, 3–18. (In Chinese)

56. Zheng, G. *Reform and Development Strategy of China' Social Security System*; People's Publishing House: Beijing, China, 2011; Volume Social Endowment Insurance. (In Chinese)

57. Xiao, X.; Peng, K. Research on the Generalized Non-Equidistance GM (1, 1) Model based on Matrix Analysis. *Grey Syst. Theory Appl.* **2011**, *1*, 87–96. (In Chinese)

58. Between 2010 and 2012, the research team surveyed 5032 rural residents. These people were randomly selected from six counties in three provinces: Chencang District, Shangnan Country, Tongxu Country, Xixia Country, Gaochun Country, and Changshu City (Six Counties for short).

59. Wang, L.; Liu, J. The Construction and Application of the Coordinating Urban and Rural Population Projection Model. *Northwest Popul.* **2009**, *30*, 24–28. (In Chinese)

60. National Bureau of Statistics of the People's Republic of China. 2010 Sixth National Population Census Data Gazette (No. 1). Available online: http://www.stats.gov.cn/tjsj/tjgb/rkpcgb/qgrkpcgb/201104/t20110428_30327.html (accessed on 28 April 2011). (In Chinese)

61. Wang, L. Robust Stability Analysis for the New Type Rural Social Endowment Insurance System with Minor Fluctuations in China. *Discret. Dyn. Nat. Soc.* **2012**, *2012*, 1–9.

62. Lei, X.; Zhang, C.; Zhao, Y. Incentive problems in China's new rural pension program. *Res. Labor Econ.* **2013**, *37*, 181–201.

63. Pei, X.; Tang, Y. Rural Old Age Support in Transitional China: Efforts between Family and State. In *Aging in China: Implications to Social Policy of a Changing Economic State*; Chen, S., Powell, J.L., Eds.; Springer: New York, NY, USA, 2012; pp. 61–81.

64. Hu, Y.Y. Social Security Fund's Annual Return at 8.4%. Available online: http://www.chinadaily.com.cn/business/2013-02/28/content_16266098.htm (accessed on 28 February 2011).

65. Samanez-Larkin, G.R.; Kuhnen, C.M.; Yoo, D.J.; Knutson, B. Variability in nucleus accumbens activity mediates age-related suboptimal financial risk taking. *J. Neurosci.* **2010**, *30*, 1426–1434.

66. Drahokoupil, J.; Domonkos, S. Averting the funding-gap crisis: East European pension reforms since 2008. *Glob. Soc. Policy* **2012**, *12*, 283–299.

67. Feng, J.; He, L.; Sato, H. Public pension and household saving: Evidence from urban China. *J. Comp. Econ.* **2011**, *39*, 470–485.

Insights into the Regional Greenhouse Gas (GHG) Emission of Industrial Processes: A Case Study of Shenyang, China

Zuoxi Liu, Huijuan Dong, Yong Geng, Chengpeng Lu and Wanxia Ren

Abstract: This paper examines the GHG emission of industrial process in Shenyang city, in the Liaoning province of China, using the 2006 IPCC greenhouse gas inventory guideline. Results show that the total GHG emissions of industrial process has increased, from 1.48 Mt in 2004 to 4.06 Mt in 2009, except for a little decrease in 2008. The cement industry, and iron and steel industries, are the main emission sources, accounting for more than 90% of the total carbon emissions. GHG emissions in 2020 are estimated based on scenario analysis. The research indicates that the cement industry, and iron and steel industries, will still be the largest emission sources, and the total carbon emissions under the business as usual (BAU) scenario will be doubled in 2020 compared with that of 2009. However, when countermeasures are taken, the GHG emission will reduce significantly. Using more clinker substitutes for blended cement, and increasing direct reduction iron process and recycled steel scraps are efficient measures in reducing GHG emission. Scenario 4, which has the highest ratio of 30/70 blended cement and the highest ratio of steel with recycled steel-EAF process, is the best one. In this scenario, the industrial process GHG emission in 2020 can almost stay the same as that of 2009. From the perspective of regions, cement industry and iron and steel industry accounted for the vast majority of GHG emission in all industries. Meanwhile, these two industries become the most potential industries for reduction of GHG emission. This study provides an insight for GHG emission of different industries at the scale of cities in China.

Reprinted from *Sustainability*. Cite as: Liu, Z.; Dong, H.; Geng, Y.; Lu, C.; Ren, W. Insights into the Regional Greenhouse Gas (GHG) Emission of Industrial Processes: A Case Study of Shenyang, China. *Sustainability* **2014**, *6*, 3669-3685.

1. Introduction

Along with the rapid industrialization and urbanization, China's high greenhouse gas emissions have become an important issue both domestically and internationally [1,2]. China's total CO_2 emissions by fossil fuel consumption were estimated to be 2.63 billion tons in 2012, which ranked China first in the world [3]. Hence, China faces increasing international pressure to curb its CO_2 emissions. The government has made the commitment of reducing CO_2 emissions per unit of GDP by 40%–45% in 2020 compared with 2005 [4], and the industries sector especially the heavy industries such as steel, cement, and chemicals production, contribute most of the GHG emissions [5].

Till now, by employing various approaches, few studies have examined sources and reduction potentials of industrial GHG emissions. For example, Liaskas *et al.*, used the algebraic disaggregation method to identify the factors influencing CO_2 emissions generated in the industrial sector of European Union countries [6], Zhou *et al.*, estimated the carbon footprint of China's Ammonia production and analyzed the potential for carbon mitigation in the industry [7], Sheinbaum *et al.*, analyzed Energy and

CO_2 emission trends of Mexico's iron and steel industry during the period 1970–2006, examining CO_2 emissions related to energy use and production process [8]. Kim and Worrell [9], and Kirschen *et al.* [10] present the analysis on energy-related carbon footprint in the iron and steel industrial sector of seven countries and electric arc furnace, respectively. With regard to the energy efficiency sector, Lee *et al.*, decomposed the changes of CO_2 emissions into eight factors from Taiwan's petrochemical industries during 1984 and 1994 [11], Lin *et al.*, identified the key factors which affecting CO_2 emission changes of industrial sectors in Taiwan by using the divisia index approach [12], Hendriks *et al.* [13] and Van Puyvelde *et al.* [14] produced papers investigating emissions and reductions from the cement sector. However, most of the studies related to industrial GHG emission mainly focuses on GHG emissions from energy consumption. Put forward by the Intergovernmental Panel on Climate Change (IPCC), GHG emission of industrial process is usually discussed as one of the main components in the greenhouse gas emission inventory of different countries or regions, but seldom has intensive studies on its reduction potentials and relative countermeasures. A comprehensive study on industrial process GHG emission is necessary and is a new point of view that can contribute in a certain degree to the climate change.

Industrial process GHG emission refers to the greenhouse gas emissions from industrial processes that chemically or physically transform materials [15]. According to the report of the National Development and Reform Commission (NDRC) of China, industrial process emissions contributed about 9% of total CO_2 emissions in 1994 [16]. Though relatively small at the total level, GHG emission of industrial process, characterized by its important, un-negligible and fast growth in emissions, has naturally become the research focus in this field. Ever since 1996, iron and steel output has firmly held the first place [17]. China's steel industry has grown rapidly on the back of a huge growth in domestic demand. Crude steel production in China reached 273 million tons in 2004, about three times the figure for 1994, and accounted for around 25.8% of global steel production [18]. In addition, China was the biggest cement producing and consumption country in this world [19]. Cement output increased from 209.7 to 1868 million tonnes of cement between 1990 and 2010 in China, and now represents over half of the world's total cement production [20]. These dramatic increases in steel and cement production are consistent with the trends in energy consumption and CO_2 emission. It should be mentioned that the amount of GHG emission from process-related accounted for growing larger portion in total GHG emission in China. In the year 2009, the process-related GHG emission from cement industry accounted for 54.1% of total cement industry in China [21]. The GHG emission of industrial process from iron and steel industry contributed more than 60% of the total GHG emissions in steel industry [22]. Therefore, it is crucial for researchers to focus on the study of GHG emission of the industrial process, including cement, steel and other industries.

The purpose of this study is to estimate the GHG emission of the industrial process and predict its reduction potentials with scenario analysis, so as to give some policy implications to local government. Thus, this study selects a city scale as a case study to study the GHG emission of industrial process, which is more accessible and representative to make scenario analysis. The paper is structured as follows: first, we give a brief description of the background information of Shenyang city and the industrial process GHG emission; then, we present the method used to assess the GHG emission of the industrial process; third, we introduce results for the case study of Shenyang city and make a general understanding of current state of industrial process GHG emission from year 2004 to 2009; fourth, we present scenario analysis based on business as usual(BAU) scenario and scenario with countermeasures to predict future industrial process GHG emission of 2020, so that proper scenarios and feasible GHG abating measures could be obtained; and finally, conclusions are drawn.

2. Background

2.1. Study Area

Shenyang, the capital city of Liaoning Province, located in the south of northeastern China and the central of Liaoning, with an administrative area of 12,881 km^2 (Figure 1). Shenyang is the biggest international metropolis of the Northeastern area of China and is known as one of the most important heavy industrial bases of China [23]. Since 2003, the central government has implemented the strategy of "revitalizing the old industrial bases in Northeastern China", and Shenyang was selected as the first city to demonstrate the strategy. It aimed to shift the industrial structure by improving resources efficiency and reducing the environmental pressure, thus later Shenyang was identified as the core of the new-industrialization zone for national demonstration [24]. The new-industrialization zone is mainly composed of equipment manufacturing industry, metallurgical industry, and petrochemical industry, which is taken as typical and representative case in blazing a trail to new industrialization and new urbanization, and expected to offer a demonstration for China's change in industrial development mode and economic development transformation. Under such circumstances, Shenyang's economy and urbanization will definitely continue to increase rapidly, and industrial sectors of cement, steel and relative material requirement will take an important role.

Figure 1. Location of Shenyang City.

2.2. Definition of Industrial Process GHG Emission

Generally, the total GHG emission of industrial production is mainly composed of two parts: energy related emissions and process related emissions. Energy related emissions are mainly from energy consumption, such as direct emissions from combustion of fossil fuel, and indirect emissions from consumption of electricity and heat. And almost all industrial sectors cause energy related emissions. While only some industrial sectors cause process related emissions, including [25]:

- Metal production—e.g., carbon dioxide and perfluorocarbon emissions from aluminum smelting; and carbon dioxide, methane and nitrous oxide emissions from iron and steel production.
- Chemical industry—nitrous oxide emissions from the production of nitric acid (largely used in production of ammonium nitrate); carbon dioxide emissions from ammonia production; and methane emissions from the production of organic polymers and other chemicals.
- Mineral products—carbon dioxide emissions from cement clinker and lime production, the use of limestone and dolomite in industrial smelting processes, soda ash use and production, magnesia production, and the use of other carbonates (sodium bicarbonate, potassium carbonate, barium carbonate, lithium carbonate and strontium carbonate).
- Food and drink production—carbon dioxide emissions from ammonia production, carbon dioxide wells, ethylene oxide production and sodium bicarbonate use.

Consumption and production of halocarbons and sulfur hexafluoride is also covered by the industrial processes sector but are not considered in this paper for lack of data.

Industrial process GHG emission means the process related emissions discussed above. And the industrial sectors selected in this paper are determined according to the 2006 IPCC Guidelines for National Greenhouse Gas Inventories [15] and the industrial condition of Shenyang. Twelve major industrial sectors are examined including cement production, glass production, refractory production, ammonia production, methanol production; graphite and carbon black, coke production, pig iron production; raw steel production, primary aluminum production, and Lead production are focused on.

3. Methodology

3.1. GHG Emission Calculation

Estimating GHG emission resulting from the industrial process is by no means straightforward. In some industrial processes, the GHG is produced by chemical reaction (e.g., the cement production, glass production *et al.*); in some industrial processes, the GHG is produced by oxidization of fossil fuels used as reducing agent to smelt metals (e.g., the metallurgy industry), whereas in some other processes, the GHG is produced by the chemical reaction during the production of certain chemicals (e.g., ammonia). The estimation of GHG emissions related to industrial process was part of the IPCC 2006, which provide internationally agreed methodologies intended for use by countries to estimate greenhouse gas inventories to report to the United Nations Framework Convention on Climate Change. The estimation of GHG emissions was mainly based on the IPCC 2006 tier 1 method.

$$E_{i,CO_2} = AD_{i,CO_2} \times EF_{i,CO_2} \tag{1}$$

$$E_{CO_2} = \sum_{i=1}^{12} E_{i,CO_2} \tag{2}$$

$$E_{i,CH_4} = AD_{i,CH_4} \times EF_{i,CH_4} \tag{3}$$

$$E_{CH_4} = \sum_{i=1}^{12} E_{i,CH_4} \tag{4}$$

$$E_i = E_{i,CO_2} + E_{i,CH_4} \times 21 \tag{5}$$

$$E = \sum_{i=1}^{12} E_i \tag{6}$$

where: $E_{[tons-(t)]}$ represents the emission amount of CO_2 or CH_4 from the industrial process. The final GHG emission for each industrial process is changed to CO_{2e} after multiplying the Global warming potential 21 [26]; AD represents the activity data that is the physical production of industrial products; $EF_{[t\,CO2\,or\,CH4/t\,product]}$ represents the emission factor of CO_2 or CH_4; i represent the index number of industrial sector.

3.2. Data Sources

We derive (i) production data from the Shenyang Statistic Yearbook (2005–2010); and (ii) process- and product-specific emission factors from IPCC 2006 and other sources [19]. Production data of the twelve industrial sectors of Shenyang City in the year 2004–2009 is shown in Table 1.

Table 1. Production data of the industrial processes in Shenyang in the year 2004–2009.

Industrial Sector	Production (t)					
	2004	2005	2006	2007	2008	2009
Cement	1,747,092	1,792,400	2,310,000	2,854,000	2,740,393	3,975,319
clinker	484,276	553,600	280,000	246,000	272,610	3,048,451
Glass	506,513	413,916	339,065	281,815	528,134	477,723
Refractory products	125,671	43,389	75,343	71,372	132,417	200,275
Ammonia	11,761	3048	3677	2774	/	1382
Methanol	8745	4861	/	/	/	/
Graphite and carbon black	/	6026	2511	1739	1402	1654
Coke	335,075	670,400	763,453	909,560	833,739	908,417
Pig iron	32,932	10,016	3288	1986	/	/
Raw steel	130,223	205,707	201,737	409,495	425,883	217,398
Primary aluminum	580	18,270	10,824	8980	700	/
Lead	3823	5222	7356	8771	2462	2139

Emission factors of carbon dioxide and methane for each industrial process are summarized in Table 2. It is difficult to get industrial process emission factors specified for China or Shenyang, so the emission factors are mainly from the default values of IPCC 2006. According to IPCC 2006, all the industrial sectors selected in this paper cause CO_2 emissions. However, most of the industrial sectors do not cause CH_4 emissions, except for methanol, carbon black, coke, pig iron, and raw steel production.

Table 2. Emission factors of the industrial process.

Industrial Sector	Emission Factor of CO_2 (t CO_2/t Product)	Emission Factor of CH_4 (10^{-3} t CH_4/t Product)
Cement	0.41 [a]	/
Clinker	0.52	/
Glass	0.12 [b]	/
Refractory products	0.20 [c]	/
Ammonia	3.27	/
Methanol	0.67	2.30
Graphite and carbon black	2.62	0.06
Coke	0.56	0.0001
Pig iron	1.35	14.50
Raw steel	1.06	0.03
Primary aluminum	1.65	/
Lead	0.52	/

[a] The value is the average emission factors of China and is obtained from reference [15]. [b] The emission factor that does not consider cullet ratio is 0.2. The cullet ratio of glass in this paper is 0.4, hence the final emission factor for glass is 0.12. [c] IPCC 2006 has mentioned the refractory products, but has not given the emission factor. So the emission factor refers to that of the glass.

4. Current Industrial Process GHG Emission for Shenyang

4.1. Total GHG Emission during 2004–2009

The GHG emissions (CO_{2e}) of industrial process from 2004 to 2009 are given in Figure 2. The total GHG footprint emission of industrial process for Shenyang is increasing year by year from 1.48 Mt in 2004 to 4.06 Mt in 2009, while decreases a little in 2008, about 2.28 Mt. It is obvious that the main component of carbon GHG footprint emission of industrial process is carbon dioxide, and methane takes up only a very small proportion. In 2004–2009, carbon dioxide accounts for 99.30%, 99.80%, 99.94%, 99.96%, 99.99%, and 99.99%, respectively, and the ratio is increasing year by year. Hence, the impact of methane can almost be neglected.

28

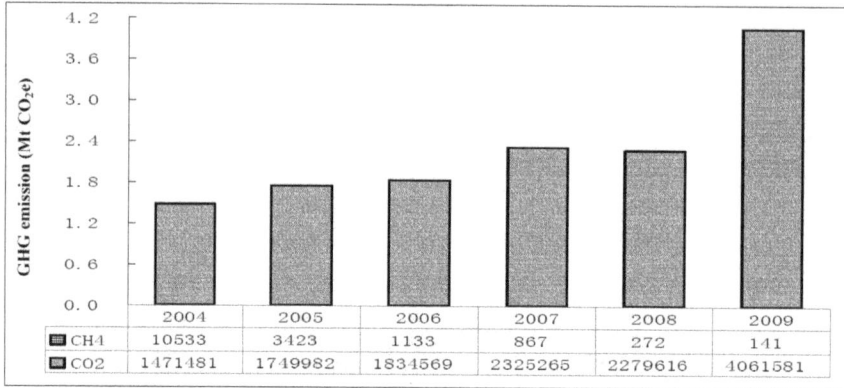

Figure 2. CarbonGHG footprintemission of industrial process from 2004 to 2009.

4.2. GHG Footprint Emissions of Twelve Industrial Sectors during 2004–2009

As shown in Figure 3, Cement, coke, raw steel, clinker, and glass production are the top five sectors that have the largest GHG emissions of industrial process in the years 2004–2009. GHG emissions of the five sectors add up accounting for about 91.41%, 95.01%, 96.69%, 97.83%, 98.57% and 98.77%, respectively from year 2004 to 2009. Among the top five emission sectors, cement production has the largest GHG emissions, accounting for about 40–50 percent from year 2004 to 2009, and is the most important sector that should be paid attention to. Coke production and raw steel production are two second important sectors that contribute to the total GHG emissions. Besides, the three major industrial sectors show the same emission trend with the total GHG emissions of industrial process, which is increasing year by year, although slightly decreases in 2008. It demonstrated that the variation of total GHG emission was mainly determined by the three sectors.

It can be seen from Figure 4 that the top five major emission sectors in 2004 are cement production, clinker production, coke production, raw steel production, and glass production orderly. In addition, the emission percentage is 46.43%, 16.73%, 12.47%, 9.18% and 6.73%, respectively. The total percentage of the five industrial sectors is 91.54%. In addition, the GHG emissions of the other six industrial processes account for only 8.46%. While GHG emission from coke production and raw steel production increased obviously in 2005. In addition, emissions from cement and clinker process begin to shrink. So the top five major sectors in 2005 are cement production, coke production, clinker production, raw steel production, and glass production orderly. Furthermore, the emission percentage is 40.54%, 21.23%, 16.28%, 12.34% and 4.68%, respectively. The sum of the five industrial sectors comes to about 95%. Emission percentage of cement, coke and raw steel production keeps increasing on the whole in year 2006–2008. However, emission percentage of clinker production keeps decreasing. The top five major emission sectors in 2006–2008 are cement production, coke production, raw steel production, clinker production, and glass production orderly. In addition, the sum of the five sectors in year 2006, 2007 and 2008 are 96.7%, 97.8% and 98.6% respectively. The percentage of other sectors becomes less and less, from 8.46% in 2004 to 1.42% in 2008. However, in 2009, the ratio of clinker increased significantly because a very large cement enterprises in Shenyang enlarged the production scale in 2009 so that the GHG emission of clinker industry was sharply increased.

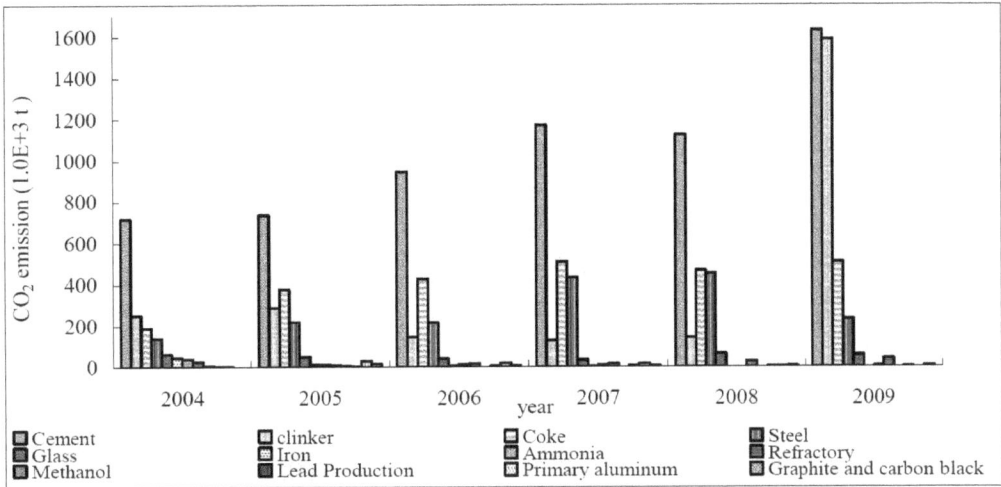

Figure 3. Comparison of GHG emissions of twelve industrial processes from 2004 to 2009.

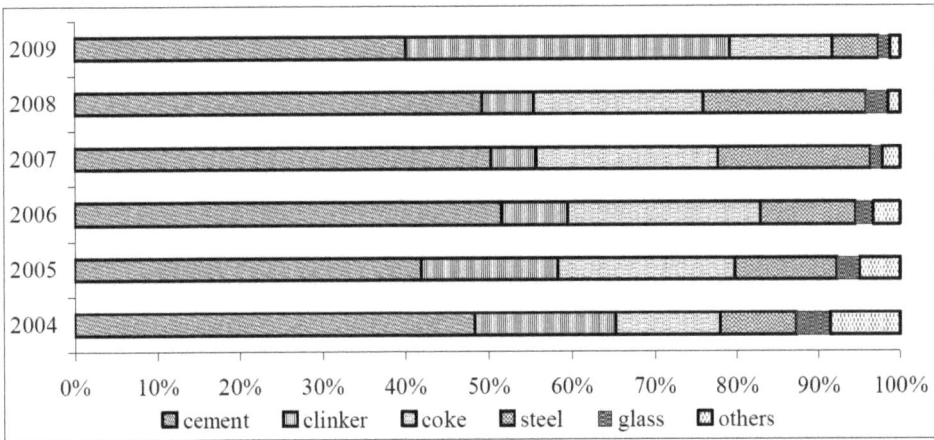

Figure 4. GHG emission profiles of major industrial processes in year 2004–2009.

4.3. Comparison of GHG Emission of Three Categories during 2004–2009

To compare the industrial process GHG emission from a more specific scale, the twelve industrial processes can be classified into three categories that are the cement industry, the iron and steel industries, and others. The cement industry consists of cement production and clinker production. The iron and steel industry consists of coke production, pig iron production, and raw steel production. Others mean the sum of the rest industrial processes. Analysis from Figure 5 demonstrated that GHG emission of industrial process is mainly from the cement industry, and iron and steel industry. From year 2004 to 2009, total GHG emissions of the two categories are 1.35 Mt, 1.63 Mt, 1.74 Mt, 2.25 Mt,

2.19 Mt and 3.96 Mt, respectively. And account for about 91%, 93%, 95%, 97%, 95%, and 97%, respectively. It is obvious that the total share of the two categories increases year by year from 2004 to 2009.

Analysis above shows that the cement industry, and iron and steel industry, contribute most of the industrial process GHG emission. It is mainly caused by the high consumption requirement of cement and steel with the quick urbanization and industrialization. To reduce GHG emission of industrial process, solutions can be taken from reducing the production of cement and steel by controlling consumption or lowering the emission factors by improving the industrial technique. However, China is undergoing transformation of economy development, and it is unpractical for Shenyang to reduce the production of cement and iron & steel.

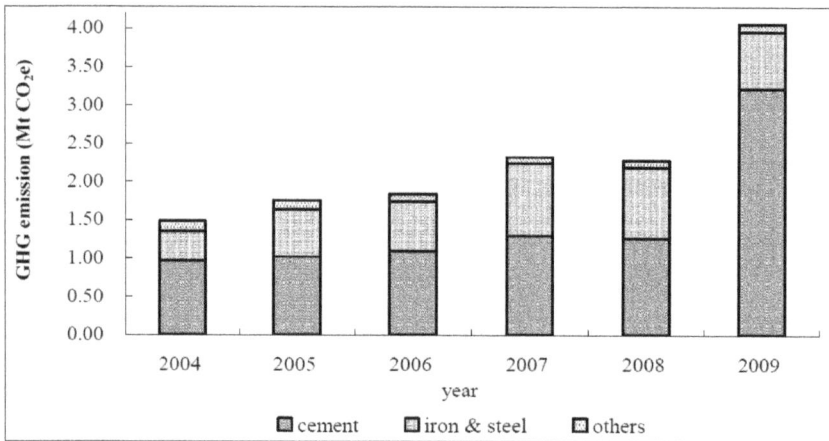

Figure 5. Comparison of GHG emission of three main industrial categories in 2004–2009.

5. Scenario Analysis

A scenario may be defined as a "hypothetical sequence of logical and plausible (but not necessarily probable) events, constructed in order to focus attention on causal processes and decision points" [27]. Scenario analysis has been applied to diverse efforts ranging from literary descriptions to model-based projections, from visionary thinking to minor adjustments to BAU projections [28]. Scenario analysis here is to predict the GHG reduction potentials and find out feasible countermeasures. The cement industry, and iron and steel industry are the main sources of industrial process GHG emission (accounting for more than 97%), so countermeasures are only focus on the two industries. However, steel and cement industry has grown rapidly due to a huge growth on economy [29]. And the growth is expected to continue over the next few years. Hence, it is more feasible to reduce the emission factors than reduce the production of both the cement, iron and steel industries. Solutions for less GHG emissions may be from the improvement of production technique, adoption of new technologies, implementation of incentive measures or other policies that can reduce GHG emission factors.

5.1. Countermeasures Introduction

For cement industry, the industrial process GHG emission accounts for about 52% of the total cement production emissions [30], and it is mainly caused by the chemical process of making clinker, so measures can be taken from the aspects of using clinker substitute, which can reduce the amount of limestone that needs to be reacted at the cement plant while maintaining the amount of cement produced [14]. The cement industry is a typical example of "scavenger" in industrial ecology [31]. The practices of utilizing different wastes in cement production process has been known for quite a long time, which is now recognized as one opportunity for reducing CO_2 emissions [32].

Mineral components (MC), including pozzolana, blast furnace slag, fly ash, and waste materials from other production processes such as steel and coal-fired power production, can be used as clinker substitute to produce blended cement [33]. The clinker fraction of different cement types can range from a high of 95%–97% for a straight Portland cement, to 25% or less for cement (Table 3) [15]. And scenarios for cement was designed based on the percentage of different blended cement types.

Table 3. Percent clinker in the cement production mix (Source: IPCC 2006).

Country Production Mix (PC/blend) [a]	Percent Additives (Pozzolana + Slag) in the Blended Cement [b]				
	10%	20%	30%	40%	75%
0/100	85	76	66	57	24
15/85	87	79	71	63	26
25/75	88	81	74	66	42
30/70	88	82	75	68	45
40/60	89	84	78	72	52
50/50	90	85	81	76	60
60/40	91	87	84	80	66
70/30	92	89	86	84	74
75/25	93	90	88	85	77
85/15	94	92	91	89	84
100/0	Straight Portland cement having 95% clinker fraction				

[a] Country production mix refers to the range of products of a country, e.g., "75/25" means 75% of total production is Portland cement (PC) and the rest is blended cement. [b] The inclusion of slag allows for a base to the blend of Portland or Portland blast furnace slag cement or both. All Portland in blended cement is assumed to be 95% clinker.

For the iron and steel industry, the industrial process GHG emission is complicated. Steel production can occur at integrated facilities from iron ore, or at secondary facilities that produce steel mainly from recycled steel scrap. Integrated facilities typically include coke production, pig iron production by blast furnaces (BF), and raw steel production by basic oxygen furnaces (BOFs), or open hearth furnaces (OHFs) some times. Raw steel is more often produced by using BOFs than by using OHFs from pig iron produced by the BFs, and secondary steelmaking usually, however, occurs in electric arc furnaces (EAFs). The iron and steel industrial process emissions come from three major processes: coke production, BF, and BOF (OHF or EAF). However, most emissions are from BF/OHF and coke production, and emissions from EAF are relatively small. Besides, direct reduced iron (DRI) process, in which coke is not needed, is a high quality, low energy consumption and low pollution process. DRI is normally used as a replacement for scrap metal in the electric arc furnace steelmaking route or used as coolant in the BOFs to improve the production of raw steel, and is encouraged by the government. It is a good substitute for BF and should be encouraged. Emission factors for separate processes are shown in Table 4. So measures can be taken from the aspect of steel production types. If steel were produced directly from recycled scrape steel, process of BF and coke will be omitted and GHG emission can be reduced significantly. It is inevitably to make steel from iron ore. In this case, GHG reduction can be achieved by using the DRI process to instead the traditional BF process.

Table 4. Emission factors of separate iron and steel production processes (Source: [15]).

Steel Production Process	coke	BF	DRI	OHF	BOF	EAF
Emission factor (t CO_2e/t product)	0.56	1.35	0.70	0.37	0.11	0.08

5.2. Scenario Building

Five scenarios are supposed, business as usual (BAU) scenario and four with countermeasures. BAU scenario, with no countermeasures, assumes that the production of industrial process increase with linear growth, and emission factors keep the same as current emission factors. Scenarios with countermeasures assume that the productions of industrial sectors are the same as BAU, but the values of emission factors are reduced through various countermeasures. Scenarios for cement industry and iron and steel industry are supposed in Table 5. The cement type is a little different from that in Table 3. It is not the ratio of PC and blended cement, but the ratio of PC and MC. For instance, "70/30" means the blended cement with 70% of PC and 30% of MC additives. BAU scenario shows that the situation of China's current cement production mix is 63% Portland cement and 37% 50/50 blended cement. From scenario 1 to scenario 4, the ratio of Portland cement becomes less and less, and ratio of high MC content blended cement increases. The BAU scenario for raw steel production is based on the default composition of steel production in IPCC 2006. For scenario 1, 50% of the steel in Shenyang is produced by the iron ore-BF-BOF production process, 20% is produced by the iron ore-DRI-BOF production process, and 30% is made from recycled steel-EAF process. The ratio of steel produced by recycled steel-EAF production process increased, and ratio of steel produced by iron ore-DRI-BOF process decreases from scenario 1 to scenario 4.

Table 5. Scenarios for cement industry, and iron and steel industry.

Sector		BAU	Scenario 1	Scenario 2	Scenario 3	Scenario 4
Blended cement type	100/0	63%	55%	40%	20%	10%
	70/30	0	10%	20%	20%	10%
	50/50	37%	25%	20%	30%	40%
	30/70	0	10%	20%	30%	40%
Steel production type	Iron ore-BF-BOF	65%	50%	30%	10%	0%
	Iron ore-DRI-BOF	5%[a]	20%	30%	40%	30%
	Recycled steel-EAF	30%	30%	40%	50%	70%

[a] In BAU scenario, 5% does not means the ratio of iron ore-DRI-BOF (EAF) production type but that of iron ore-BF-OHF type. That is to say raw steel is produced using OHFs from pig iron produced by the BFs.

Emission factors for different cement type are calculated based on the clinker content. Portland cement is assumed to be 95% clinker. Values calculated for cement type are: % PC*95%*0.52. While for different steel production type, the emission factors are the sum of emission factors of corresponding separate processes in Table 4. The final results are demonstrated in Table 6.

Table 6. Emission factors of different cement and steel types.

Blended Cement Type (PC/MC)	Emission Factor (t CO_2e/t cement)	Steel Production Type	Emission Factor (t CO_2e/t steel)
100/0	0.49	Iron ore-BF-BOF	1.46
70/30	0.35	Iron ore-DRI-BOF	0.81
50/50	0.25	Recycled steel-EAF	0.08
30/70	0.15	Coke	0.56

5.3. Results and Discussions

The forecast of GHG emission of BAU scenario is shown in Table 7. Results show that GHG emission of industrial sectors such as ammonia, methanol, carbon black, pig iron, primary aluminum and lead disappeared by linear regression in 2020. Only six major industrial sectors exist, of which cement, clinker, raw steel, and coke are still the largest emission sectors, accounting for about 98% of the total GHG emissions. Cement, iron and steel industries are still the largest emission sources. However, the total GHG emissions under this forecast will increase obviously, reaching to about 8.28 Mt that is about two times of 4.06 Mt in year 2009.

GHG emissions of the four scenarios with countermeasures are shown in Table 8. It demonstrates that all the four scenarios have impact on reducing industrial process GHG emission compared with BAU scenario. The total GHG emissions of cement, iron and steel industries for scenario 1 to 4 are 85%, 74%, 60% and 49% of the BAU scenario, respectively. Scenario 4 is the best one, and its total GHG emissions can almost keep the same with that of 2009.Comparing cement, iron and steel industries, it is more sensitive for the iron and steel industry than for the cement industry to keep the GHG emissions. Take scenario 4 for an example, the GHG emission of cement and clinker is predicted to be 1.98 Mt and 1.70 Mt respectively, which are still larger than 1.63 Mt and

1.59 Mt in year 2009 (or are larger than half that of the BAU scenario). While for raw steel and coke production process, the GHG emissions are to be controlled at 0.23 Mt and 0.1 Mt respectively, which are smaller than 0.23 Mt and 0.51 Mt in year 2009 (or are far smaller than that of the BAU scenario). However, because of the big difference of base numbers, the absolute values for cement industry are bigger than that for the iron and steel industry.

Table 7. Estimated industrial production and GHG emissions in 2020.

Industrial Sectors	Production (t)	GHG Emission (t CO_2e)
Cement	8,173,954	3,351,321.21
Clinker	5,421,091	2,818,967.30
Glass	479,088	57,490.58
Refractory	353,443	70,688.68
Ammonia	/	/
Methanol	/	/
Graphite and carbon black	/	/
Coke	2,087,867	1,169,209.97
Pig iron	/	/
Raw steel	768,107	814,677.71
Primary aluminum	/	/
Lead	/	/
Total	/	8,282,355.45

Table 8. Estimated GHG emissions of different scenarios in 2020 (t CO_2e).

	Sector	BAU	Scenario 1	Scenario 2	Scenario 3	Scenario 4
	100/0		2,220,863.30	1,615,173.31	807,586.66	403,793.33
	70/30		282,818.81	565,637.62	565,637.62	282,818.81
Cement	50/50		504,741.66	403,793.33	605,689.99	807,586.66
industry	30/70		120,974.52	241,949.04	362,923.56	483,898.08
	Sum	3,351,321.21	3,129,398.29	2,826,553.29	2,341,837.82	1,978,096.87
	Clinker [a]	2,818,967.30	2,695,408.69	2,434,562.69	2,017,209.08	1,704,193.88
	Iron ore-BF-BOF		560,718.11	336,430.87	112,143.62	0.00
Iron and	Iron ore-DRI-BOF		124,433.33	186,650.00	248,866.67	186,650.00
steel	Recycled steel-EAF		18,434.57	24,579.42	30,724.28	43,013.99
industry	Sum	814,677.71	703,586.01	547,660.29	391,734.57	229,663.99
	Coke [b]	1,169,209.97	376,485.59	265,176.81	153,868.02	98,213.63
	Total	8,154,176.19	6,904,878.58	6,073,953.08	4,904,649.49	4,010,168.37

[a] The content of clinker in different types of cements is different. So consumption of clinker and emissions from clinker will change with different scenarios. GHG emission of clinker in different scenarios is calculated according to the ratio of different scenario to BAU scenario. [b] Coke are used only in the iron ore-BF-BOF production type, so for different steel production type, the consumption of coke varied accordingly. However, not all the coke obtained from the Statistic Yearbook is used for iron and steel sector. It can also be used for other uses. Here 85 percent of the coke production is assumed for iron and steel production and 15 percent is for other uses.

This study discussed GHG reduction measures from the aspect of technology. For cement industry, scenarios focus mainly on clinker substitute. It is one of the four technological measures summarized by the World Business Council for Sustainable Development [34], and shows larger potentials than the other three measures [35]. The efficiency of clinker substitute in GHG reduction is manifested in this paper. It is also useful for energy related emission reduction and can benefit relative industries that produce MIC wastes. Average ratio of PC/S is 85/15 in China, and large potential for blended cement can be expected [36]. However, MIC resources are in decline and most of the world production of suitable blast furnace slag and fly ash are already destined for use in the cement industry [37]. As for the iron and steel industry, DRI process is a promising technique for GHG reduction in industrial process. GHG emission was reduced not only during the process itself, but also in the coke production process for less coke needed. Besides, production type of recycled steel-EAF is effective in reducing the GHG emission, due to the high proportion of recycled scrap and the use of electrical energy [10]. Increasing the ratio of DRI and recycled iron and steel will be efficient measures in reducing GHG emission, only if corresponding technique is available [38].

For the rapidly developing regions, industrial CO_2 emission is affected by both industrial production changes and process technology [39]. Hence, there is more GHG reduction potential if industrial activity/production were considered. Such measures may include new additives that can reduce the amount of cement requirement, other low carbon building materials that can replace cement, adjustment measures that can control the population and construction, *etc.* However, industrial activity/production is not discussed in this study for two reasons. Firstly, it is complicate to consider the influencing factors and difficult to quantify the factors for building prediction model. Secondly, the final result of the measures is the reduction of industrial production, and this can be easily figured out based on the results obtained in this paper.

6. Conclusions

Industrial process GHG emissions increase quickly with fast urbanization and industrialization. Intensive study of the industrial process GHG emission should be investigated to respond to climate change. This paper simulates industrial process GHG emission and its reduction potentials by taking feasible technological measures, taking Shenyang as a case. Twelve industrial sectors that cause industrial process emissions are selected for study. Results indicate that industrial process GHG emission shows an increasing trend from year 2004 to 2009, although decreasing a little in 2008. The cement, iron the steel sectors are the main sources of industrial process GHG emission, occupying more than 90 percent.

Based on scenario analysis, one BAU and four with different countermeasures, the industrial process GHG emission of Shenyang in 2020 is predicted. The results show that if the industrial sectors go under current trend (BAU scenario), the total GHG emissions will reach 8.28 Mt in 2020, about two times of that of 2009. However, the four other scenarios with countermeasures prove to be effective in reducing industrial process GHG emission. And scenario 4 is the best one. It can almost keep the GHG emission at the same level as that of 2009. It can be demonstrated from scenario analysis that using clinker substitute for blended cement and increasing the ratio of DRI process and recycled steel will be efficient measures for Shenyang.

Acknowledgments

This work is supported by Natural Science Foundation of China (71325006, 41101126, 71033004 and 71303230), 100 Talents Program of the Chinese Academy of Sciences (2008-318), Ministry of Science and Technology of China (2011BAJ06B01; 2011DFA91810).

Author Contributions

Liu and Dong drafted the paper and contributed to data collection and calculation; Dong, Geng, Lu and Ren contributed to data analysis.

Conflicts of Interest

The authors declare no conflict of interest.

References

1. Guan, D.; Liu, Z.; Geng, Y.; Lindner, S.; Hubacek, K. The gigatonne gap in China's CO_2 inventories. *Nat. Clim. Chang.* **2012**, *2*, 672–675.
2. Xue, B.; Ren, W. China's uncertain CO_2 emissions. *Nat. Clim. Chang.* **2012**, *2*, Article 762.
3. Boden, T.; Andres, J.; Marland, G. *Global, Regional, and National Fossil-Fuel CO_2 Emissions*; Carbon Dioxide Information Analysis Center: Oak Ridge, TN, USA, 2013.
4. Ma, Z.; Xue, B.; Geng, Y.; Ren, W.; Fujita, T.; Zhang, Z.; de Oliveira, J.P.; Jacques, D.A.; Xi, F. Co-benefits analysis on climate change and environmental effects of wind-power: A case study from Xinjiang, China. *Renew. Energ.* **2013**, *57*, 35–42.
5. Geng, Y.; Zhao, H.; Liu, Z.; Xue, B.; Fujita, T.; Xi, F. Exploring driving factors of energy-related CO_2 emissions in Chinese provinces: A case of Liaoning. *Energy Policy* **2013**, *60*, 820–826.
6. Liaskas, K.; Mavrotas, G.; Mandaraka, M.; Diakoulaki, D. Decomposition of industrial CO_2 emissions: The case of European Union. *Energy Econ.* **2000**, *22*, 383–394.
7. Zhou, W.; Zhu, B.; Li, Q.; Ma, T.; Hu, S.; Griffy-Brown, C. CO_2 emissions and mitigation potential in China's ammonia industry. *Energy Policy* **2010**, *38*, 3701–3709.
8. Sheinbaum, C.; Ozawa, L.; Castillo, D. Using logarithmic mean divisia index to analyze changes in energy use and carbon dioxide emissions in Mexico's iron and steel industry. *Energy Econ.* **2010**, *32*, 1337–1344.
9. Kim, Y.; Worrell, E. International comparison of CO_2 emission trends in the iron and steel industry. *Energy Policy* **2002**, *30*, 827–838.
10. Kirschen, M.; Risonarta, V.; Pfeifer, H. Energy efficiency and the influence of gas burners to the energy related carbon dioxide emissions of electric arc furnaces in steel industry. *Energy* **2009**, *34*, 1065–1072.

11. Lee, C.; Lin, S. Structural decomposition of CO_2 emissions from Taiwan's petrochemical industries. *Energy Policy* **2001**, *29*, 237–244.

12. Lin, S.; Lu, I.; Lewis, C. Identifying key factors and strategies for reducing industrial CO_2 emissions from a non-Kyoto protocol member's (Taiwan) perspective. *Energy Policy* **2006**, *34*, 1499–1507.

13. Hendriks, C.; Worrell, E.; de Jager, D.; Blok, K.; Riemer, P. Emission reduction of greenhouse gases from the cement industry. In Proceedings of the Fourth International Conference on Greenhouse Gas Control Technologies, Interlaken, Switzerland, 1998; pp. 939–944.

14. Van Puyvelde, D. CCS opportunities in the Australian Industrial Processes sector. *Energ. Procedia* **2009**, *1*, 109–116.

15. Intergovernmental Panel on Climate Change (IPCC). *Guidelines for National Greenhouse Gas Inventories*; IPCC: Hayama, Japan, 2006.

16. Chen, H.M. Analysis on Embodied CO_2 Emissions Including Industrial Process Emission. *China Popul. Resour. Environ.* **2009**, *19*, 25–30.

17. Hu, X.; Ping, H.; Xie, C.; Hu, X. Globalization and China's iron and steel industry: Modeling China's demand for steel importation. *J. Chin. Econ. For. Trade Stud.* **2008**, *1*, 62–74.

18. IISI. *Steel Statistical Year Book*; International Iron and Steel Institute: Brussels, Belgium, 2004.

19. Cai, W.; Wang, C.; Wang, K.; Zhang, Y.; Chen, J. Scenario analysis on CO_2 emissions reduction potential in China's electricity sector. *Energ. Pol.* **2007**, *35*, 6445–6456.

20. MIIT. *Cement industry 12th Five-Year Development Plan*; Ministry of Industry and Information Technology of the People's Republic of China: Beijing, China, 2011.

21. Wang, Y.; Zhu, Q.; Geng, Y. Trajectory and driving factors for GHG emissions in the Chinese cement industry. *J. Clean. Product.* **2013**, *53*, 252–260.

22. Tian, Y.; Zhu, Q.; Geng, Y. An analysis of energy-related greenhouse gas emissions in the Chinese iron and steel industry. *Energy Policy* **2013**, *56*, 352–361.

23. Ren, W.; Xue, B.; Geng, Y.; Sun, L.; Ma, Z.; Zhang, Y.; Mitchell, B.; Zhang, L. Inventorying heavy metal pollution in redeveloped brownfield and its policy contribution: Case study from Tiexi district, China. *Land Use Policy* **2014**, *38*, 138–146.

24. Ren, W.; Geng, Y.; Xue, B.; Fujita, T.; Ma, Z.; Jiang, P. Pursuing co-benefits in China's old industrial base: A case of Shenyang. *Urban Clim.* **2012**, *1*, 55–64.

25. Department of Climate Change and Energy Efficiency (DCCEE). *Industrial Process Emissions Projections 2010*; DCCEE: Caberra, Australian, 2010.

26. *Revised 1996 IPCC Guidelines for National Greenhouse Gas Inventories*; Intergovernmental Panel on Climate Change: Brussels, Belgium, 1997.

27. Shiftan, Y.; Kaplan, S.; Hakkert, S. Scenario building as a tool for planning a sustainable transportation system. *Transp. Res. Part D* **2003**, *8*, 323–342.

28. Swart, R.; Raskin, P.; Robinson, J. The problem of the future: sustainability science and scenario analysis. *Global Environ. Chang.* **2007**, *14*, 137–146.

29. Wang, K.; Wang, C.; Lu, X.; Chen, J. Scenario analysis on CO_2 emissions reduction potential in China's iron and steel industry. *Energy Policy* **2007**, *35*, 2320–2335.

30. Baumert, K.; Herzog, T.; Pershing, J. *Navigating the Numbers: Greenhouse Gas Data and International Climate Policy*; The World Resources Institute: Washington, DC, USA, 2005.

31. Reijnders, L. The cement industry as a scavenger in industrial ecology and the management of hazardous substances. *J. Ind. Ecol.* **2008**, *11*, 15–25.

32. Hashimoto, S.; Fujita, T.; Geng, Y.; Nagasawa, E. Realizing CO_2 emission reduction through industrial symbiosis: A cement production case study for Kawasaki. *Resour. Conservat. Recycl.* **2010**, *54*, 704–710.

33. World Business Council for Sustainable Development (WBCSD). *CO_2 Accounting and Reporting Standard for the Cement Industry*; WBCSD: Geneva, Switzerland, 2005.

34. WBCSD/IEA. *Cement Technology Roadmap 2009: Carbon Emissions Reductions up to 2050*; International Energy Agency: Paris, France, 2009.

35. Lei, Y.; Zhang, Q.; Nielsen, C.; He, K. An inventory of primary air pollutants and CO_2 emissions from cement production in China, 1990–2020. *Atmos. Environ.* **2011**, *45*, 147–154.

36. Worrell, E.; Price, L.; Martin, N.; Hendriks, C.; Meida, L. Carbon dioxide emissions from the global cement industry energy environment. *Annu. Rev. Energy Environ.* **2001**, *26*, 303–329.

37. Tyrer, M.; Cheeseman, C.; Greaves, R.; Claisse, P.; Ganjian, E.; Kay, M.; Churchman-Davies, J. Potential for carbon dioxide reduction from cement industry through increased use of industrial pozzolans. *Adv. Appl. Ceram.* **2010**, *109*, 275–279.

38. Zeng, S.; Lan, Y.; Huang, J. Mitigation paths for Chinese iron and steel industry to tackle global climate change. *Int. J. Greenh. Gas Control.* **2009**, *3*, 675–682.

39. Akashi, O.; Hanaoka, T.; Matsuoka, Y.; Kainuma, M. A projection for global CO_2 emissions from the industrial sector through 2030 based on activity level and technology changes. *Energy* **2011**, *36*, 1855–1867.

Scenario-Based Analysis on Water Resources Implication of Coal Power in Western China

Jiahai Yuan, Qi Lei, Minpeng Xiong, Jingsheng Guo and Changhong Zhao

Abstract: Currently, 58% of coal-fired power generation capacity is located in eastern China, where the demand for electricity is strong. Serious air pollution in China, in eastern regions in particular, has compelled the Chinese government to impose a ban on the new construction of pulverized coal power plants in eastern regions. Meanwhile, rapid economic growth is thirsty for electric power supply. As a response, China planned to build large-scale coal power bases in six western provinces, including Inner Mongolia, Shanxi, Shaanxi, Xinjiang, Ningxia and Gansu. In this paper, the water resource implication of the coal power base planning is addressed. We find that, in a business-as-usual (BAU) scenario, water consumption for coal power generation in these six provinces will increase from 1130 million m^3 in 2012 to 2085 million m^3 in 2020, experiencing nearly a double growth. Such a surge will exert great pressure on water supply and lead to serious water crisis in these already water-starved regions. A strong implication is that the Chinese Government must add water resource constraint as a critical point in its overall sustainable development plan, in addition to energy supply and environment protection. An integrated energy-water resource plan with regionalized environmental carrying capacity as constraints should be developed to settle this puzzle. Several measures are proposed to cope with it, including downsizing coal power in western regions, raising the technical threshold of new coal power plants and implementing retrofitting to the inefficient cooling system, and reengineering the generation process to waterless or recycled means.

Reprinted from *Sustainability*. Cite as: Yuan, J.; Lei, Q.; Xiong, M.; Guo, J.; Zhao, C. Scenario-Based Analysis on Water Resources Implication of Coal Power in Western China. *Sustainability* **2014**, *6*, 7155-7180.

1. Introduction

The extensive development mode in China has led to rapid economic growth in a short time, but has been followed by a lot of environmental problems, especially air pollution, in recent years. Taking the increasingly serious air pollution into consideration, the Chinese government announced the ban on new energy-and-pollution intensive projects in eastern regions, among which coal-fired power plants are strictly banned [1].

Meanwhile, China's rapid economic growth has brought growing electricity demand. The national total electricity consumption has ballooned from 2481 trillion Watt hour (TWh) in 2006 to 4200 TWh by 2010, with an average annual growth of 11.1% during the 11th Five-year-plan (FYP) period (2006–2010). Then in 2012, it reached 4950 TWh [2]. In 2020, the total electricity consumption is expected to reach around 8000 TWh, with an average annual growth rate of 6.8% [3].

In order to meet the growing energy demand, five comprehensive energy bases located in Shanxi, Ordos, Xinjiang region, and eastern and southwestern Inner Mongolia will be built, according to the 12th FYP [4]. The 12th Energy Development plan also intended to develop 14 large coal mine bases and 16 coal power bases, and the total size of thermal power capacity is expected to reach more than 600 GW in these bases [5,6].

However, the water resource constraint in the coal power bases and the implication of the coal power bases on regional water resource is not fully considered in the planning. This paper is an attempt to address these two interweaved issues. The structure of the paper is as follows. Section 1 presents the background and purpose of the study. Section 2 will briefly review the water–energy nexus and provide background information on water resource and consumption in China's coal power base provinces. Section 3 will propose the model for analyzing water usage of thermal power generation. Section 4 will analyze the demand and impact of the coal power planning on water resource in China's coal power base provinces. Section 5 will provide analysis on water saving measures and an alternative scenario and Section 6 concludes the paper.

2. Literature Review and Research Background

2.1. Energy-Water Nexus

Globally, 80% of electricity generation comes from thermo-electric power stations (such as fossil fuels and nuclear), all of which require cooling for efficient and safe operation [7]. Most of the cooling system is provided by water abstractions from, and thermal discharges to, the natural environment, including rivers, tidal estuaries and coasts. The production of energy requires large quantities of water in processes such as thermal plant cooling systems or raw materials extraction. Analysis on whether water resources in a given region can support energy production will be a critical issue in the near future. The energy-water nexus has been a topic of increasing importance in recent years. In 2003 and 2006, numerous power plants in Europe had to be throttled in summer due to water shortages and high water temperatures caused by a hot and dry summer [8]. The agriculture sector currently has the highest water demand at the global scale, followed by the industry-energy sector that is responsible for 20% of the total water withdrawals [9]. In the U.S., the energy sector is expected to be the fastest growing water consuming sector, being responsible for 85% of the increase in domestic water consumption during 2005–2030 [10]. In some MENA (Middle Eastern and Northern Africa) countries, the interdependencies are already being manifested. For instance, in Saudi Arabia almost all of the natural gas currently produced is consumed domestically, primarily in the petrochemical industry and in seawater desalination [11]. Joint consideration of both water and energy domains can identify new options for increasing overall resource use efficiencies [12]. Gleick calculated the water consumption of different forms of energy [13]. Morgan Bazilian et al. (2011) considered the energy, water and food nexus, primarily from a developing country perspective [14]. Hagen Koch et al. described how to integrate the calculation of water demand of power plants into water resource management model [15]. Alexander et al. [16] studied the relationship between water and energy in an interactive lifecycle framework. Benjamin et al. [17] highlighted the most likely locations of severe water shortages in 22 counties because of thermoelectric capacity additions, and identified an assortment of technologies and policies that could respond to these electricity–water tradeoffs. Edward et al. [18] presented a model that quantifies current water use of the UK electricity sector, disaggregated by generation type, cooling method and cooling source. Also, it tested six decarbonisation pathways for the UK by combining projections of cooling methods and cooling sources for future thermoelectric generation. Aurelie et al. [19] proposed a model to assess optimal "water-energy" mix considering opportunities for water reuse and non-conventional water use in the water-scarce Middle East region. Kuishuang et al. [20] applied an integrated hybrid LCA approach to eight different electricity generation technologies in China to calculate their total life-cycle CO_2 emissions and water consumption throughout national supply chains.

China is a drought-hit country. Though total freshwater resources reach 2.8 trillion m^3, accounting for 6% of global water resources, ranking fourth in the world after Brazil, Russia and Canada, per capita water resource is just 2200 m^3, a quarter of the world average, or 1/5 of the United States level [21]. Energy and water have become major factors limiting sustainable development in China. Energy efficiency and optimization of water management are critical for the healthy growth of the Chinese economy [22]. In addition to energy shortages, China is also confronted with numerous water resource challenges, including shortage, pollution and aquatic environment deterioration.

It is noteworthy that the development of China's coal power base is uncoordinated with the distribution of water resources. The coal-rich regions often face water scarcity, while regions with plentiful water resources mostly face coal shortage [23]. According to the 12th Energy Development Plan [5], coal and energy supply in China is relying on the west coal-rich regions, but these areas are mostly located in the dry arid and semi-arid regions. That is to say, water resource conditions will inevitably become an important factor in the development of coal power bases. Because of limited freshwater resources, decision-making has to take the most valuable use of this limited resource into consideration. In water-stressed areas of the coal power bases, power plants will increasingly compete with other water users. Large-scale coal power bases in arid region of northwest will lead to severe water crisis, and the conflicts between coal resources and water are most significant in the following six provinces (autonomous regions), including Inner Mongolia, Shanxi, Shaanxi, Xinjiang, Ningxia and Gansu. To the best scope of our knowledge, only a book [24] explicitly addressed this issue, but it only raised a question and did not provide a solution to it. It is rightly the study scope of this paper.

2.2. Water Resources in China's Western Coal Power Base Provinces

According to the China Statistical Yearbook 2013, the data of water resources and water use of the main coal power bases is reported in Figure 1. In terms of total water resources, Xinjiang is the richest while Ningxia is the poorest region in these provinces. In particular, the water use in Ningxia is much more than its total water resources, indicating that water is scarce. Though the water resource in Xinjiang Uygur Autonomous Region seems rich, the water resources in Hami (coal power base) is merely 1.696 billion m^3, accounting for only 2.13% of its total water resource endowment [24].

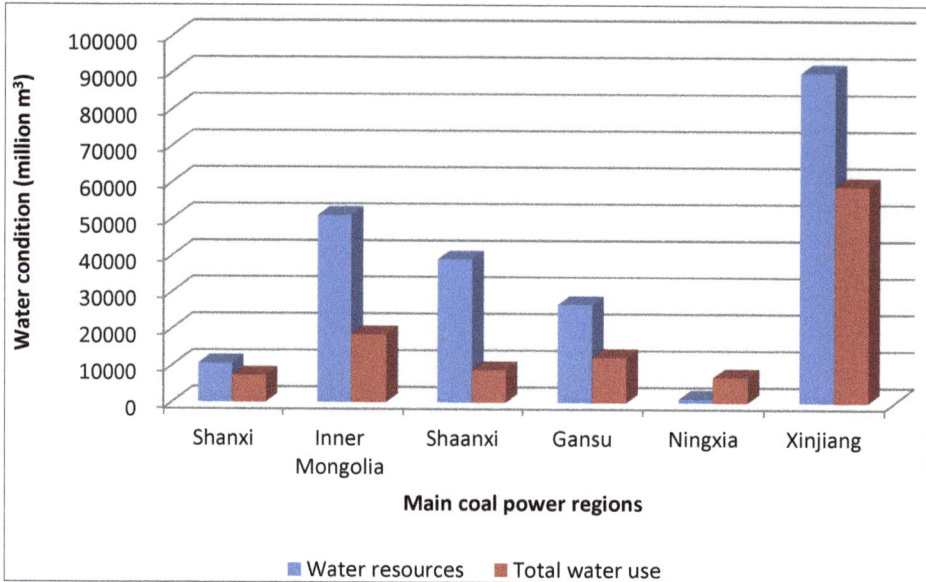

Figure 1. The water resources and water use in China's coal-power base provinces.

As is shown in Figure 1, the water resource in Inner Mongolia Autonomous Region is second to that of Xinjiang, but its coal power bases' water resources are not as rich. With the exception of Hulunbeir coal power base, other bases all have water shortage issues at varying degrees. For example, water resource in Jungar coal power base account for just 0.67% of Inner Mongolia water endowment [25]. For Xilingol League coal power base, the figure is 5.8%, for Ordos coal power base, 5.4% [26], and for Holingol coal power base, 0.09% [27]. The coal power bases located in the other four provinces all have water shortage issues. Especially, the condition of water shortage in Ningxia is the most serious, whose water resource ranked the last in this country.

2.3. Water Use in the Coal Power Base Provinces of Western China

The consumption of water resources can be divided into the following categories: agricultural water use, household water use, ecological water use and industrial water use. The electricity sector is one of the largest water consumers in China after agriculture [28]. Thus, the electricity sector can be a contributor to water scarcity which has already occurred in many parts of the country, in particular in Northern China [29]. The water use of the main coal power base provinces in 2012 is shown in Figure 2 [30].

From Figure 2, we can easily see that agricultural water use absolutely takes up the most part of whole water use, and industrial use is only second to agricultural use. Though total water use in Shanxi and Shaanxi is not as much as that in Xinjiang, their industrial water use has accounted for much larger share. Because of water shortage and agricultural water demand, Ningxia does not seem to have enough water to support industry development, including coal power bases. Considering total water resources and water use, the potential of developing coal power bases in Shanxi, Gansu and Ningxia does not look optimistic.

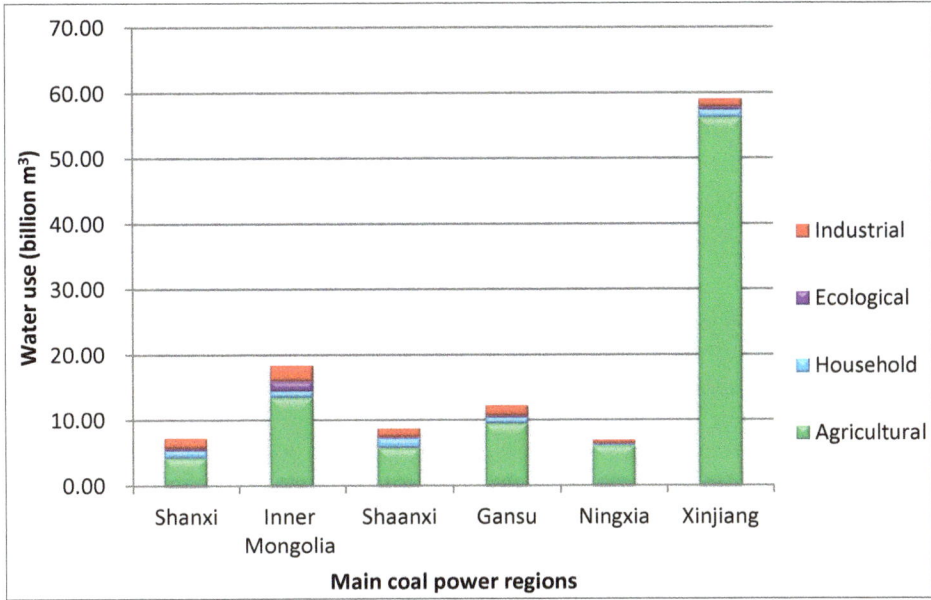

Figure 2. Water use of main coal power base provinces in 2012.

2.4. Originality and Novelty of the Study

The energy-water nexus has been a topic of increasing importance in recent years. Indeed, water is needed for the energy sector (fuel production and electricity generation), and energy is used to clean, desalinate and transport water. This relationship is referred to as the water-energy nexus, a concept originally formulated by Gleick in the 1990s with his seminar study [13]. Current studies and research mainly focus on water issues and energy consumption related to water management, and coupled water and energy challenges have been described for the energy production sector, *i.e.*, for the United States [31], Australia [32], Spain [33], the UK [18], Middle East [19], and China [22]. However, though few studies have indicated that China could face water shortages resulting from the addition of thermoelectric power plants, no precise estimate or policy study have been conducted. The goal of the present study is to reveal the potential water issues in China's coal power bases, and to provide a solution for it.

3. Methods and Data

According to the US Geological Survey's (USGS) water use survey data, Each kilowatt-hour (kWh) of thermoelectric generation requires the withdrawal of approximately 25 gallons of water (weighted-average for all thermoelectric power generation), which is primarily used for cooling purposes. Power plants also use water for operation of flue gas desulfurization (FGD) devices, ash handling, wastewater treatment, and wash water [34].

When discussing water and thermoelectric generation, it is necessary to distinguish between water withdrawal and water consumption. Water withdrawal represents the total water taken from a source but then returned to it, while water consumption represents the amount of water withdrawal that is not returned to the source. Water withdrawal is defined as the removal of water from any source or reservoir for human use. The water withdrawal constitutes the conveyance losses, consumptive use and return flow. Water consumption is defined as the amount of water extracted from a source that is no longer available for use, because it has evaporated, transpired, been incorporated into products and crops, consumed by man or livestock, ejected into the sea, or otherwise removed from freshwater resources [33]. The industry chain of large coal bases includes coal mining industry, thermal power industry and coal chemical industry. The focus of this paper is on the water consumption of thermal power industry and thus the terminology of "water consumption" is used in our analysis.

3.1. The Water Link of Thermal Power Plant

The composition of water consumption in thermal power plants are as follows: (a) water recharge of the cooling system; (b) ash and slag removal system; (c) boiler water feed system; (d) auxiliary cooling system; (e) desulfurization system; (f) water use of coal yard; and (g) domestic water use [35].

Thermal plants use fossil fuels to generate electricity or heat, and their water needs can vary drastically depending on the type of plant, type of refrigeration system, type of fuel and region [36]. The amount of water withdrawal and consumption depends on the type of technology used at a given plant. According to Meldrum *et al.* [37], water used for cooling purposes dominated the life cycle water use of electricity generation.

Large quantities of cooling water are required for thermoelectric power plants to support the generation of electricity. There are basically three types of cooling system designs: wet recirculating (open-loop and closed-loop) and dry recirculating (air-cooled). Open-loop (once-through) systems do not consume large quantities of water but require large quantities of water withdrawal in the cooling process. Closed-loop systems require smaller withdrawal of water, but most of the water withdrawal is lost in evaporation [13]. Air-cooled systems, also referred to as dry recirculating cooling systems, use either direct or indirect air-cooled steam condensers. In a direct air-cooled steam condenser, the turbine exhaust steam flows through air condenser tubes that are cooled directly by conductive heat transfer using a high flow rate of ambient air that is blown by fans across the outside surface of the tubes. Therefore, cooling water is not used in the direct air-cooled system. In an indirect air-cooled steam condenser system, a conventional water-cooled surface condenser is used to condense the steam, but an air-cooled closed heat exchanger is used to conductively transfer the heat from water to ambient air [38].

3.2. Water Consumption Rate in Coal Power Plants

China Electricity Council (CEC) is responsible for energy efficiency benchmarking in the power industry and releases detailed energy efficiency and water efficiency statistics [39]. By collating and analyzing the data, we calculated the water consumption and coal consumption indexes for power generation in the 600 MW units. Table 1 reports the water consumption rates while Table 2 reports the heat rates of the coal power units with different cooling system (Interested readers may refer to the Appendix for detailed information).

Table 1. Water consumption rates in China's 600 MW coal power units, 2012.

Types of Cooling System	Counted Units	Proportion (%)	Comprehensive Water Consumption Rate (m³/MWh)		
			Optimal Value	The Top 30% Average	Average
Closed-loop	128	39.26%	0.23	1.27	1.83
Open-loop	134	41.10%	0.02	0.18	0.29
Air-cooled	64	19.63%	0.18	0.23	0.31

Table 2. Heat rates in China's 600 MW coal power units, 2012.

Types of Cooling System	Counted Units	Heat Rates (gce/kWh)		
		Optimal Value	The Top 30% Average	Average
Closed-loop	128	285.00	296.41	307.80
Open-loop	134	275.85	290.28	303.89
Air-cooled	64	306.90	320.86	331.18

As shown in Table 2, on the average, air-cooled unit performs worst in energy efficiency, with about 25 grams coal equivalent gce/kWh more than water-cooled units; while open-loop water cooling system is with lowest heat rate. Combining the information reported in Tables 1 and 2, we can easily see that the open-loop system is the most efficient, because of the lowest heat rate and the lowest comprehensive water consumption rate. However, the pre-condition for open-loop water-cooled system is the existence of large body of fresh water like rivers or lakes adjacent to the power plant. Also, the direct pollution of freshwater also indicates the infeasibility of open-loop water-cooled system.

3.3. Method

According to the data on installed capacity and comprehensive water consumption rates of thermal power plants by [39], we can calculate the water consumption of thermal power plants in these coal power bases located in western provinces. The water consumption of power plants is calculated by:

$$Q = \sum_{i=1}^{3} MW_i \times H_i \times R_i \div 10^6 \tag{1}$$

where

Q	water consumption [10^6 m³/a]
MW	installed capacity of thermal power plants [MW]
H	Annual operation hours of the thermal power plants
R	comprehensive water consumption rate [m³/MWh]
i = 1	closed-loop cooling system
i = 2	open-loop cooling system
i = 3	air-cooled cooling system

The methodology for water consumption estimate and scenario study is presented in Figure 3. In its simplest form, water consumption is calculated by multiplying the electricity generated from a certain technology and the consumption factor for that technology, which is subject to various assumptions in our study. Because of the scope of our study, we limit our study in the above-mentioned six western coal power base provinces.

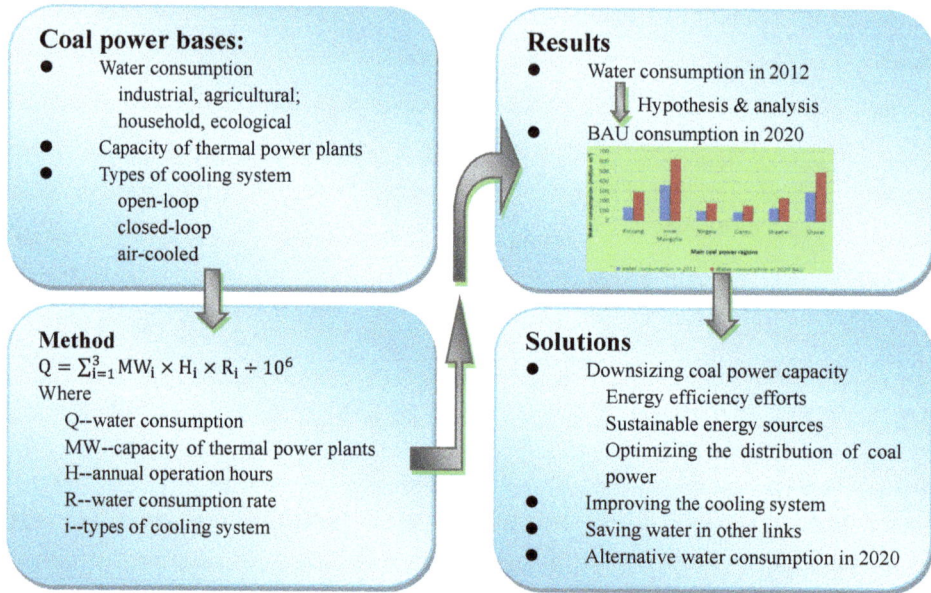

Coal power bases:
- Water consumption
  industrial, agricultural;
  household, ecological
- Capacity of thermal power plants
- Types of cooling system
  open-loop
  closed-loop
  air-cooled

Results
- Water consumption in 2012
  Hypothesis & analysis
- BAU consumption in 2020

Method
$$Q = \sum_{i=1}^{3} MW_i \times H_i \times R_i \div 10^6$$
Where
  Q--water consumption
  MW--capacity of thermal power plants
  H--annual operation hours
  R--water consumption rate
  i--types of cooling system

Solutions
- Downsizing coal power capacity
  Energy efficiency efforts
  Sustainable energy sources
  Optimizing the distribution of coal
  power
- Improving the cooling system
- Saving water in other links
- Alternative water consumption in 2020

Figure 3. Methodology framework of our study.

4. Results

4.1. Current Water Consumption Status

Annual Development Report on China's power industry 2013 by CEC has counted the whole-diameter capacity in every province by the end of 2012 [2]. According to this report, by the end of 2012, the installed coal power capacity is 50,110 MW in Shanxi, 60,190 MW in Inner Mongolia, 22,270 MW in Shaanxi, 22,570 MW in Xinjiang; 16,400 MW in Ningxia and 15,510 MW in Gansu.

According to the 2012 basic information and scoring statistics on nationwide 600 MW thermal power units announced by CEC in 2013, we can estimate the distribution of cooling systems in these coal power provinces. There are approximately 75% of units adopting air-cooled system and other 25% of units adopting closed-loop cooling system, as shown in Table 3. Open-loop system, constrained by resource endowment in these western regions, is rarely deployed. According to Equation (1), we can estimate the water consumption by thermal power plants in these provinces, and the results are shown in Table 4.

Table 3. Statistics on different cooling systems in China's coal power base provinces.

Region	Count of Different Cooling System		
	Open-loop	Closed-loop	Air-cooled
Shanxi	0	4	18
Inner Mongolia	2	12	20
Shaanxi	0	2	13
Gansu	0	0	2
Ningxia	0	0	8
Total	2	18	61

Table 4. Water consumption by thermal power plants in coal power base provinces, 2012.

Province	Coal power Capacity/ MW	Types of Cooling System	Water Consumption Rate (m³/MWh)	Installed Capacity /MW	Water Consumption (10⁶ m³/a)	Total Water Consumption (10⁶ m³/a)
Xinjiang	22,570	Closed-loop	1.83	5640	90.414	136.389
		Air-cooled	0.31	16,930	45.975	
Inner Mongolia	60,190	Closed-loop	1.83	15,050	241.264	363.846
		Air-cooled	0.31	45,140	122.582	
Ningxia	16,400	Closed-loop	1.83	4100	65.726	99.128
		Air-cooled	0.31	12,300	33.402	
Gansu	15,510	Closed-loop	1.83	3870	62.039	93.649
		Air-cooled	0.31	11,640	31.610	
Shaanxi	22,270	Closed-loop	1.83	5570	89.292	134.642
		Air-cooled	0.31	16,700	45.351	
Shanxi	50,110	Closed-loop	1.83	12,530	200.866	302.918
		Air-cooled	0.31	37,580	102.052	
Aggregate	187,050				1130.572	1130.572

Then according to the statistical data on industrial water consumption, we can estimate the proportion of thermal power generation in the overall industrial water use in these provinces. The results are shown in Figure 4.

It is found that in both Ningxia and Shanxi provinces, coal power accounts for as high as 20% of industrial water consumption. For Inner Mongolia, coal power consumes 15% of the industrial water use. For Xinjiang and Shaanxi provinces, the percent is 10%. Only in Gansu province the percentage is less than 10%. Because of the scarcity of water resource in these provinces, we assume that further increase in industrial water supply is impossible or is much likely to bring conflicts between industrial water use and other purposes (ecological use or household use, for example) [18]. Therefore, we assume that further big increase in coal power/industrial water use percent is definitely unacceptable in the perspective of sustainable development in these provinces.

48

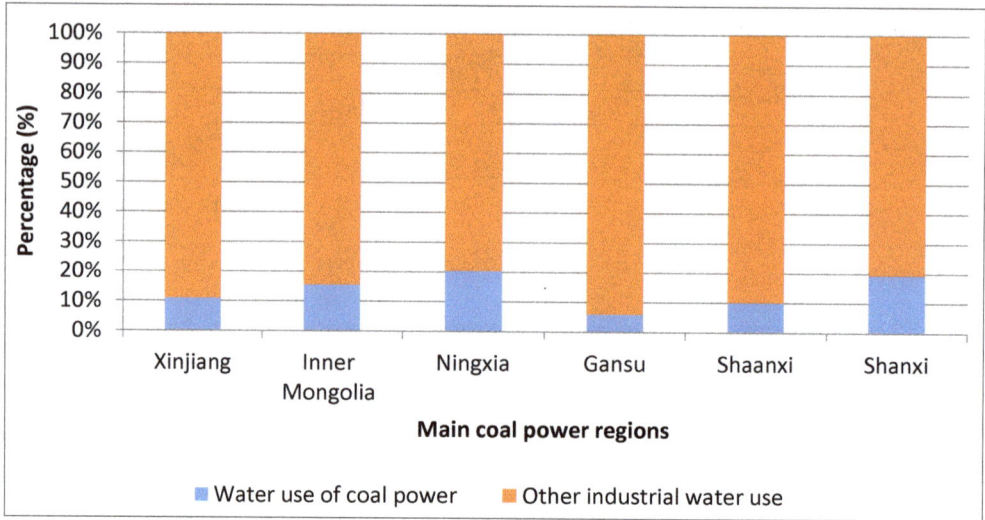

Figure 4. Coal power *vs.* industrial water use in the coal power base provinces, 2012.

4.2. BAU Water Consumption Scenario in 2020

According to the "Power Development Strategic Research Report 2013" by CEC, the capacity of China's coal power plants in 2020 is expected to reach 1100 GW, and coal power bases will account for 55% of total newly-constructed plants. The BAU power demand and power planning is shown in Table 5.

Table 5. BAU (business-as-usual) power demand and power planning.

	Capacity (GW)	Electricity Production (TWh)
Hydropower	360	1260
Pumped storage	60	48
Coal power	1050	5250
Gas power	100	300
Nuclear power	58	406
Wind power	200	400
Solar power	70	112
Biomass energy	15	67.5
Total	1913	7795.5
Balancing loss	-	90.5
Electricity demand	-	7705

Assume constant water consumption rates, we can estimate the business-as-usual (BAU) scenario of water consumption by coal power plants in these coal power base provinces in 2020 (Table 6). According to the BAU scenario, it is estimated that water consumption for coal power generation will double during 2012–2020, provided that the comprehensive water consumption rates and the composition of cooling system are unchanged (Figure 5). It is worthwhile pointing out that all these coal power bases are puzzled by water shortage, and a double growth in water demand will certainly break the balance of water ecology system in these regions.

Table 6. BAU Water consumption by thermal power plants in coal power base provinces, 2020.

Province	Total Capacity /MW	Types of Cooling System	Water Consumption Rate (m³/MWh)	Installed Capacity /MW	Water Consumption (10⁶ m³/a)	Total water Consumption (10⁶ m³/a)
Xinjiang	47,970	Closed-loop	1.83	11,990	192.209	289.917
		Air-cooled	0.31	35,980	97.707	
Inner Mongolia	102,900	Closed-loop	1.83	25,730	412.472	622.035
		Air-cooled	0.31	77,170	209.563	
Ningxia	29,670	Closed-loop	1.83	7420	118.949	179.371
		Air-cooled	0.31	22,250	60.422	
Gansu	26,300	Closed-loop	1.83	6580	105.483	159.034
		Air-cooled	0.31	19,720	53.552	
Shaanxi	39,520	Closed-loop	1.83	9880	158.384	238.875
		Air-cooled	0.31	29,640	80.490	
Shanxi	82,950	Closed-loop	1.83	20,740	332.479	501.416
		Air-cooled	0.31	62,210	168.937	
Total	329,310			329,310	1990.648	1990.648

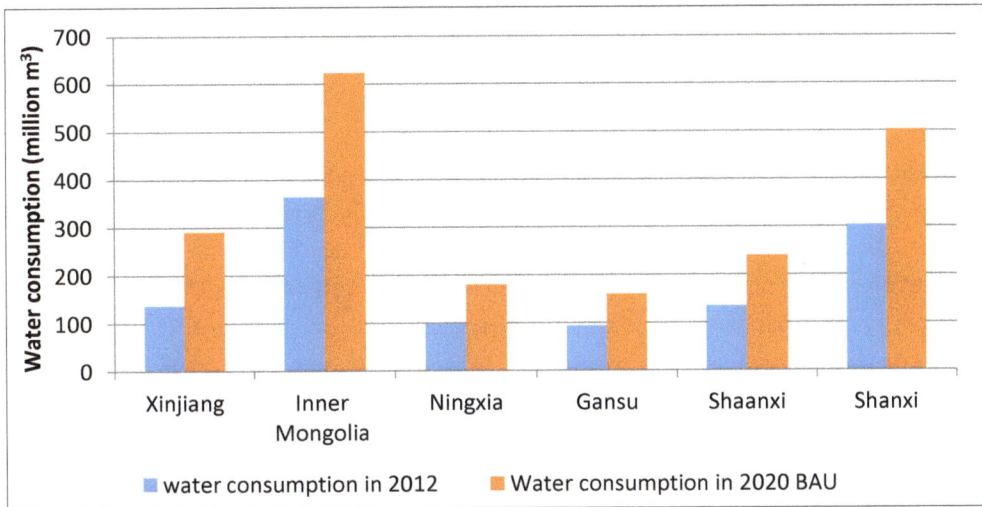

Figure 5. 2020 BAU *vs.* 2012 water consumption by coal power in coal power base provinces.

Furthermore, current water consumption by the power generation is substantial in volume and critical to its operation, but pressures of population growth, hydrological variability and climate change will complicate the issue. Water consumption in these provinces is always constrained by available supply and consumption in other industries is relative stable in the past. Hence, we hypothesize that water consumption in other industrial sectors will hold constant during 2012–2020, and then we can update the percentage of coal power/industrial water use for these provinces in 2020 (Figure 6). According to the estimate, with the exception of Gansu province, in the other five provinces, the percentages of coal power/industrial water use will be largely enlarged. In Shanxi and Ningxia provinces, coal power is expected to consume 30% of industrial water use in the BAU scenario; while in Xinjiang, Inner Mongolia and Shaanxi, coal power is expected to consume around 20% of industrial water use. In other words, the BAU scenario envisions higher possibility of water resource conflicts in these provinces.

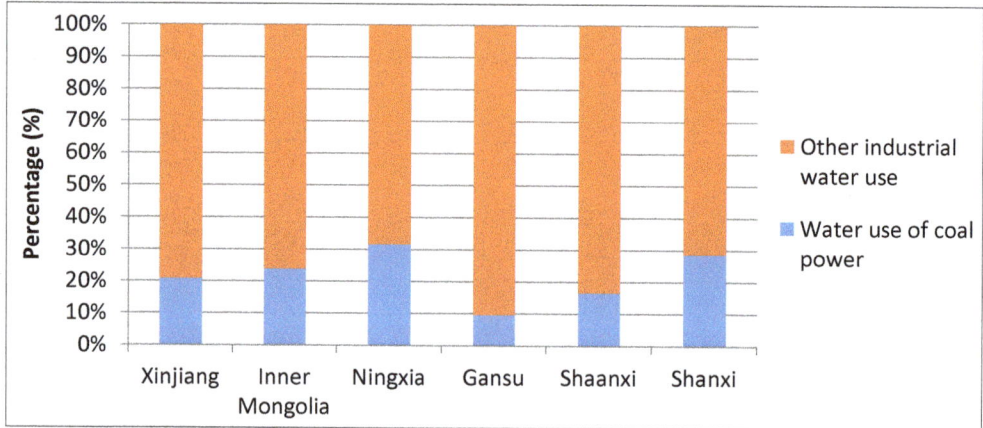

Figure 6. Percentage of coal power/industrial water use in the BUA scenario.

5. Water Conservation Measures and Alternative Scenario

Our projected result in these provinces has clearly indicated that serious water crisis will be likely to erupt in the BAU scenario. A strong implication is that the Chinese Government must take water resource constraint into consideration as a critical point in its overall sustainable development plan, in addition to energy supply and environment protection. The point is that an integrated energy-water resource plan with regionalized environmental carrying capacity as constraints is a desideratum to settle this puzzle. Hence, in this section, firstly possible measures to cut down (or avoid) water consumption will be discussed. Then, an alternative scenario will be proposed to guide the sustainable energy development in these coal power base provinces.

5.1. Water Conservation Measures

5.1.1. Downsizing Coal Power Capacity

The most effective measure to cope with water resource crisis caused by power generation is cutting down the scale of coal power capacity in western provinces. In other words, an alternative power plan is needed. The following strategies can be employed in developing the power sector plan:

- Reducing electric power demand by active energy efficiency efforts. Energy efficiency has been regarded as the fifth energy source besides coal, oil, gas and hydropower. Worldwide experience has also clearly demonstrated the efficacy of energy efficiency in optimizing energy system. We estimate that with active energy efficiency efforts, the demand growth for electric power can be slowed down by 2%–3% and results in energy conservation at 200–300 TWh annually. The energy efficiency potential into 2020 by various active efforts is shown in Table 7.

Table 7. The energy efficiency potential of active efforts into 2020.

Technology/Sector	Energy Efficiency Potential (TWh)
Green lighting	108.51
High efficiency motor	130.00
Energy-saving transformer	8.87
Frequency converter	97.70
High efficiency appliance	442.00
Ground source heat pump technology	18.00
Total	805.08

- Optimizing generation mix by deploying clean and sustainable energy sources. On one hand, in China the hydropower development is yet to reach its economically developable resource limit. On the other hand, China is endowed with abundant renewable energy sources like wind power and solar power, and active support of renewable energy development can make a big difference.

- Optimizing the regional distribution of coal power. With the development of zero emission technology, coal power can generate electricity at emission levels (SO_2, NOx and dust) closer to or even lower than gas power [40]. Hence, the government can lift the ban on the new construction of coal power plants in east regions and plan to build some coal power plants with cutting-edge pollutant control technologies in the load centers. The benefits are twofold. First, China's potential water resource crisis in western can be partly relieved. Second, the demand on long-distance electric power transmission—another important concern of decision-makers—can also be reduced.

5.1.2. Saving Water by Improving the Cooling System

The cooling system consumes most of water in the power generation process. Therefore, another priority of water conservation is to improve the cooling system in coal power plants. In the existing coal power fleet, these plants with advanced closed-loop water cooling system perform as effectively as those with air-cooled system in terms of the comprehensive water consumption rate while enjoying super energy efficiency advantage. Here, an important measure is to raise the technical threshold of new coal power plants and require that all the newly-constructed coal power plants perform better in their water consumption than the existing records. For existing plants, technical retrofitting can be implemented in those plants with below than average water efficiency performance.

5.1.3. Saving Water in Other Links

The demand on water quality in the other links of power plants is not as strict as in the cooling system. For example, treated wastewater, instead of fresh water, can be used in the desulfurization system and the coal yard. Use of nontraditional water sources, such as secondary-treated municipal wastewater, provides an option to reduce freshwater usage in thermal power production [41]. In the subsequent subsection, concrete water conservation measures are proposed from many links in the generation process.

- Ash and slag removal system: Dry-type ash and slag removal system, which consumes no water at all, is a mature option. On the other hand, with this process, the produced ash and slag can be reclaimed as building material [41].
- Cooling system of auxiliary systems: The cooling water for the auxiliary systems can be processed in a centralized way. Similarly, a small-scale wet recirculating or dry recirculating cooling system can be employed for the cooling of the pooled water.
- Desulfurization system: Dry desulfurization process consumes much less water than wet process. For the wet desulfurization process, the cyclic utilization of processing water is a feasible option. Other option is to use the reclaimed water in the cooling system for the auxiliary systems in desulfurization process.
- Water use of coal yard: The water quality demand in the coal yard is much lower. Retreated industrial wastewater or domestic wastewater can be collected for this purpose. Besides, the wastewater produced in coal yard can also be reutilized to further cut down water consumption.

5.2. An Alternative Water Consumption Scenario

In the alternative water consumption scenario, with energy efficiency efforts, electricity demand is projected to drop from 7705.5 TWh to 7560 TWh. In addition, with more clean and sustainable energy sources, the demand for thermal power is reduced by almost 100 GW, as shown in Table 8.

Table 8. Alternative power demand and power planning.

	Capacity (GW)	**Power Production (TWh)**
Hydropower	360	1260
Pumped storage	70	56
Coal power	959.91	4799.6
Gas power-centralized	70	350
Gas power-distributed	50	125
Nuclear power	60	420
Wind power	230	460
Solar power-centralized	40	64
Solar power-distributed	60	72
Biomass energy	14	63
Total	1913.91	7594.9
Balancing loss	-	34.9
Electricity demand	-	7560

Due to technological advances and structural optimization, the comprehensive water consumption rate could be cut down by a large extent in the alternative scenario. Our estimate is that the water consumption rate could be cut down by 30% from the existing level. In addition, we project that more air-cooled units would be built. The details of the alternative scenario are reported in Table 9.

Table 9. Alternative water consumption in coal power base provinces.

Province	Total Capacity /MW	Types of Cooling System	Water Consumption Rate (m³/MWh)	Installed Capacity /MW	Water Consumption (10⁶ m³/a)	Total Water Consumption (10⁶ m³/a)
Xinjiang	45,050	Closed-loop	1.27	9010	100.238	172.851
		Air-cooled	0.23	36,040	72.613	
Inner Mongolia	100,000	Closed-loop	1.27	20,000	222.504	383.688
		Air-cooled	0.23	80,000	161.184	
Ningxia	27,580	Closed-loop	1.27	5520	61.411	105.858
		Air-cooled	0.23	22,060	44.446	
Gansu	24,390	Closed-loop	1.27	4880	54.291	93.600
		Air-cooled	0.23	19,510	39.309	
Shaanxi	38,140	Closed-loop	1.27	7630	84.885	146.357
		Air-cooled	0.23	30,510	61.472	
Shanxi	80,950	Closed-loop	1.27	16,190	180.117	310.595
		Air-cooled	0.23	64,760	130.478	
Aggregate	316,110			316,110	1212.949	1212.949

In the alternative scenario, planned coal-fired generation capacity in these provinces would be 13 GW lower than in the BAU scenario. Also, the comprehensive water consumption rate of the closed-loop cooling system would be lowered to 1.27 m³/MWh. The water consumption by coal power in these provinces will be 1212 million m³, with only a slight increase relative to 2012 level. In this way, the envisioned water crisis caused by the electric power industry in these water-deprived provinces could possibly be avoided. However, we should not be too optimistic about it. Great risk exists in the implementation of water conservation measures and their actual effects.

6. Concluding Remarks

Coal power is an inevitable choice for meeting the increasing electric power demand in China. Considering the ever worsening atmospheric pollution, the Chinese Government has to adhere to the path of developing large-scale coal power bases in western provinces. However, the analysis presented in this paper clearly indicates that there will be furious water-energy conflicts in the development of coal bases, and water resource constraints will seriously restrict their development. An integrated planning of energy and water resources which takes regionalized environmental carrying capacity as the constraint is a final resort to this sustainable development puzzle. Several concrete water conservation measures are proposed to address the water crisis in China's coal power bases.

54

Certainly, our study is suffering from some limitations. For example, the perspective employed in the study is water resource constraint. Other important factors, such as the deployment cost of new cooling system and the retrofitting cost of existing cooling system, the economic appraisal of water conservation in links other than cooling system, are not addressed in the study. Water issue is only projected to worsen in the future, as a consequence of climate change. Therefore, further and more in-depth analysis on the issues presented here is needed to provide guidance for Chinese government and other stakeholders.

Acknowledgments

The authors would like to thank the anonymous reviewers for their useful comments and suggestions. The work reported in the paper is funded by Beijing Higher Education Young Elite Teacher Project (YETP0707) and the Fundamental Research Funds for the Central Universities.

Author Contributions

Jiahai Yuan contributed to the research idea and the framework of this study. Other authors contributed equally to the study.

Appendix

Table A1. Energy efficiency and water efficiency statistics on coal-fired power plants in China, 2012.

Plant No.	Capacity (MW)	Comprehensive Water Consumption Rate (m³/MWh)	Types of Cooling System	Heat Rates (gce/kWh)
1	640	1.91	Closed-loop	298.29
2	640	2.10	Closed-loop	297.71
3	1000	0.30	Open-loop	275.85
4	630	1.31	Closed-loop	296.2
5	640	1.98	Closed-loop	297.85
6	630	0.35	Open-loop	296.77
7	640	1.99	Closed-loop	298.36
8	600	0.23	Open-loop	306.28
9	630	1.31	Closed-loop	297.33
10	640	1.92	Closed-loop	298.15
11	1000	0.30	Open-loop	276.44
12	630	2.20	Closed-loop	296.78
13	600	0.23	Open-loop	308.21
14	500	1.83	Closed-loop	306.05
15	600	2.00	Closed-loop	307.3
16	660	0.25	Air-cooled	310.21
17	600	0.29	Open-loop	298.97
18	500	1.60	Closed-loop	324.06
19	700	2.10	Closed-loop	307.61

Table A1. *Cont.*

Plant No.	Capacity (MW)	Comprehensive Water Consumption Rate (m³/MWh)	Types of Cooling System	Heat Rates (gce/kWh)
20	600	2.00	Closed-loop	309.35
21	600	0.24	Air-cooled	326.41
22	600	1.72	Closed-loop	302.35
23	600	2.00	Closed-loop	298.51
24	600	1.70	Closed-loop	302.12
25	600	0.24	Air-cooled	326.7
26	600	1.57	Closed-loop	310.03
27	1000	0.23	Closed-loop	287.18
28	1000	0.10	Open-loop	282.
29	600	0.23	Air-cooled	330.13
30	600	0.47	Air-cooled	318.14
31	600	0.25	Open-loop	310.5
32	600	0.23	Air-cooled	330.05
33	900	0.43	Open-loop	299.82
34	500	1.83	Closed-loop	306.28
35	630	1.47	Closed-loop	300.58
36	630	0.35	Open-loop	298.53
37	600	0.24	Air-cooled	327.53
38	630	0.35	Open-loop	300.3
39	600	0.33	Air-cooled	329.62
40	660	0.32	Open-loop	288.2
41	660	1.73	Closed-loop	298.01
42	630	1.99	Closed-loop	305.3
43	700	2.10	Closed-loop	311.39
44	600	2.83	Closed-loop	310.97
45	660	0.19	Air-cooled	323.55
46	660	0.43	Open-loop	291.51
47	670	2.07	Closed-loop	301.11
48	640	1.88	Closed-loop	305.12
49	670	2.07	Closed-loop	302.07
50	600	0.26	Air-cooled	328.14
51	680	0.19	Open-loop	298.6
52	600	0.26	Air-cooled	327.11
53	660	1.31	Closed-loop	291.34
54	600	1.90	Closed-loop	307.57
55	600	1.87	Closed-loop	302.45
56	680	0.11	Open-loop	289.98
57	600	0.24	Air-cooled	331.9
58	630	1.85	Closed-loop	301.08
59	1000	2.01	Closed-loop	285.71
60	600	2.10	Closed-loop	315.48
61	1000	2.01	Closed-loop	286.11

56

Table A1. *Cont.*

Plant No.	Capacity (MW)	Comprehensive Water Consumption Rate (m³/MWh)	Types of Cooling System	Heat Rates (gce/kWh)
62	660	0.41	Open-loop	311.3
63	600	2.03	Closed-loop	301.49
64	630	0.42	Air-cooled	323.42
65	600	0.24	Open-loop	311.66
66	630	0.42	Air-cooled	323.39
67	600	2.81	Closed-loop	319.58
68	600	1.85	Closed-loop	300.86
69	1000	0.26	Open-loop	285.58
70	630	2.26	Closed-loop	304.13
71	600	0.34	Open-loop	300.71
72	660	0.43	Open-loop	291.51
73	600	0.29	Open-loop	312.11
74	600	0.24	Open-loop	299.99
75	600	0.20	Open-loop	310.03
76	900	0.43	Open-loop	301.01
77	1000	0.27	Open-loop	285.55
78	600	0.24	Air-cooled	330.73
79	650	0.33	Closed-loop	303.07
80	660	0.17	Open-loop	294.85
81	600	2.46	Closed-loop	304.26
82	630	1.99	Closed-loop	306.8
83	600	0.41	Air-cooled	330.38
84	600	2.83	Closed-loop	311.91
85	600	0.29	Open-loop	313.48
86	600	0.24	Open-loop	301.48
87	600	0.24	Air-cooled	331.15
88	600	2.46	Closed-loop	304.87
89	600	1.78	Closed-loop	302.49
90	600	1.57	Closed-loop	314.35
91	660	1.62	Closed-loop	297.16
92	600	1.85	Closed-loop	308.1
93	1000	1.65	Closed-loop	288.77
94	630	2.20	Closed-loop	301.75
95	600	1.75	Closed-loop	315.73
96	660	0.19	Air-cooled	323.41
97	630	0.41	Open-loop	302.36
98	600	0.49	Air-cooled	320.25
99	650	0.28	Open-loop	302.93
100	600	0.24	Open-loop	300.39
101	600	1.76	Closed-loop	306.17
102	600	0.20	Open-loop	313.97
103	600	2.01	Closed-loop	305.2

Table A1. *Cont.*

Plant No.	Capacity (MW)	Comprehensive Water Consumption Rate (m³/MWh)	Types of Cooling System	Heat Rates (gce/kWh)
104	600	1.95	Closed-loop	320.04
105	1000	0.31	Air-cooled	306.9
106	600	1.57	Closed-loop	314.62
107	600	0.29	Open-loop	302.52
108	630	1.97	Closed-loop	303.66
109	600	2.90	Closed-loop	319.42
110	600	1.57	Closed-loop	314.97
111	1000	0.32	Open-loop	283.93
112	660	0.41	Open-loop	315.05
113	1000	0.81	Closed-loop	285.
114	600	1.75	Closed-loop	316.69
115	600	0.19	Open-loop	313.92
116	600	1.76	Closed-loop	306.39
117	600	0.24	Open-loop	301.66
118	600	2.05	Closed-loop	318.
119	660	0.32	Open-loop	293.12
120	640	2.10	Closed-loop	302.07
121	600	2.03	Closed-loop	303.19
122	660	0.36	Air-cooled	314.82
123	600	1.76	Closed-loop	305.96
124	1000	0.19	Open-loop	289.61
125	600	1.83	Closed-loop	306.03
126	660	0.26	Open-loop	298.98
127	660	0.31	Open-loop	292.89
128	600	0.36	Open-loop	305.13
129	1000	0.23	Closed-loop	291.46
130	600	0.41	Air-cooled	332.7
131	600	2.46	Closed-loop	305.95
132	600	0.27	Air-cooled	331.91
133	630	0.35	Open-loop	303.38
134	600	0.24	Open-loop	304.19
135	600	1.75	Closed-loop	319.14
136	630	0.14	Open-loop	306.29
137	630	1.67	Closed-loop	305.48
138	600	0.23	Air-cooled	337.29
139	600	0.34	Open-loop	303.17
140	600	2.46	Closed-loop	305.71
141	680	0.19	Open-loop	303.58
142	600	0.24	Open-loop	311.64
143	660	0.18	Open-loop	298.47
144	600	0.41	Air-cooled	333.87

58

Table A1. Cont.

Plant No.	Capacity (MW)	Comprehensive Water Consumption Rate (m³/MWh)	Types of Cooling System	Heat Rates (gce/kWh)
145	1000	0.16	Open-loop	286.6
146	680	0.11	Open-loop	294.31
147	600	2.00	Closed-loop	318.28
148	600	1.80	Closed-loop	307.34
149	600	0.29	Closed-loop	313.94
150	650	0.33	Closed-loop	306.63
151	600	0.24	Open-loop	304.92
152	600	2.16	Closed-loop	311.51
153	600	1.83	Closed-loop	305.74
154	600	0.29	Open-loop	303.69
155	1000	0.29	Closed-loop	290.67
157	650	0.30	Open-loop	305.25
158	660	1.03	Closed-loop	324.
159	660	0.28	Open-loop	292.09
160	660	1.73	Closed-loop	302.49
161	1000	0.16	Open-loop	288.99
162	1000	1.65	Closed-loop	289.5
163	1000	0.28	Open-loop	289.07
164	600	1.60	Closed-loop	309.03
165	600	2.03	Closed-loop	310.84
166	660	2.05	Closed-loop	309.78
167	630	0.34	Open-loop	308.04
168	600	1.84	Closed-loop	307.46
169	600	0.29	Open-loop	303.06
170	600	1.76	Closed-loop	306.29
171	660	0.24	Open-loop	299.56
172	1000	0.28	Open-loop	285.09
173	600	2.49	Closed-loop	309.85
174	660	0.24	Open-loop	296.98
175	600	0.39	Air-cooled	338.06
176	660	0.02	Open-loop	294.52
177	600	0.41	Air-cooled	334.8
178	1000	0.28	Open-loop	289.2
179	1000	0.89	Closed-loop	285.05
180	660	2.05	Closed-loop	309.78
181	630	0.34	Open-loop	316.79
182	1000	0.31	Air-cooled	308.01
183	1000	0.28	Open-loop	289.02
184	600	0.36	Open-loop	309.38
185	660	0.31	Air-cooled	320.86
186	660	0.18	Open-loop	299.94
187	600	2.00	Closed-loop	306.58

Table A1. *Cont.*

Plant No.	Capacity (MW)	Comprehensive Water Consumption Rate (m³/MWh)	Types of Cooling System	Heat Rates (gce/kWh)
188	630	0.43	Open-loop	312.91
189	1000	0.40	Open-loop	289.51
190	630	0.39	Open-loop	309.73
191	500	1.60	Closed-loop	327.8
192	600	0.23	Air-cooled	335.44
193	600	0.24	Open-loop	304.24
194	1000	0.19	Open-loop	291.34
195	660	0.02	Open-loop	295.42
196	600	0.31	Air-cooled	337.53
197	600	0.27	Open-loop	309.85
198	600	2.13	Closed-loop	321.41
199	660	0.18	Open-loop	301.81
200	1000	0.39	Open-loop	292.2
201	660	1.03	Closed-loop	324.7
202	600	1.75	Closed-loop	309.03
204	660	1.65	Closed-loop	300.1
205	600	2.13	Closed-loop	321.41
206	600	1.84	Closed-loop	308.99
207	660	0.17	Open-loop	306.56
208	630	0.34	Open-loop	308.87
209	600	2.20	Closed-loop	318.28
210	1000	0.40	Open-loop	288.15
211	660	0.41	Open-loop	314.32
212	1000	0.21	Open-loop	284.56
213	600	1.75	Closed-loop	320.92
214	600	0.40	Air-cooled	330.38
215	660	0.28	Open-loop	292.8
216	660	0.31	Air-cooled	321.15
217	650	0.32	Open-loop	307.82
218	600	1.71	Closed-loop	320.74
219	600	0.33	Air-cooled	338.41
220	600	1.78	Closed-loop	307.04
221	660	1.74	Closed-loop	305.11
222	1000	0.32	Open-loop	293.25
223	600	0.39	Air-cooled	338.4
224	680	0.39	Open-loop	302.96
225	660	0.28	Open-loop	302.73
226	660	0.36	Air-cooled	328.9
227	660	0.35	Air-cooled	333.58
228	600	0.18	Open-loop	297.56
229	660	0.31	Air-cooled	325.03
230	600	2.07	Closed-loop	320.33

Table A1. *Cont.*

Plant No.	Capacity (MW)	Comprehensive Water Consumption Rate (m³/MWh)	Types of Cooling System	Heat Rates (gce/kWh)
231	630	2.26	Closed-loop	308.05
232	600	2.07	Closed-loop	313.43
233	660	0.31	Air-cooled	328.35
234	630	0.63	Open-loop	306.7
235	660	0.34	Air-cooled	329.91
236	660	1.30	Closed-loop	291.34
237	660	1.71	Closed-loop	305.72
238	630	0.14	Open-loop	306.36
239	600	1.96	Closed-loop	320.28
240	600	0.31	Air-cooled	338.79
241	600	2.02	Closed-loop	308.06
242	630	0.41	Open-loop	303.68
243	600	1.81	Closed-loop	313.89
244	660	0.24	Open-loop	310.58
245	600	3.83	Closed-loop	327.47
246	600	0.27	Air-cooled	339.18
247	600	0.25	Open-loop	322.2
248	600	0.28	Open-loop	306.39
249	600	0.22	Air-cooled	337.27
250	660	0.17	Open-loop	308.22
251	600	0.26	Open-loop	322.76
252	600	0.22	Air-cooled	338.28
253	630	0.39	Open-loop	319.89
254	630	0.63	Open-loop	307.89
255	600	0.28	Open-loop	313.82
256	600	0.19	Open-loop	306.97
257	660	1.55	Closed-loop	308.27
258	660	0.24	Open-loop	311.67
259	600	1.86	Closed-loop	315.92
260	660	0.31	Air-cooled	328.35
261	1000	0.31	Open-loop	284.56
262	1000	0.19	Open-loop	295.69
263	600	0.40	Open-loop	322.3
264	600	2.03	Closed-loop	310.84
265	600	0.40	Air-cooled	330.15
266	600	2.05	Closed-loop	317.36
267	600	0.44	Air-cooled	344.06
268	660	0.50	Open-loop	305.45
269	1000	0.39	Open-loop	294.82
270	660	0.30	Open-loop	291.2
271	660	0.31	Air-cooled	324.62
272	600	0.31	Air-cooled	326.92

Table A1. *Cont.*

Plant No.	Capacity (MW)	Comprehensive Water Consumption Rate (m³/MWh)	Types of Cooling System	Heat Rates (gce/kWh)
273	600	0.27	Air-cooled	338.92
274	600	0.40	Open-loop	318.71
275	600	0.10	Open-loop	313.33
276	600	0.20	Open-loop	310.75
277	600	0.44	Air-cooled	344.06
278	600	2.10	Closed-loop	316.79
279	700	0.41	Open-loop	321.73
280	700	0.49	Open-loop	310.74
281	660	0.38	Air-cooled	328.51
282	600	2.08	Closed-loop	323.87
283	600	1.86	Closed-loop	316.44
284	660	0.55	Air-cooled	325.81
285	600	0.40	Open-loop	318.23
286	600	0.20	Open-loop	310.77
287	660	0.45	Open-loop	303.24
288	600	2.07	Closed-loop	318.
289	600	0.27	Air-cooled	338.82
290	600	2.16	Closed-loop	311.67
291	600	0.40	Open-loop	324.25
292	600	3.83	Closed-loop	326.53
293	600	0.19	Open-loop	306.98
294	600	0.26	Open-loop	317.15
295	600	0.26	Open-loop	318.05
296	600	0.31	Air-cooled	339.33
297	600	0.38	Open-loop	313.76
298	600	0.24	Open-loop	303.91
299	1000	1.87	Closed-loop	291.08
300	1000	0.32	Open-loop	298.1
301	600	1.71	Closed-loop	322.59
302	600	0.25	Open-loop	322.2
303	600	1.61	Closed-loop	332.22
304	600	0.31	Air-cooled	326.75
305	630	0.39	Open-loop	309.04
306	600	0.25	Open-loop	322.2
307	600	0.28	Open-loop	308.48
308	630	0.42	Open-loop	317.64
309	600	0.23	Open-loop	309.43
310	600	1.79	Closed-loop	331.2
311	700	0.41	Open-loop	322.14
313	600	0.76	Closed-loop	303.75
314	600	0.26	Open-loop	320.52
315	660	1.71	Closed-loop	304.94

Table A1. *Cont.*

Plant No.	Capacity (MW)	Comprehensive Water Consumption Rate (m³/MWh)	Types of Cooling System	Heat Rates (gce/kWh)
316	600	0.27	Air-cooled	339.58
317	600	0.76	Closed-loop	305.18
318	600	0.18	Air-cooled	346.8
319	600	1.78	Closed-loop	316.43
320	600	0.23	Open-loop	314.1
321	600	0.35	Air-cooled	349.18
323	800	0.51	Open-loop	355.23
324	600	0.28	Air-cooled	341.09
327	600	0.25	Open-loop	325.51
328	700	0.50	Open-loop	312.49
330	600	0.35	Air-cooled	351.6
331	600	0.28	Air-cooled	339.29
333	600	0.27	Air-cooled	333.16
334	600	0.18	Air-cooled	346.8

Conflicts of Interest

The authors declare no conflict of interest.

References

1. Li Keqiang: Industry in the East Has to Move to the West, and a Ban on New Thermal Power in Beijing, Tianjin and Other Regions. Available online: http://news.bjx.com.cn/html/20140626/521911.shtml (accessed on 26 June 2014).
2. China Electricity Council (CEC). *Annual Development Report on China's Power Industry*; China Market Press: China, Beijing, 2013.
3. China Electricity Council (CEC). Power Development Strategic Research Report, 2013. Available online: http://www.doc88.com/p-8015468257297.html (accessed on 20 August 2014).
4. The Fifth Session of the Seventeenth CPC Central Committee. Economic and Social Development 12th FYP, 2011. Available online: http://www.gov.cn/2011lh/content_1825838.htm (accessed on 20 August 2014).
5. The State Council. The 12th Energy Development Plan, 2013. Available online: http://wenku.baidu.com/link?url=cVcFR_cw1iXEDQ4XkOiRSlb23mSeZ12d6GpRzNF7QkDh zFpW8w6Rq0ikPcOSofYj-0v-V6IH2a9zj8g1H4nVba6Wp_FKpo8n3KGxp188fbK (accessed on 20 August 2014).
6. National Energy Administration. The 12th Coal Industry Development Plan, 2012. Available online: http://finance.qq.com/a/20120322/005082.htm (accessed on 20 August 2014).
7. International Energy Agency. IEA Energy Statistics—Electricity for World. Available online: http://www.iea.org/stats/electricity-data.asp?COUNTRY_CODE=29 (accessed on 20 August 2014).

8. Rebetez, M.; Dupont, O.; Girond, M. An Analysis if the July 2006 Heatwave Extent in Europe Compared to the Record Year of 2003. Available online: http://china.springerlink.com/content/y5rk72801647k547/?p=a484fa0be21d4cad8d361b2b6c22eaca&pi=0 (accessed on 20 August 2014).

9. UN-Water. *Coping with Water Scarcity: An Action Framework for Agriculture and Food Security*; Food and Agriculture Organization of the United Nations: Rome, Italy, 2012.

10. Carter, N.T. *Energy's Water Demand: Trends, Vulnerabilities, and Management*; Diane Publishing Company: Collingdale, PA, USA, 2010.

11. Kajenthira, A.; Siddiqi, A.; Anadon, L.D. A new case for promoting wastewater reuse in Saudi Arabia: Bringing energy into the water equation. *J. Environ. Manag.* **2012**, *102*, 184–192.

12. Siddiqi, A.; Kajenthira, A.; Anadón, L.D. Bridging decision networks for integrated water and energy planning. *Energy Strategy Rev.* **2013**, *2*, 46–58.

13. Gleick, P.H. Water and energy. *Annu. Rev. Energy Environ.* **1994**, *19*, 267–299.

14. Bazilian, M.; Rogner, H.; Howells, M.; Hermann, S.; Arent, D.; Gielen, D.; Steduto, P.; Mueller, A.; Komor, P.; Tol, R.S.J.; *et al.* Considering the energy, water and food nexus: Towards an integrated modelling approach. *Energy Policy* **2011**, *39*, 7896–7906.

15. Koch, H.; Vögele, S. Dynamic modelling of water demand, water availability and adaptation strategies for power plants to global change. *Ecol. Econ.* **2009**, *68*, 2031–2039.

16. Alexander, T.D.; Melissa, M.B. The Regional Energy & Water Supply Scenarios (REWSS) model. Part I: Framework, procedure, and validation. *Sustain. Energy Technol. Assess.* **2014**, *7*, 227–236.

17. Benjamin, K.S.; Kelly, E.S. Identifying future electricity—Water tradeoffs in the United States. *Energy Policy* **2009**, *37*, 2763–2773.

18. Edward, A.B.; Jim, W.H.; Jaime, M.A. Electricity generation and cooling water use: UK pathways to 2050. *Glob. Environ. Chang.* **2014**, *25*, 16–30.

19. Aurelie, D.; Edi, A.; Stephanie, B.; Sandrine, S.; Nadia, M. Water modeling in an energy optimization framework—The water-scarce middle east context. *Appl. Energy* **2013**, *101*, 268–279.

20. Kuishuang, F.; Klaus, H.; Yim, L.S.; Xin, L. The energy and water nexus in Chinese electricity production—A hybrid life cycle analysis. *Renew. Sustain. Energy Rev.* **2014**, *39*, 342–355.

21. Huang, X. On the Status of Water in China. *Dossier* **2013**, *3*, 185.

22. Alun, G.; Fei, T.; Yu, W. China energy-water nexus: Assessing the water-saving synergy effects of energy-saving policies during the eleventh Five-year Plan. *Energy Convers. Manag.* **2014**, *85*, 630–637.

23. Zhang, Z.; Bai, J.; Zhen, H. Distributing water resources optimumly and promoting the development of coal power bases. *Electric Power* **2007**, *40*, 20–24.

24. Institute of Geographical Sciences and Natural Resource Research, Chinese Academy of Sciences, Land Key Laboratory of Water Cycle and Surface Process. *Thirsty Coal*; China Environmental Science Press: Beijing, China, 2012.

25. Wei, X; Liu, F. A brief analysis of protection and sustainable use of water resources in Inner Mongolia Jungar Banner. *Shanxi Architect.* **2010**, *36*, 362–363.

26. Tuo, Y. Saving water and promoting sustainable use of water resources in Inner Mongolia. *Inner Mongolia Water* **2009**, *3*, 87–88.

27. Water Resources in Holingola. Available online: http://www.hlgls.gov.cn/userlist/admin/newshow-7304.html (accessed on 06 June 2013).

28. National Bureau of Statistics of China. *China Economic Census Yearbook 2008*; China Statistics Press: Beijing, China, 2008.

29. Jiang, Y. China's water scarcity. *J. Environ. Manag.* **2009**, *90*, 3185–3196.

30. National Bureau of Statistics of China. *China Statistical Yearbook 2013*; China Statistics Press: Beijing, China, 2013.

31. Department of Energy. Energy Demands on Water Resources—Report to Congress on the Interdependency of Energy and Water. Available online: http://climateknowledge.org/figures/Rood_Climate_Change_AOSS480_Documents/DoE_Energy_Water_Resources_2006.pdf (accessed on 20 August 2014).

32. Cammerman, N. Integrated Water Resource Management and the Water, Energy, Climate Change Nexus. Available online: http://www.watercentre.org/resources/publications/attachments/ncammerman (accessed on 20 August 2014).

33. Rio Carillo, A.M.; Frei, C. Water: A key resource in energy production. *Energy Policy* **2009**, *37*, 4303–4312.

34. US Geological Survey (USGS). Estimated Use of Water in the United States in 2000. Available online: http://pubs.usgs.gov/circ/2004/circ1268/index.html (accessed on 20 August 2014).

35. Cheng, Z.; Wen, Z. Water resources utilization of thermal power plant. *Manag. J.* **2011**, *17*, 375.

36. Electric Power Research Institute (EPRI). US Water Consumption for Power Production—The Next Half Century. Available online: http://www.circleofblue.org/waternews/wp-content/uploads/2010/08/EPRI-Volume-3.pdf (accessed on 20 August 2014).

37. Meldrum, J.; Nettles-Anderson, S.; Heath, G.; Macknick, J. Life cycle water use for electricity generation: A review and harmonization of literature estimates. *Environ. Res. Lett.* **2013**, doi:10.1088/1748–9326/8/1/015031.

38. Thomas, J.F., III; Timothy, J.S.; Gary, J.S., Jr.; Andrea, M.; Michael, N.; Brian, S.; James, T.M.; Lynn, M. Water: A critical resource in the thermoelectric power industry. *Energy* **2008**, *33*, 1–11.

39. China Electricity Council (CEC). Notification of Energy Efficiency Benchmarking and Competition Data of 2012 National 600MW Thermal Power Units. Available online: http://kjfw.cec.org.cn/kejifuwu/2013-04-07/99877.html (accessed on 20 August 2014).

40. Wang, H.; Nakata, T. Analysis of the market penetration of clean coal technologies and its impacts in China's electricity sector. *Energy Policy* **2009**, *37*, 338–351.
41. Li, H.; Chien, S.H.; Hsieh, M.K.; Dzombak, D.A.; Vidic, R.D. Escalating water demands for energy production and the potential for use of treated municipal wastewater. *Environ. Sci. Technol.* **2011**, *45*, 195–200.

Water Quality Changes during Rapid Urbanization in the Shenzhen River Catchment: An Integrated View of Socio-Economic and Infrastructure Development

Hua-peng Qin, Qiong Su, Soon-Thiam Khu and Nv Tang

Abstract: Surface water quality deterioration is a serious problem in many rapidly urbanizing catchments in developing countries. There is currently a lack of studies that quantify water quality variation (deterioration or otherwise) due to both socio-economic and infrastructure development in a catchment. This paper investigates the causes of water quality changes over the rapid urbanization period of 1985–2009 in the Shenzhen River catchment, China and examines the changes in relation to infrastructure development and socio-economic policies. The results indicate that the water quality deteriorated rapidly during the earlier urbanization stages before gradually improving over recent years, and that rapid increases in domestic discharge were the major causes of water quality deterioration. Although construction of additional wastewater infrastructure can significantly improve water quality, it was unable to dispose all of the wastewater in the catchment. However, it was found that socio-economic measures can significantly improve water quality by decreasing pollutant load per gross regional production (GRP) or increasing labor productivity. Our findings suggest that sustainable development during urbanization is possible, provided that: (1) the wastewater infrastructure should be constructed timely and revitalized regularly in line with urbanization, and wastewater treatment facilities should be upgraded to improve their nitrogen and phosphorus removal efficiencies; (2) administrative regulation policies, economic incentives and financial policies should be implemented to encourage industries to prevent or reduce the pollution at the source; (3) the environmental awareness and education level of local population should be increased; (4) planners from various sectors should consult each other and adapt an integrated planning approach for socio-economic and wastewater infrastructure development.

Reprinted from *Sustainability*. Cite as: Qin, H.; Su, Q.; Khu, S.-T.; Tang, N. Water Quality Changes during Rapid Urbanization in the Shenzhen River Catchment: An Integrated View of Socio-Economic and Infrastructure Development. *Sustainability* **2014**, *6*, 7433-7451.

1. Introduction

Many catchments in developing countries are experiencing rapid urbanization [1,2]. These catchments are faced with the challenge of maintaining (or improving) water quality while allowing for economic growth and population expansion [3]. Understanding the characteristics and mechanisms of water quality changes in the catchment can help policy makers evaluate the effectiveness of water management measures, avoid repetition of past mistakes and create a sustainable development environment.

Many studies have investigated and characterized the changes in water quality during urbanization. Ren *et al.* [4] found that urbanization from 1947 to 1996 in Shanghai corresponded to a rapid degradation of water quality in the Huangpu River. A separate study in Shanghai found that water quality in the city center had deteriorated from the early 1980s to the early 1990s but has been improving since the 1990s [3]. Chang [5] reported that with rapid urbanization and economic development, the water quality of the Han River in Korea declined gradually during the 1960s and 1970s but improved after control measures were put in place in the 1980s. Kannel *et al.* [6] detected increasing phosphorus contents from 1999 to 2003 in the urban areas of the Bagmati River catchment, Nepal.

Studies have also been performed to determine the causes of water quality change in urbanized catchments [4,7–9]. It is well known that population growth and industrial activities are drivers of water quality change. Groppo *et al.* [10] found that population growth and increases in untreated sewage were the main causes of water quality deterioration in rivers in the State of Sao Paulo, Brazil. Ma *et al.* [11] found that sharp increases in industrial pollution and domestic discharge were the major causes of water quality deterioration in the Shiyang River, Northwest China. Suitable water management measures, however, may also be included in the urbanization process to mediate water quality deterioration [12] such as installing wastewater infrastructure (e.g., wastewater treatment plant (WWTP) and urban drainage system) to increase wastewater treatment capacity. Socio-economic policies such as industrial structure regulation, economic incentive and finance policies, command and control, and legal measures are able to improve water consumption efficiencies and reduce pollutant load generation [13,14].

Although many policies and measures have been enacted, water quality degradation continues to be a serious problem in some rapidly urbanizing areas. For the past three decades, surface water deterioration has coincided with rapid economic growth in China despite the implementation of modern water management measures [15]. Many researchers found that such deterioration is due to the complex interaction between engineered infrastructures and social, economic, legal, and political issues during rapid urbanization [16,17]. However, most research addressed water quality change in urbanized areas rather than catchments in the process of rapid urbanization [6,10]. Furthermore, even fewer studies have investigated the integrated impact of socio-economic and infrastructure development on changes in water quality [16]. Current understandings of the impact of urbanization on water quality are mainly anecdotal. Further research is required to generate an integrated view of water quality changes, pollution generation in the socio-economic system and pollution control measures in the management system.

Using the Shenzhen River catchment in China as a case study, this paper will examine the water quality changes in the catchment during the rapid urbanization period of 1985 to 2009 based on measured water quality data and estimated pollutant loads. The objectives of this study are: (1) to characterize the water quality changes and better understand the factors that determine the changes in the catchment; (2) to evaluate the effects of infrastructure development and socio-economic policies on water quality changes; and (3) to propose possible solutions on water quality changes in the rapid urbanization catchment.

2. Material and Methods

2.1. Study Area

The Shenzhen River is located in a rapidly urbanizing coastal region of Southeast China and forms the administrative border between Shenzhen City and Hong Kong (Figure 1). The total catchment area of the Shenzhen River is 312 km². The river is a typical tidal river and is 14 km long. The main river drains southwest into Deep Bay, which joins the Pearl River estuary on its seaward side. Three sampling stations, S1, S2 and S3, are located in the upper, middle and lower streams of the river, respectively (Figure 1). The catchment has a mild, subtropical maritime climate with a mean annual temperature of 22.4 °C and a mean annual precipitation of 1933 mm.

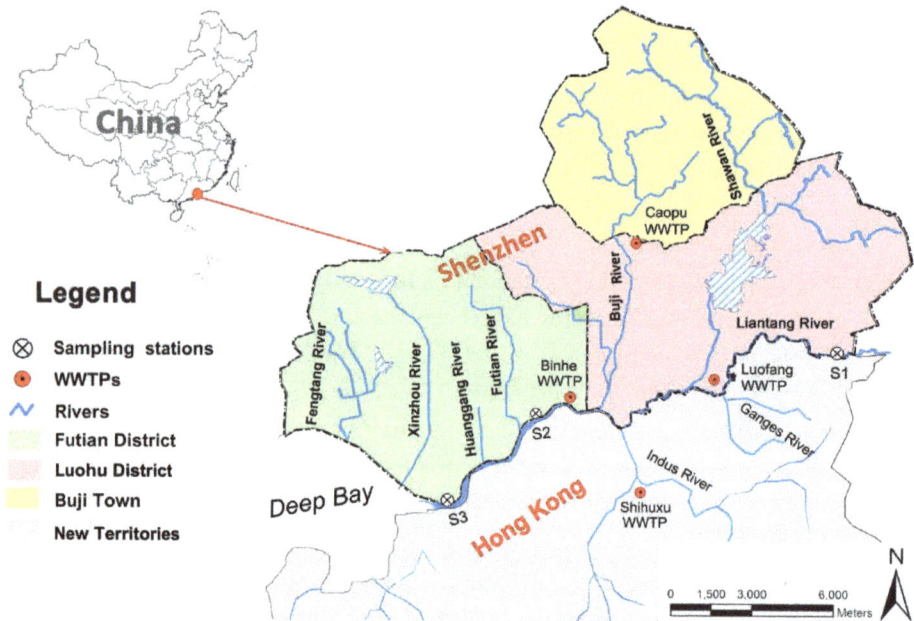

Figure 1. Overview of the Shenzhen River catchment.

The northern catchment, with an area of 188 km², includes three administrative areas of Shenzhen: Luohu District, Futian District, and Buji Town. Over the past three decades, this area has experienced rapid urbanization. Between 1980 and 2009, the area's gross regional product (GRP) increased from 0.4–116.3 billion yuan (using the 1990 price), total population increased from 0.16 million to 2.69 million and the ratio of built-up to developable land increased from 12.0%–91.9%. In contrast, the southern catchment of the Shenzhen River belongs to the Northern New Territories of Hong Kong. Most of the area within the southern catchment is rural with low population densities.

2.2. Water Quality Data

The water quality in the Shenzhen River has been measured by the Environmental Monitoring Station of Shenzhen since 1985. The data set measured from 1985–2009 was used to characterize the water quality change during rapid urbanization. The water quality was measured bi-monthly from 1985–2002 and monthly after 2002. To consider the effects of tides, two water samples were collected during ebb and flood tides. Ten water quality parameters are considered in this study: BOD_5, NH_3-N, TP, petroleum, volatile phenol (V-ArOH), hexavalent chromium (Cr^{6+}), mercury (Hg), lead (Pb), copper (Cu) and cadmium (Cd). These parameters are the target indicators in China's environmental monitoring system. The river is classified as a water body for landscape requirements. The maximum permitted concentrations (MPCs) for various water quality parameters are prescribed by the Environmental Quality Standard for Surface Water in China (GB 3838-2002). For example, MPCs for BOD_5, NH_3-N and TP are 10 mg/L, 2 mg/L and 0.4 mg/L, respectively. In order to obtain an overall understanding of the combined effect of the changes in the target indicator values a composite indicator—Water Pollution Index (WPI)—was used to assess the comprehensive status of the water quality of the Shenzhen River. The index is defined as follows [18,19]:

$$WPI = \frac{1}{n}\sum_{i=1}^{n} C_i / S_i \qquad (1)$$

where n represents the number of water quality parameters; C_i is the average measured concentration of the ith parameter, mg/L; and S_i is the MPC for the ith parameter. The water quality can be classified into six types by WPI: very pure, pure, moderately polluted, polluted, impure and heavily impure with corresponding WPI values of below 0.3, 0.3–1.0, 1.0–2.0, 2.0–4.0, 4.0–6.0 and over 6.0, respectively [18]. Furthermore, to show the pollutants that most significantly reduce the overall water quality (in the sense of violating their objective levels), the contribution of each polluting substance to the summed pollution index of all pollutants is calculated as follows:

$$K_i = \frac{C_i / S_i}{\sum_{n=1}^{i} C_i / S_i} \qquad (2)$$

where K_i is the contribution of each substance to the summed pollution index.

2.3. Methods to Estimate the Pollutant Load

The pollutant load discharged into the river is the direct driving factor of water quality change during urbanization. In the Shenzhen River catchment, however, the pollutant load has not been directly measured and therefore a simple method was proposed to estimate this discharge.

Reasons for change in pollutant load discharge during rapid urbanization are usually varied and complex. They are either directly or indirectly linked to social-economic development and engineering interventions. In this study, a cause diagram is used to break down broad items into increasingly finer levels of detail, and is also useful for identifying the causal relationships between items of interest.

Based on cause-and-effect analysis, a diagram was constructed to trace the causes of pollutant load changes in the Shenzhen River catchment. As shown in Figure 2, the pollutant load discharge into the river is related to the difference between the total pollutant load generated within the catchment and the pollutant load removed by wastewater infrastructure. Pollution is often categorized into point source and non-point source. However, water quality during wet periods in the Shenzhen River is much better than that during dry periods [20]. According to estimates by Liu and Lu [21], non-point source pollutant loads are much smaller than the pollutant loads from industrial and domestic sources in the catchment and therefore only point source pollution is considered in this study.

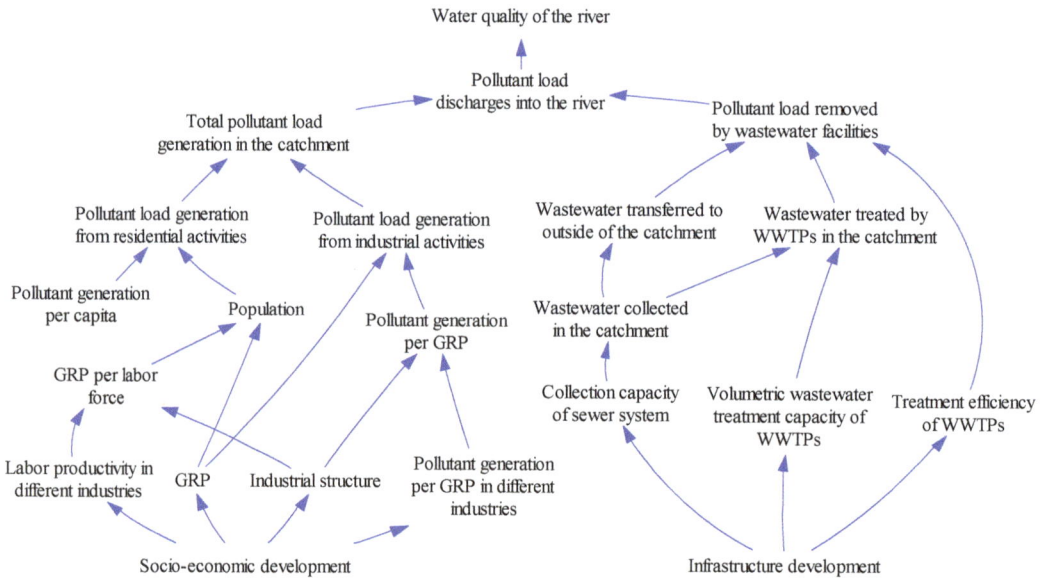

Figure 2. Cause diagram of water quality changes in the Shenzhen River catchment.

The pollutant load generation is affected by socio-economic development. Industrial wastewater is discharged by manufacturing and production processes and commercial enterprises. Pollution from industrial activities is estimated based on GRP and pollutant load per unit GRP. The respective equation can be expressed as follows:

$$LG_{e,t} = GRP_t * LOAD_{GRP,t} \qquad (3)$$

where $LG_{e,t}$ and $LOAD_{GRP,t}$ denote industrial pollutant load (tons) and the pollutant load per unit GRP in year t, respectively. $LOAD_{GRP,t}$ is a function of industrial structure and pollutant per GRP in different industries. The industry structures in the study area can be categorized into primary (e.g., agriculture), secondary (e.g., manufacturing) and tertiary (e.g., services) industries. Each industry type can be further sub-divided according to the primary factor in the production process. $LOAD_{GRP,t}$ can be expressed as follows:

$$LOAD_{GRP,t} = \sum_i (PROP_{i,t} * LOAD_{GRPi,t}) \tag{4}$$

where $PROP_{i,t}$ and $LOAD_{GRPi,t}$ are GRP proportion and the pollutant load per unit GRP of industry i in year t, respectively. Therefore, pollutant load generation from industrial activities can be estimated by GRP, industrial structure and pollutant load per unit GRP of different industries. Furthermore, water pollutant loads from tertiary industries are usually identified as domestic pollution and are thus included in the estimation of pollutant generation in residential activities (Equation (5)). And thus Equations (3) and (4) were only applied to estimate pollutant load in the secondary industry in this study.

Domestic wastewater comes from residential sources, including toilets, sinks, baths and laundry. Pollution from residential activities is estimated according to the population size and pollutant load per capita. The respective equation can be expressed as follows:

$$LG_{r,t} = Pop_t * LOAD_{cap,t} \tag{5}$$

where $LG_{r,t}$ denotes residential pollutant load (tons) in year t; $LOAD_{cap,t}$ denotes the pollutant load per capita, which is assumed to be constant in this study.

To understand the changes of pollution from residential activities, population growth during rapid urbanization should be further analyzed. Fertility, mortality and migration are principal determinants of population growth. However, migration accounts for a large percentage of total population in rapidly urbanizing areas (e.g., more than 80% in Shenzhen City, China in 2007). Population growth in an urbanizing catchment is mainly determined by the labor demand of economic development [13] and expressed as

$$Pop_t = R_{pop,t} * GRP_t / LABOR_{GRP,t} \tag{6}$$

where $R_{pop,t}$ is the ratio of a population to the number of labors at year t in a catchment. $LABOR_{GRP,t}$ is the GRP per labor force, which is a measurement of labor force productivity. $LABOR_{GRP,t}$ is a function of industrial structure and labor productivity for different industries, and it can be expressed as follows:

$$LABOR_{GRP,t} = 1 / \sum_i (PROP_{i,t} / LABOR_{GRPi,t}) \tag{7}$$

where $LABOR_{GRPi,t}$ denotes GRP per labor force of industry i in year t.

However, with the development of wastewater infrastructure, the capacity for removal of the pollutant load has steadily increased. The pollutant load removed by wastewater infrastructure can be determined by the wastewater volume collected in the catchment and the pollutant removal rate of the WWTPs. In addition, some wastewater is transported out of the catchment and then discharged into the sea. Thus, the pollutant load removed by wastewater treatment or transfer facilities can be calculated by the following expression:

$$LR_t = (V_{p,t} * \alpha_t + V_{s,t}) * C_w \tag{8}$$

where LR_t is the pollutant load removed by wastewater infrastructure in year t; $V_{p,t}$ and $V_{s,t}$ are the wastewater volume treated by WWTPs in the catchment and the volume transported out of the catchment in year t, respectively; α_t is the WWTPs' pollutant removal rate in year t; and C_w is the averaged pollutant concentration of municipal wastewater.

The GRP, population, industrial structure and labor productivity in different industries are obtained from the Statistical Year Book of Shenzhen (1985–2009) [22]. The pollutant load per capita, pollutant load per unit GRP of different industries, and the status of wastewater infrastructure are obtained from published government reports [23–25].

3. Results

3.1. Temporal Trend of Water Quality

Water quality measured at the sampling station S3 (Figure 1) was used to calculate the WPI as it represents the effect of pollution discharges in the whole catchment on the water quality of the river. As shown in Figure 3, the annual average WPI varied between 1.6 and 2.2 from 1988–1995, decreased after 1996, and reached 1.5 in 2009.

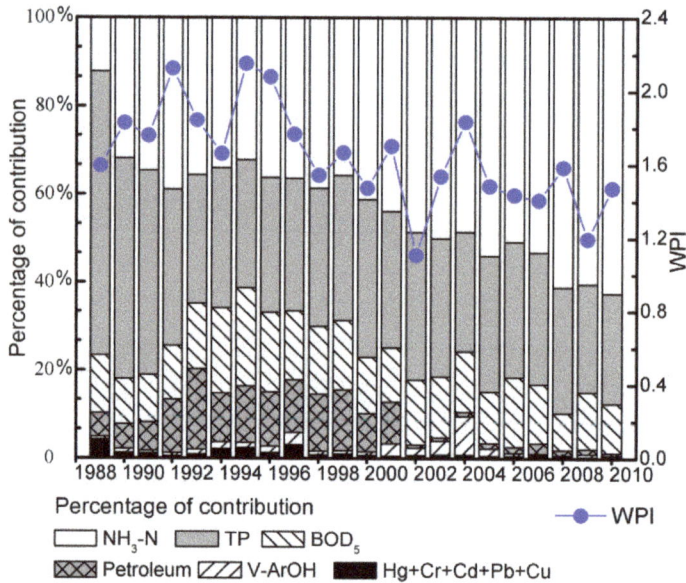

Figure 3. Annual average WPI (Water Pollution Index) and relative contributions of each pollutant to WPI in the Shenzhen River.

The contribution of each polluting substance to the WPI was calculated according to Equation (2). The results indicated that the main pollutants in the river were NH$_3$-N, TP and BOD$_5$ and that different pollutants had different variation trends (Figure 3): the contribution of BOD$_5$ to the WPI initially increased from 13% in 1988 to 22% in 1994 but then declined after 1995; the contribution of NH$_3$-N gradually increased from 12% in 1988 to 63% in 2009; the contribution of TP decreased from 64% in 1988 to 25% in 2009.

Figure 3 also indicates that the industrial effluent pollutants, e.g., petroleum and heavy metals, significantly affected the water quality. The contribution of petroleum varied between 5.3% and 18.3% before 2000, sharply decreased to 0.6% in 2001, and then stayed at a low level. Heavy metals had some effects on WPI before 1997, whereas V-ArOH had some effects on WPI before 2005. However, the effects of heavy metals and V-ArOH have become negligible in recent years.

3.2. Water Quality vs. Pollutant Load Discharges

Since BOD_5, NH_3-N and TP are the main pollutants in the river, their variations were further investigated. As observed in Figure 4, the annual average concentrations of the three pollutants followed different temporal trends: BOD_5 increased for 1985–1994 and then gradually decreased, reaching a concentration of 16.6 mg/L in 2009; NH_3-N increased for 1985–1991, remained relatively stable for 1992–1999, and then gradually increased; while TP decreased continuously after 1988.

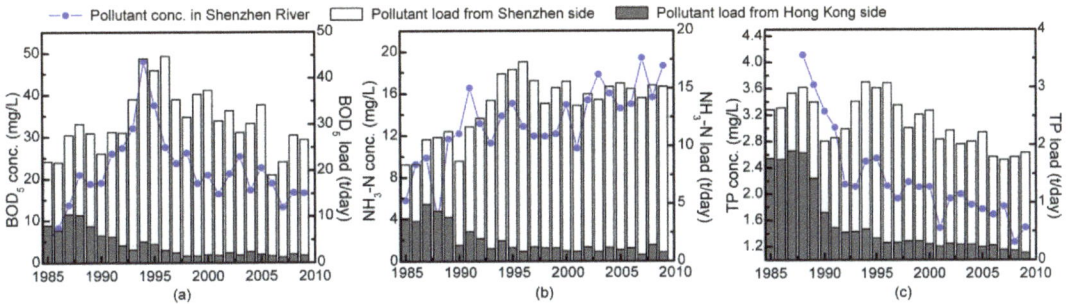

Figure 4. Water quality and pollutant loads discharged in the Shenzhen River. (**a**) BOD_5; (**b**) NH_3-N; (**c**) TP.

The temporal changes in water quality were mainly caused by the changes of the pollutant load discharges into the river from Shenzhen and Hong Kong. The indiscriminate discharge of livestock waste from the Hong Kong side was an important pollution source of the Shenzhen River before 1989 [24]. With the implementation of the Livestock Waste Control Scheme (LWCS) in 1987, many farmers on the Hong Kong side have ceased keeping livestock or have installed waste treatment facilities. Accordingly, the loads of NH_3-N, TP and BOD_5 from Hong Kong rapidly decreased after 1989 (Figure 4). From Shenzhen, the loads of all three pollutants increased for 1985–1996; after which the TP and BOD_5 loads then decreased, while the NH_3-N loads remained relatively stable at approximately 13–15 t/day after 1997.

The total pollutant load discharges can be estimated by summing the pollutant loads from the Shenzhen and Hong Kong sides. As observed in Figure 4, the total loads of the three main pollutants increased for 1985–1996; and then the BOD_5 and TP loads decreased, while the NH_3-N loads fluctuated at 13–16 t/day, for 1997–2009. Pearson's correlation analysis revealed that the pollutant concentration in the river was significantly correlated with the pollutant load discharges, with correlation coefficients of 0.743 ($p < 0.01$), 0.571 ($p < 0.01$) and 0.514 ($p < 0.05$) for BOD_5, NH_3-N and TP, respectively.

3.3. Pollutant Load Generation vs. Removal

Since the loads from Shenzhen accounted for 95%–99% of the total pollutant load discharges into the river after 1995, the pollutant load discharges on the Shenzhen side were further related to the pollutant load generation and removal. While the total pollutant load generation on the Shenzhen side increased over the study period, slowing after 2006 (Figure 5), the total pollutant load removal on the Shenzhen side slowly increased until 1996, rapidly increased from 1997–2006, and stabilized after 2006. Therefore, from 1985–1996, the pollutant load discharges from Shenzhen increased as the pollutant load generation increased faster than the pollutant load removal. From 1997–2009, the BOD_5 and TP load discharges decreased as the load removal of the two pollutants increased faster than the load generation, while the NH_3-N load discharges followed a slightly increasing trend as the NH_3-N load generation increased slightly faster than the load removal.

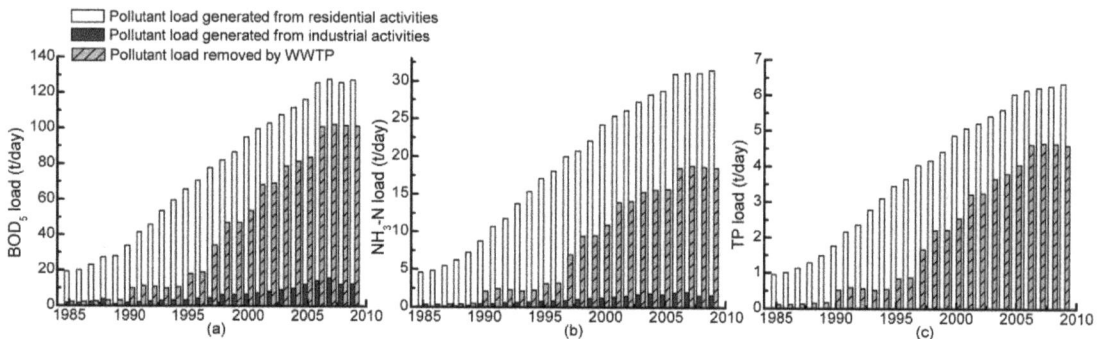

Figure 5. Pollutant load generation and removal on the Shenzhen side. (**a**) BOD_5; (**b**) NH_3-N; (**c**) TP.

The main sources of water pollution in the catchment include industrial and domestic wastewater. Only the BOD_5 and NH_3-N sources were further analyzed in this study for lack of TP load generation data from industrial activities. As shown in Figure 5, NH_3-N and BOD_5 load generation from residential activities increased dramatically from 1985–2005, slowing after 2006. In contrast, NH_3-N and BOD_5 load generation from industrial activities increased slightly from 1985–2007 and decreased slightly after 2008. Figure 5 also indicates that the pollutant load generation from residential activities was the main source of pollution in the catchment.

3.4. Capacity for Pollutant Load Removal

The pollutant load removal was attributed to the capacity and efficiency of the wastewater facilities in the catchment. Only a small amount of wastewater was treated by WWTPs or transferred outside the catchment before 1995, but the amount increased rapidly from 1996–2006 (Figure 6a). In addition, the pollutant removal efficiency of WWTPs increased significantly after 1995 (Figure 6b). However, WWTPs have different removal efficiencies for different pollutants. For example, the removal efficiencies of BOD_5 and TP have increased to approximately 90% and 80%, respectively, while the removal efficiency of NH_3-N remains lower than 60% (Figure 6b). These differences resulted in different temporal trends in load discharges of BOD_5, TP and NH_3-N.

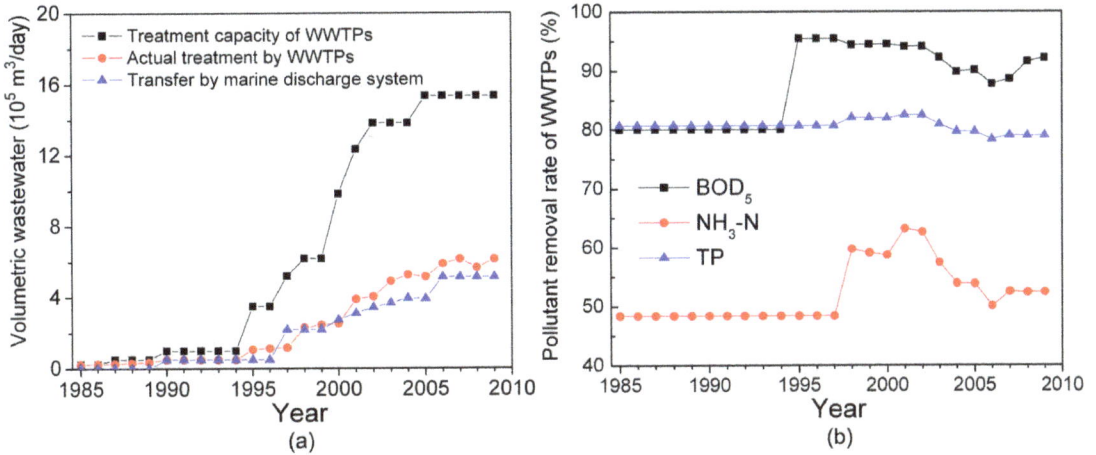

Figure 6. Capacity of wastewater facilities in the catchment. (**a**) Wastewater collection, transfer and treatment; (**b**) Pollutant removal efficiency.

4. Discussion

4.1. Water Pollution Characteristics during Urbanization

Water pollution in the Shenzhen River has undergone a deterioration stage and improvement stage. As shown in Figure 5, the period of 1985–1995 is identified as a deterioration stage as the pollutant load generation increased faster than the pollutant load removal, while the period of 1996–2009 is identified as an improvement stage as the load removal increased faster than the load generation.

In both stages, water quality management in the catchment was under the pressure of rapid increases in both population and economy. For example, the average annual growth rate of GRP in the catchment for 1985–1995 and 1996–2009 was 30.3% and 12.6%, respectively; and the average annual growth rate of population for 1985–1995 and 1996–2009 was 14.3% and 4.5%, respectively. The pollutant load generation from residential activities was the main source of pollution in the catchment during rapid urbanization (Figure 5). For the period of 1996–2009, industry-derived petroleum and heavy metal pollution had been well controlled and domestic-derived water quality parameters (e.g., BOD_5, NH_3-N and TP) also improved greatly, however still had a high rate of non-compliance with water quality objectives.

4.2. Effects of Socio-Economic Measures

4.2.1. Socio-Economic Measures

Over the past three decades many socio-economic measures have been implemented in the Shenzhen River catchment. Three measures have had particularly significant effects on water quality improvement (Table 1):

(a) "Deadline requirements for pollution control" may suspend the operations of or close an enterprise or institution that has caused severe pollution to water bodies but has failed to adhere to the specified reduction goals by the deadline [26,27].

(b) "Centralized control of waste" was first instituted in China in 1999, requiring all levels of government to generate economies of scale and improve efficiency in waste disposal [14,26]. Based on this principle, ecological industrial parks are encouraged, in which businesses cooperate with each other and the local community in an attempt to reduce waste and pollution and efficiently share resources (such as information, materials, water, energy, infrastructure, and natural resources).

(c) To accelerate the transformation of economic development and promote industrial restructuring and upgrading, the local government has compiled the "catalog of industrial structure adjustment" annually since 1993 [28], in which the industries are categorized into three groups: encouraged, restricted, and prohibited. The prohibited industries include printing and dyeing, tinning, plating, and eight other labor-intensive industries. According to these catalogs, the industries with high labor productivity and low pollution emission were encouraged to develop, e.g., the cultural industry, the electronic information industry, the biotechnology and pharmaceutical industry, and the advanced material and new energy industry.

Table 1. Pollution control measures in the Shenzhen River catchment [23].

Measures	Name	Year	Description
Socio-economic measure	Deadline requirements for pollution control	1998	A total of 173 companies that had severely polluted water bodies were required to reduce pollution by the deadline, and 43 were ordered to close down.
	Centralized control of waste	1999	Focusing on petroleum discharge control, 336 companies were examined.
		2004	Focusing on both V-ArOH and petroleum discharge control.
		2000	An electroplating park was established, and all wastewater from the electroplating industry can now be effectively collected and treated in the park before discharge.
	Catalog of industrial structure adjustment	1993–2009	The industries were categorized into three groups: encouraged, restricted, and prohibited. Industries with high labor productivity and low pollution emission were encouraged to develop.
WWTP construction & technology improvement	Binhe WWTP	1985–1995	Wastewater treatment capacity: 2.5×10^4 m^3/day in 1985, 3.0×10^5 m^3/day in 1995; actual wastewater treatment: 1.1×10^5 m^3/day in 1995.

Table 1. *Cont.*

Measures	Name	Year	Description
	Luofang WWTP	1998–2002	Wastewater treatment capacity: 1.0×10^5 m³/day in 1998, 3.5×10^5 m³/day in 2002; actual wastewater treatment: 2.2×10^5 m³/day in 2002.
	Caopu WWTP	2003	Wastewater treatment capacity: 1.5×10^5 m³/day; actual wastewater treatment: 5.0×10^4 m³/day.
Wastewater transfer and marine discharge system	Stage I, II & III	1990–2001	Wastewater transfer capacity: 5.0×10^4 m³/day in 1990, 7.4×10^5 m³/day in 2001; actual wastewater transfer 2.4×10^5 m³/day in 2001.
Sewer system improvement	For Binhe WWTP collection area	1999	Wastewater collection capacity: 1.5×10^5 m³/day.
	Wastewater transfer and marine discharge system collection area	2003	Wastewater collection capacity: 2.8×10^5 m³/day.
	For Luofang WWTP collection area	2004	Wastewater collection capacity: 2.6×10^5 m³/day.
	For Caopu WWTP collection area	2006	Wastewater collection capacity: 1.5×10^5 m³/day.

4.2.2. Effects of "Deadline Requirements" and "Centralized Control"

In the 1980s, many industries with high pollution emission, e.g., paper and paper-board, chemical raw materials and products, dyeing and metalwork were introduced into the Shenzhen River catchment without appropriate pollution control regulations/measures. These industries contributed organic, petroleum and heavy metal pollution to the river.

The "deadline requirements for pollution control" for petroleum and V-ArOH were implemented in Shenzhen in 1999 and 2004, respectively. The effect of these measures can be seen in a decrease in pollutant levels one to two years after implementation, and the contribution of petroleum and V-ArOH to the WPI has become negligible since 2001 and 2005, respectively (Figure 3).

The electroplating industry is a dominant industry type in Shenzhen and typically causes serious heavy metal pollution. To solve this problem, in 2000 an electroplating industry park was established in Shenzhen and the related factories relocated to the park, allowing industry wastewater to be effectively collected and treated in the park before discharge. The implementation of this "centralized control of waste" measure may be one of the reasons for the low concentration level of heavy metal in the recent decade (Figure 3).

4.2.3. Effects of Industrial Structure Adjustment

With the implementation of the "catalog of industrial structure adjustment" the proportion of secondary industry decreased. Although the average annual growth rate of GRP in the catchment was as high as 18.4%, the growth rate of the GRP in the secondary industry was only 14.9% (Figure 7a). In particular, the proportion of GRP from the secondary industry decreased dramatically from 49.4% in 1990 to 16.6% in 2009. Furthermore, the decrease in GRP proportion of the labor intensive secondary industry resulted in the decrease in the pollutant generation per GRP in the secondary industry. The secondary industry can be further subdivided into labor-, technology- and capital-intensive secondary industries according to the dominant production process. As shown in Figure 7b, the GRP of the technology-intensive secondary industry increased faster than that of the labor-intensive industry, which caused the GRP proportion of technology-intensive industry to increase while that of labor-intensive industry to decrease from 1990–2009. As the pollutant generation per GRP in the technology-intensive industry is much lower than that in the labor-intensive secondary industry, the changes in the secondary industry structure decreased the pollutant generation per GRP in the secondary industry. Therefore, the industrial structure adjustment greatly reduced the growth rate of the pollutant load generation from the secondary industry (Figure 5).

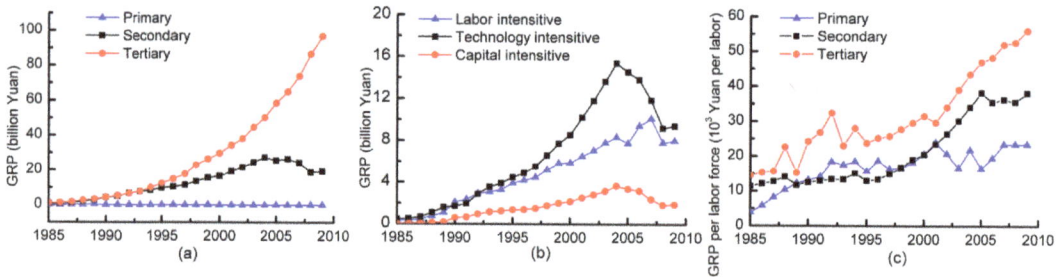

Figure 7. GRP and labor productivity in various industries. (**a**) GRP in various industries; (**b**) GRP in various secondary industries; (**c**) Labor productivity in various industries.

Furthermore, industrial structure adjustment can also greatly increase the labor productivity in the catchment and decrease the growth rates of the population and the relevant residential pollutant load. The average annual growth rate of population (8.8%) for 1985–2009 was much lower than that of GRP (20.4%), indicating an increase in the labor productivity during the urbanization process. This trend can be explained by the increase in labor productivity in various industries over the last three decades (Figure 7c). In addition, between 1990 and 2009, the proportion of the secondary industry decreased, while that of the tertiary industry increased. As the labor productivity in the tertiary industry was higher than that in the secondary industry (Figure 7c), the increase of the tertiary industry proportion resulted in an improvement of labor productivity.

4.3. Effects of Wastewater Infrastructure Development

4.3.1. Wastewater Infrastructure

The following wastewater infrastructure has been developed in the northern catchment: WWTPs, wastewater transfer and marine discharge system improvement and sewer system (Table 1 and Figure 6):

(a) Before 1994, Binhe WWTP was the only WWTP in the catchment (Figure 1). This WWTP was equipped with secondary treatment; its BOD_5, NH_3-N and TP removal rates were 80%, 48% and 81%, respectively; its volumetric treatment capacity was less than 5.0×10^4 m^3/day. Two WWTPs, Luofang WWTP with secondary treatment and Caopu WWTP with advanced primary treatment (Figure 1), were constructed and then upgraded between 1996 and 2003. In 2005, the total volumetric treatment capacity in the northern catchment rapidly increased to 1.5×10^6 m^3/day, while the BOD_5, NH_3-N and TP removal rates increased to 92%, 57% and 81%, respectively.

(b) A wastewater transfer and marine discharge system was constructed in 1990 and 5.0×10^4 m^3/day of wastewater was transferred outside of the Shenzhen River catchment and discharged into the sea through a long-distance discharge pipe. The system was gradually enlarged between 1997 and 2001, and its transfer capacity increased to 7.4×10^5 m^3/day by 2001.

(c) A large amount of wastewater in the catchment could not be collected and conveyed to the WWTPs due to non-existent or low-efficiency sewer systems during urbanization. Several projects had been carried out to retrofit or improve the sewer systems in the catchment over 1999–2006, and the amount of wastewater actually collected by the sewer system increased from 4.4×10^5 m^3/day in 1999 to 9.4×10^5 m^3/day in 2006.

4.3.2. Limitations of Wastewater Infrastructure

Although the wastewater infrastructure had significant effects on water quality improvement during urbanization, it was unable to effectively collect and dispose of all wastewater in the catchment. This inability was caused by the following:

(1) The construction of wastewater facilities lagged behind population and economic growth in the early stages of urbanization (before 1995). In the study area, socio-economic planning and wastewater facilities planning are performed by the Shenzhen Development and Reform Commission (SZDRC) and Shenzhen Municipal Water Affairs Bureau (SZWAB), respectively. The socio-economic planners of SZDRC are not expected to fully apprehend the capacity limitations of existing/future wastewater facilities in their decision-making process and usually assume that facilities construction can match the pace of socio-economic growth, while wastewater facilities planners of SZWAB do not fully account for the extent of rapid socio-economic development in decision making. Facilities planners usually estimate certain socio-economic growth potentials in their planning, however the facilities planners may over- or under-estimate the growth and fail to make timely adjustment in facilities development. Therefore, wastewater generation and treatment capacity may be

mismatched during rapid urbanization [29]. These mismatches caused water quality deterioration in the early stage of urbanization.

(2) The sewer network construction lagged behind the construction of WWTPs in the catchment. The wastewater treatment capacity of WWTPs increased much faster than the amount of wastewater actually collected by the sewer system in the catchment (Figure 6a). One reason is that the local government focused on improving wastewater treatment capacity by constructing WWTPs. However, constructing a sewer network is more difficult and requires more time to construct, and due to the delay in construction the wastewater collection capacity increased slower than the treatment capacity.

(3) The existing sewer networks were operated at a low efficiency level. Most of the early developed areas in the catchment are densely populated, and overcrowded multi-story buildings are usually prevalent in these areas. While the buildings provide cheap accommodation for the massive number of workers immigrating from other Chinese cities or villages, they usually have poorly installed pipelines which result in the mixing of sewage flows with rainfall runoff. In addition, to fully utilize the indoor living space, local residents alter balconies in the buildings to re-equip them for toilet, kitchen or laundry use, leading to sewage discharge into the river via rainwater pipes. Due to the poor environmental management and environmentally harmful behavior, wastewater cannot be efficiently collected into the existing sewer systems.

(4) The WWTPs have relatively low removal efficiencies for nutrient pollutants. The existing WWTPs in the Shenzhen River are equipped with secondary or advanced primary treatment technology. The WWTPs have high removal rates for organic matter, such as BOD_5, but are not as effective in removing nutrient substances, such as nitrogen and phosphorus, and unfortunately these nutrient substances are predominant pollutants in densely populated catchments such as the Shenzhen River catchment. For example, the NH_3-N pollutant removal rate of the WWTPs in the catchment was only 55% in 2009. Therefore, although BOD_5 has significantly decreased, NH_3-N has remained at a high level since 1995.

4.4. Solutions on Water Quality Changes

Although the water quality of the Shenzhen River has improved greatly since 1996, some domestic-derived water quality parameters still have a high rate of non-compliance with water quality objectives. For example, NH_3-N and TP of the river in 2009 were three and eight times higher, respectively, than the corresponding maximum permitted concentrations prescribed by the Environmental Quality Standard for Surface Water in China (GB 3838-2002). To satisfy the water quality requirement, further multifaceted approaches including wastewater infrastructure investment, socio-economic policies regulation and increasing environmental awareness, are required and discussed below.

4.4.1. Wastewater Infrastructure Construction

Wastewater infrastructure construction lagged behind the population and economic growth during rapid urbanization of the area, and the wastewater infrastructure still needs to be further improved in the future. The sewer networks should be constructed timely and revitalized regularly in line with urbanization. To reduce the nitrogen and phosphorus discharge, wastewater treatment facilities should be upgraded to improve their nitrogen and phosphorus removal efficiencies.

In addition to the centralized wastewater infrastructure, the construction of decentralized small WWTPs should also be promoted. These infrastructures have lower cost, higher ecological value and additional public benefits and are considered sustainable approaches for water environment. For example, Organica Food Chain Reactor is a decentralized and "living" WWTP, which can maximize the decomposition of contaminants such as nitrogen and phosphorus, and minimize space and energy *in situ* [30].

4.4.2. Socio-Economic Policies Regulation

The social-economic measures (e.g., "deadline requirements for pollution control", "centralized control of pollutant") have made great progress in addressing industry-derived petroleum and heavy metal pollution in the catchment. Local government is also reliant on economic measures such as industrial structure adjustment, which largely reduced the water consumption and wastewater generation in economic and residential activities. However, socio-economic policies still need to be promoted in the future to obtain sustainable development of economy and environment. Besides the administrative regulation policies (e.g., industrial structure adjustment), economic incentives and financial policies (e.g., water tariff adjustment, trade of pollutant emission rights) should be implemented to encourage industries to prevent or reduce pollution at the source.

4.4.3. Increasing Environmental Awareness

As mentioned in Section 4.3.2, the lack of environmental awareness of local residents is one of the main reasons for low operation efficiency of the existing sewer networks. In order to increase the environmental awareness, the local government should increase the public education level, disclose environmental information and promote public participation in environmental protection. The enterprises should be encouraged to take part in preventive approaches such as ISO 14000, environmental labeling, and cleaner production. In addition, technology support for the creation of neighborhood-based water purification installations should also be promoted to help residents and enterprises effectively reduce pollutant generation/emission and reuse wastewater.

4.4.4. Integrated Measures on the Water Environment

Both socio-economic measures and infrastructure construction are necessary to improve water quality in a rapidly urbanizing catchment. To achieve economic and environmental sustainability in the catchment, planners and policy makers across different sectors must consult with each other and jointly make decisions on integrated planning for socio-economic development and wastewater facilities improvement.

5. Conclusions

In this paper, we characterized the water quality changes and identified the factors that determine the changes in the Shenzhen River catchment during rapid urbanization. In the early stage of urbanization (1985–1995) water quality deteriorated rapidly due to the construction of wastewater infrastructure lagged behind the population and economic growth. Although the population and economy continued to grow in the second stage of urbanization (1996–2009), water quality gradually improved due to the implementation of comprehensive measures for pollution control, and rapid increases in domestic pollution discharge were identified as the major causes of water quality deterioration. Industry-derived petroleum and heavy metal pollution were well controlled; however, the domestic-derived water quality parameters (e.g., BOD_5, NH_3-N and TP) still had a high rate of non-compliance with water quality objectives.

Although the wastewater infrastructure had significant effects on water quality improvement during urbanization, some wastewater could not be efficiently collected and treated before discharge due to the delay in sewer system construction, poor environmental management and environmentally harmful behaviors of residents in the early developed areas. Results also indicate that socio-economic measures had significant effects on water quality improvement. The industry-related pollutants, e.g., heavy metal, petroleum and V-ArOH, have been well controlled by enforcing "deadline requirements for pollution control" and "centralized control of pollutant" in the catchment. In addition, industrial structure adjustment can not only directly reduce pollutant generation from secondary industry by decreasing pollutant load per GRP in the industries but also indirectly reduce pollutant generation from residential activities by increasing labor productivity.

To avoid repeating past mistakes and to institute a sustainable development regime, we suggest that: (1) the wastewater infrastructure should be constructed timely and revitalized regularly in line with the urbanization, and wastewater treatment facilities should be upgraded to improve their nitrogen and phosphorus removal efficiencies; (2) both administrative regulation policies, economic incentives and financial policies should be implemented to encourage industries to prevent or reduce the pollution at the source; (3) the environmental awareness and the education level of local population should be increased; (4) planners and policy makers across different sectors must consult with each other and jointly make decisions on integrated planning for socio-economic development and wastewater facilities improvement.

Acknowledgments

This research was supported by the European Community's Seventh Framework Programme under grant agreement NO. PIIF-GA-2008-220448 and Project of Shenzhen Municipal Development and Reform Commission of China (301200800174-01).

Author Contributions

Hua-peng Qin contributed to the development of the idea and participated in all phases. Qiong Su conducted the data collection/analysis and manuscript preparation. Soon-Thiam Khu helped perform the analysis with constructive discussions. Nv Tang helped improve the figures and manuscript. All authors have read and approved the final manuscript.

Conflicts of Interest

The authors declare no conflict of interest.

References

1. Lehmann, S. Can rapid urbanization ever lead to low carbon cities? The case of Shanghai in comparison to Potsdamer Platz Berlin. *Sustain. Cities Soc.* **2012**, *3*, 1–12.
2. Rana, M.M.P. Urbanization and sustainability: Challenges and stragety for sustainable development. *Environ. Dev. Sustain.* **2011**, *13*, 237–256.
3. Zhao, S.; Da, L.; Tang, Z.; Fang, H.; Song, K.; Fang, J. Ecological consequences of rapid urban expansion: Shanghai, China. *Front. Ecol. Environ.* **2006**, *4*, 341–346.
4. Ren, W.W.; Zhong, Y.; Meligrana, J.; Anderson, B.; Watt, W.E.; Chen, J.K.; Leung, H.L. Urbanization, landuse, and water quality in Shanghai 1947–1996. *Environ. Int.* **2003**, *29*, 649–659.
5. Chang, H. Spatial and temporal variations of water quality in the Han River and its tributaries, Seoul, Korea, 1993–2002. *Water Air Soil Pollut.* **2005**, *161*, 267–284.
6. Kannel, P.R.; Lee, S.; Kanel, S.R.; Khan, S.P.; Lee, Y.S. Spatial-temporal variation and comparative assessment of water qualities of urban river system: A case study of the River Bagmati (Nepal). *Environ. Monitor. Assess.* **2007**, *129*, 433–459.
7. Boeder, M.; Chang, H. Multi-scale analysis of oxygen demand trends in an urbanizing Oregon watershed, USA. *J. Environ. Manag.* **2008**, *87*, 567–581.
8. Ferrier, R.C.; Edwards, A.C.; Hirst, D.; Littlewood, I.G.; Watts, C.D.; Morris, R. Water quality of Scottish rivers: Spatial and temporal trends. *Sci. Total Environ.* **2001**, *265*, 327–342.
9. He, H.M.; Zhou, J.; Wu, Y.J.; Zhang, W.C.; Xie, X.P. Modelling the response of surface water quality to the urbanization. *J. Environ. Manag.* **2008**, *86*, 731–749.
10. Groppo, J.D.; de Moraes, J.M.; Beduschi, C.E.; Genovez, A.M.; Martinelli, L.A. Trend analysis of water quality in some rivers with different degrees of development within the São Paulo State, Brazil. *River Res. Appl.* **2008**, *24*, 1056–1067.
11. Ma, J.Z.; Ding, Z.Y.; Wei, G.X.; Zhao, H.; Huang, T.M. Sources of water pollution and evolution of water quality in the Wuwei basin of Shiyang river, Northwest China. *J. Environ. Manag.* **2009**, *90*, 1168–1177.
12. Duh, J.D.; Shandas, V.; Chang, H.; George, L.A. Rates of urbanisation and the resiliency of air and water quality. *Sci. Total Environ.* **2008**, *400*, 238–256.
13. Qin, H.P.; Su, Q.; Khu, S.T. An integrated model for water management in a rapidly urbanizing catchment. *Environ. Model. Softw.* **2011**, *26*, 1502–1514.

14. Zhang, K.M.; Wen, Z.G. Review and challenges of policies of environmental protection and sustainable development in China. *J. Environ. Manag.* **2008**, *88*, 1249–1261.

15. Shao, W. Effectiveness of water protection policy in China: A case study of Jiaxing. *Sci. Total Environ.* **2010**, *408*, 690–701.

16. Su, S.L.; Li, D.; Zhang, Q.; Xiao, R.; Huang, F.; Wu, J.P. Temporal trend and source apportionment of water pollution in different functional zones of Qiantang River, China. *Water Res.* **2011**, *45*, 1781–1795.

17. Weng, Q. A historical perspective of river basin management in the Pearl River Delta of China. *J. Environ. Manag.* **2007**, *85*, 1048–1062.

18. Miljašević, D.; Milanović, A.; Brankov, J.; Radovanović, M. Water quality assessment of the Borska Reka river using the WPI (water pollution index) method. *Arch. Biol. Sci.* **2011**, *63*, 819–824.

19. Nikolaidis, C.; Mandalos, P.; Vantarakis, A. Impact of intensive agricultural practices on drinking water quality in the EVROS Region (NE GREECE) by GIS analysis. *Environ. Monitor. Assess.* **2008**, *143*, 43–50.

20. Su, Q.; Qin, H.P. Environmental and ecological impacts of water supplement schemes in a heavily polluted estuary. *Sci. Total Environ.* **2014**, *472*, 704–711.

21. Liu, N.; Lu, R.F. *Water Environmental Management in Shenzhen River and Deep Bay*, 1st ed.; China Water Power Press: Beijing, China, 2006. (in Chinese)

22. Census and Statistical Bureau of Shenzhen (CSBSZ). *Statistical Year Book of Shenzhen 1985–2009*; China Statistical Press: Beijing, China, 2010. (in Chinese)

23. Environmental Protection Bureau of Shenzhen (EPBSZ). *Water Quality Report of Shenzhen 1985–2009*; EPBSZ: Shenzhen, China, 2010. (in Chinese)

24. Environmental Protection Department of Hong Kong (EPDHK). River Water Quality in Hong Kong 1985–2009. Available online: http://www.gov.hk/en/residents/environment/water/riverwater.htm (accessed on 18 October 2014).

25. Environmental Protection Department of Hong Kong (EPDHK). *Guideline for Estimating Sewage for Infrastructure Planning*; EPDHK: Hong Kong, China, 2005.

26. MacBean, A. China's Environment: Problems and Policies. *World Econ.* **2007**, *30*, 292–307.

27. Zhang, K.M.; Wen, Z.G.; Peng, L.Y. Environmental policies in China: Evolvement, feature and evaluation. *China Popul. Resour. Environ.* **2007**, *17*, 1–7.

28. Development and Reform Commission of Shenzhen (DRCSZ). Catalogue of industrial structure adjustment 1993–2009. Available online: http://www.szpb.gov.cn/fgzl/cydxml (accessed on 18 October 2014).

29. Qin, H.P.; Su, Q.; Khu, S.T. Assessment of environmental improvement measures using a novel integrated model: A case study of the Shenzhen River catchment, China. *J. Environ. Manag.* **2013**, *114*, 486–495.

30. Organica Food Chain Reactor. Lower Infrastructure Cost: Treating Wastewater at the Source Eliminates the Need for Expensive Infrastructure. Available online: http://www.organicawater.com/solutions/advantages/lower-infrastructure-costs (accessed on 29 September 2014).

Factor Analysis of Residential Energy Consumption at the Provincial Level in China

Weibin Lin, Bin Chen, Shichao Luo and Li Liang

Abstract: This paper analyzes the differences in the amount and the structure of residential energy consumption at the provincial level in China and identifies the hidden factors behind such differences. The econometrical analysis reveals that population, economic development level, energy resource endowment and climatic conditions are the main factors driving residential energy consumption; while the regional differences in energy consumption per capita and the consumption structure can be mainly illustrated by various economic development levels, energy resource endowments and climatic conditions. Economic development level has a significant positive impact on the proportion of gasoline consumption, whereas its impact on the proportion of electricity consumption is not notable; energy resource endowment and climatic condition indirectly affect both the proportion of electricity consumption and that of gasoline consumption, primarily through their impacts on the proportions of coal consumption and heat consumption.

Reprinted from *Sustainability*. Cite as: Lin, W.; Chen, B.; Luo, S.; Liang, L. Factor Analysis of Residential Energy Consumption at the Provincial Level in China. *Sustainability* **2014**, *6*, 7710-7724.

1. Introduction

The residential energy consumption of China in 2012 was about 400 million tons of coal equivalent, which approximately equals the total amount of energy consumption of Brazil in the same year and comprises 11% of the year's total energy consumption. China is a fast developing country with a vast size, and there are great differences in both the amount and structure of residential energy consumption at the provincial level. However, few studies focus on China's residential energy consumption, especially at the provincial level. Thus, there is a need for the identification of critical factors resulting in the regional differences in residential energy consumption.

From a macro point of view, national and regional energy consumptions are often handled considering the impact of income, climate, energy prices, population and other factors on residential energy consumption [1]. Nesbakken [2] studied the impact of energy prices on residential heating equipment and energy consumption during the period 1993–1995 in Norway and points out that price has a significant impact and that high-income households are more sensitive to energy prices than low-income families. Using the panel data of 50 U.S. metropolitan areas from 1997 to 2007, Alberini [3] identified the factors that affect resident's demand for electricity and natural gas and holds that, in either the long or short term, energy price is a major factor, and there is a strong correlation between demand and price, as well. Lenzen *et al.* [4] depicted the residential energy consumption changes in Australia, Denmark, Brazil, India and Japan and find that such changes do not match what is supported by the Kuznets hypothesis, concluding that there is no consistent relationship between energy demand and residential income. Results from Sarak *et al.*, Isaac *et al.* and Zhu *et al.* revealed that fuel consumption is affected by heating degree days, and climate warming in the future would lead to less residential fuel, but more electricity consumption [5–7]. Zhang examined the relationship between the unit energy consumption and heating degree-days for China, Japan, Canada and the United States [8]. Pachauri and Jiang [9] compared the household energy transitions in China and India and find that residential energy consumption both in aggregate and per capita terms in China is twice that in India, while Indians derive a slightly larger fraction of their total household energy needs from liquid fuels and grids than Chinese with comparable incomes. Besides, Nakagami *et al.* [10] surveyed the residential energy consumption and its indicators in 18 countries and demonstrate that household energy consumption per capita shows a trend toward saturation in Western countries, but it will continue to rise in Asian countries [10].

Residential energy consumption has also been studied primarily through questionnaire surveys. Brounen [11] studied residential energy consumption behavior from the perspective of resource conservation. The sample data of more than 300,000 households in the Netherlands reveal that residential natural gas consumption is mainly determined by residence features, such as construction year, building type, *etc.*; while electricity consumption is determined by residential factors, such as income, age structure of family members, *etc.* Using the U.S. 2009 residential energy consumption survey (RECS) micro-data, Tsoand and Guan [12] confirmed the statistically significant impacts of division groups, single-family detached housing, house size, usage of space heating equipment, household size and usage of air-conditioners on residential energy consumption [12]. Heinonen and Junnila [13] employed the household budget survey data of Finland and demonstrate that behavioral differences seem significant between different housing modes and that each housing mode appears to be less energy-intensive in rural areas.

Regarding the residential energy consumption issues of China, Chen [14] conducted a co-integration analysis and argued that actual expenditure is a dominant factor among the factors the affecting residential energy demand and that urbanization and changes in energy consumption structure have little effect on residential energy consumption. Nie *et al.* [15] undertook a decomposition analysis of changes in energy consumption by Chinese households and argued that the increase in energy-using appliances is the biggest contributor followed by floor space per capita, while population is the most stable factor and energy mix is the least important factor. Zhao *et al.* [16] and Qin [17] decomposed the factors that have an impact on residential energy consumption by using the logarithmic mean Divisia index and conclude that population, household income, energy efficiency and structure directly affect residential energy consumption, and especially, the factor of income contributes the most. Particularly, Zhao *et al.* [16] also pointed out that the current energy structure is undergoing an intensive change promoting the usage of high-quality and cleaner energy, such as electricity and natural gas. Chen *et al.* [18] analyzed the data of residential energy consumption in Hangzhou and find that the resident age structure has more influence than income. Fu *et al.* [19] conducted a micro-demographic analysis on residential energy consumption in China and indicated that population change, urbanization and aging are sensitive, while population age is not, except for the 60+ age group. Golley [20] extended the notion of household energy consumption by including indirect energy requirements, then examined the extent of variation in total energy requirements and emissions across households with different income levels in China and, finally, identified that, while richer households do indeed emit more per capita, poorer households tend to be more emissions-intensive.

With regards to rural areas of China, Xu [21] analyzed the residential energy consumption and its structural changes and suggested that income level, merchantability, energy quality and renewability determine the residential energy consumption level of rural households and the consumption structure in China. Lou [22] and Zhang *et al.* [23] studied rural residents' selection behavior in energy consumption, and the results reveal that household wealth, resources availability and the level of education and other household characteristics are the principal factors that determine the level and the structural change of China's energy consumption. Li *et al.* [24] discussed the current status and the regional differences of residential energy consumption in rural China and indicated that significant regional differences exist in the level of residential energy consumption of rural areas, with a gradually decreasing trend along the north-southwest axis of China. Lun *et al.* [25] reported the findings of field surveys in forest villages in Weichang County as a case study of rural energy consumption in northern China and find that local climate, family size and household income have strong influences on rural residential energy consumption. Suo [26] analyzed the residential energy consumption of rural Beijing, implying that the population is a major influence factor and that the energy-saving transformation can effectively reduce the energy consumption quantity.

As mentioned above, there is still a lack of research on regional differences of the residential energy consumption structure and on the factors causing such differences. This paper thereby uses provincial-sectional data of China to explore the regional differences in residential energy consumption and adds climate condition as an explanatory and supplementary factor to the other factors mentioned in previous studies. Meanwhile, it examines the impacts on energy consumption structure brought by the economic development level, energy resource endowments and climatic condition. Therefore, this paper would fill the gap of the lack of a database and help with understanding Chinese residents' energy consumption status.

This paper is organized as follows: Section 2 describes the data and methodology; Section 3 shows the empirical results and some discussions; and Section 4 summarizes the main findings.

2. Data and Methodology

2.1. Data

This study uses panel data of residential energy consumption of China's 28 provinces in 2011. The data of energy consumption are from the *China Energy Statistical Yearbook 2012*. The electricity and gasoline consumptions are derived from the regional energy balance table of the *China Energy Statistical Yearbook 2012*. The data of population, GDP, urbanization rate, average temperature and energy resource endowment are obtained from the *China Statistical Yearbook 2011* and *China Statistical Yearbook 2012*. The mid-year population, the average of year-end population, is treated as population in this paper. GDP per capita is the ratio of gross domestic product in the 2011 price to the mid-year population. Given that this paper is only concerned with the case in year 2011, the GDP figure is taken directly without price adjustment. The urbanization rate is the percentage of residents living in urban areas for more than 6 months in a year. Since coal plays a primary role in China's energy production and consumption, regional coal reserves are used to represent the region's energy resource endowment. Due to the failure of finding an indicator of average temperature in a province, the average temperatures of the province in January and in August are approximated by the average temperatures of the provincial capital city in these two months, respectively.

2.2. Differences in the Amount of Residential Energy Consumption

In 2011, (given that most provincial data of residential energy consumption in 2012 were missing, the inter-provincial cross-section data in 2011 were used instead in the analysis of regional differences in residential energy consumption and its influencing factors), the national (referring to mainland China, excluding Hong Kong, Macao and Taiwan) average of residential energy consumption in 28 provinces (including municipalities and autonomous regions), except Zhejiang, Tibet and Ningxia, is 13.96 million tons of standard coal equivalent (tce, hereafter). The residential energy consumptions of the following seven provinces are more than 20 million tce: Guangdong (36.85 million tons), Henan (25.36 million tons), Hebei (24.99 million tons), Sichuan (23.95 million tons), Shandong (21.17 million tons), Hunan (20.98 million tons) and Heilongjiang (2011 million tons). The residential energy consumption of the nine provinces, namely Hainan (1.3 million tons), Qinghai (2.01 million tons), Chongqing (5.13 million tons), Gansu (6.62 million tons), Jiangxi (7.1 million tons), Tianjin (7.56 million tons), Jilin (7.96 million tons), Guangxi (8.79 million tons) and Xinjiang (9.73 million tons), however, are less than 10 million tons. As shown in Figure 1, there are huge differences in residential energy consumption in different provinces.

According to the description above, there are indeed huge differences in residential energy consumption in different provinces. Then, which factors lead to such differences?

Population is one of the factors resulting in such differences. Those provinces with residential energy consumptions over 20 million tce, except Heilongjiang, are the most populated provinces in China; while the provinces with residential energy consumptions less than 500 tons, such as Hainan and Qinghai etc., have a population of less than 10 million. The correlation coefficient between the provincial data of residential energy consumption and that of population is 0.83, showing that regional difference in residential energy consumption is highly correlated with population.

Besides population, residential energy consumption per capita is another influencing factor. Figure 1 illustrates that there are huge differences in residential energy consumption per capita in different regions and that large differences also exist in the rank of areas according to both per capita and total residential energy consumption. For instance, the residential energy consumption of Guangdong ranks the first, but the level of residential energy consumption per capita is just slightly above the national average level, which is 343 kg tce. Values for Inner Mongolia (773 kg), Beijing (656 kg), Tianjin (570 kg) and Heilongjiang (525 kg) are all higher than 500 kg tce, while those of Hainan (149 kg), Jiangxi (159 kg), Yunnan (178 kg), Anhui (184 kg) and Guangxi (190 kg) are less than 200 kg tce.

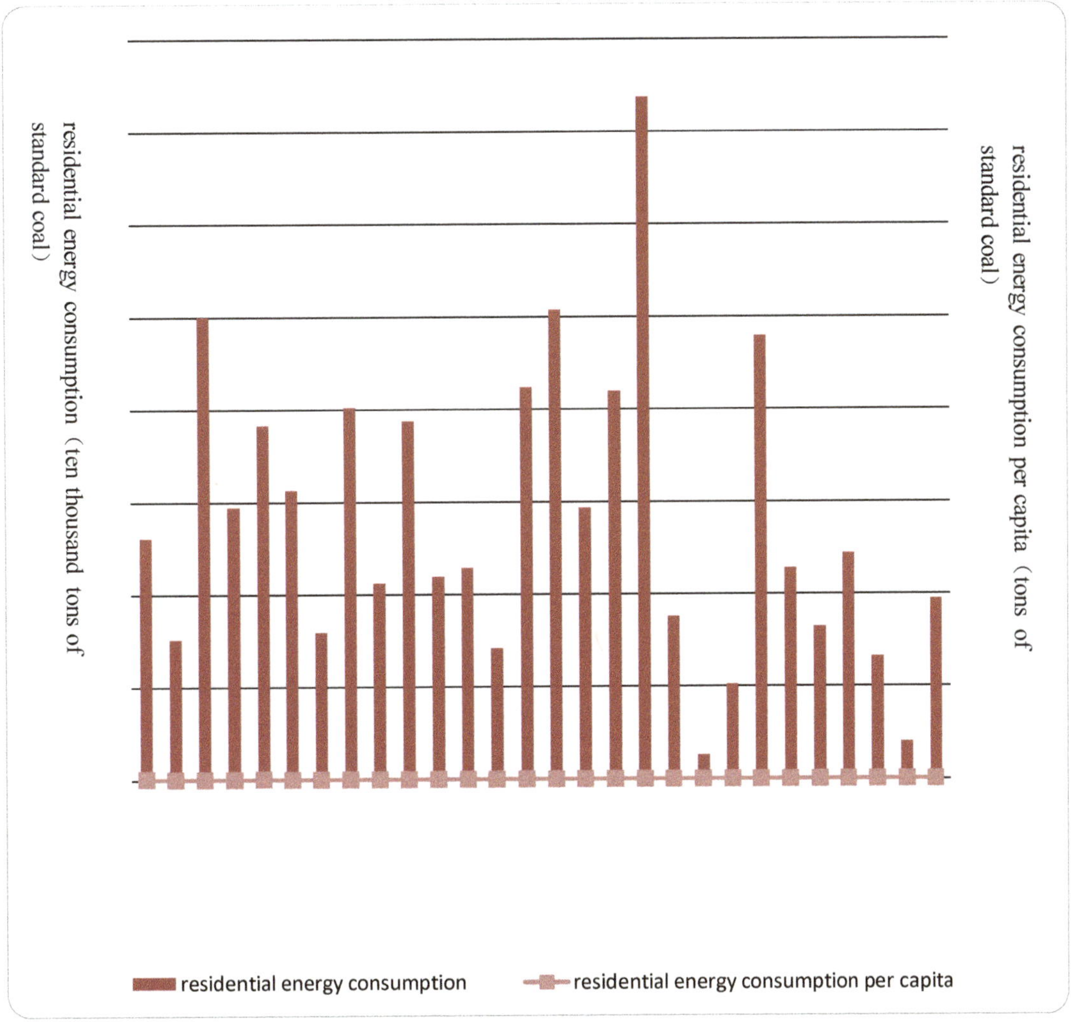

Figure 1. Provincial data of residential energy consumption (2011).

Economic development level may contribute to such differences. The higher the economic development level of a region is, the higher the income and living standards are. Therefore, more electricity, gasoline and other energy commodities would be consumed. For example, Beijing, Tianjin and Shanghai have higher residential energy consumption per capita, because of strong economies, while Hainan, Jiangxi, Yunnan, Anhui and Guangxi have much lower ones, due to relatively weak economies. Using GDP per capita, an indicator of the economic development level, the correlation coefficient between GDP per capita and residential energy consumption per capita is calculated to be 0.62, implying that there is a strong connection between them (see Figure 2).

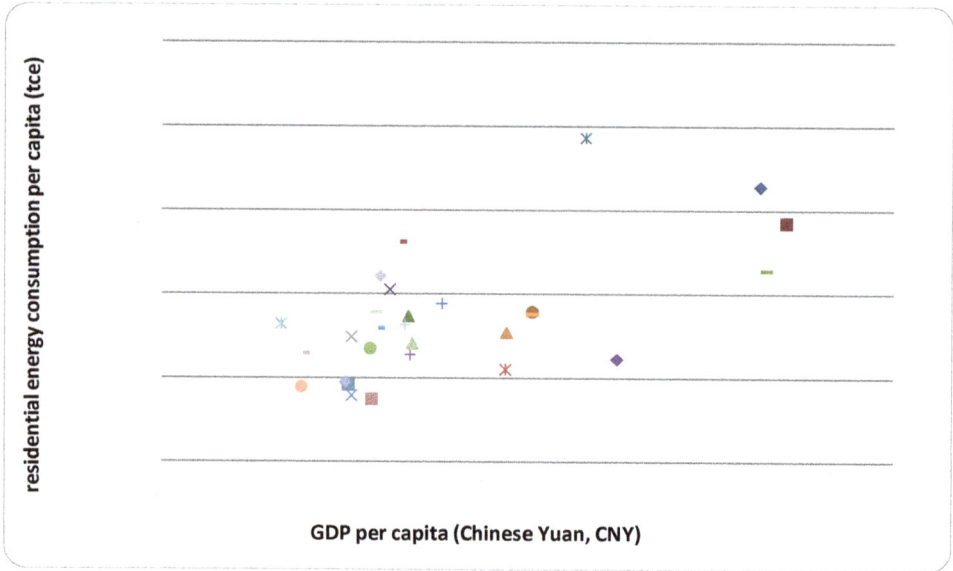

Figure 2. Residential energy consumption per capita and GDP per capita (2011).

Urbanization also contributes to the differences, because urban and rural residents differ greatly in many ways regarding energy consumption. The energy supply systems in urban areas are relatively perfect, with electricity supply, heat supply and natural gas supply providing easy access to clean, convenient and comprehensive energy services for local residents. In contrast, rural residents still primarily rely on traditional ways to access and consume energy. With rapid urbanization and rural residents moving to cities and towns, coal, wood and other traditional energy sources would be replaced by cleaner energies, such as electricity, natural gas, *etc.* In 2011, residential energy consumption per capita in China's rural areas was 0.23 tce per capita, which is 43.5% lower than that of 0.33 tce per capita in urban areas. Statistical tests conducted on both the residential energy consumption per capita and urbanization rate in various regions in 2011 demonstrate that there is a significantly positive correlation between these two factors, with the correlation coefficient being 0.57.

Moreover, the price or cost of the energy can be an explanatory factor that affects residential energy demand. Increases in price lead to decreases in demand, with other conditions unchanged. If the price of energy rises, the economic burden of residential energy consumption will increase, and consequently, the consumption will decrease.

China is a vast country with resources varying significantly in different areas, which results in the differences in the costs of energy. Normally, in a region where energy resources are abundant, such as Shanxi, Inner Mongolia, Xinjiang, *etc.*, the cost of energy will be lower, which leads to higher energy consumption per capita. In this paper, coal reserves are used to represent the differences in the costs of energy among various regions.

Furthermore, climatic condition is another important factor that has an impact on residential energy consumption via the energy demand for heating and cooling. In summer, the areas with hot weather have great demand for air conditioning, while in winter, the regions with cold weather have heavy demand for space heating. Take Heilongjiang for example: its energy consumption per capita ranks the fourth in China, which is largely determined by its cold weather.

A residential energy demand equation can be established based on the analysis above:

$$REC = \beta_0 + \beta_1 POP + \beta_2 GDPP + \beta_3 URB + \beta_4 RES + \beta_5 TEMP1 + \beta_6 TEMP8 + \varepsilon \tag{1}$$

where *REC* is the residential energy consumption; *POP* denotes population; *GDPP* represents GDP per capita, a measure of the economic development level of a region; *URB* is the urbanization rate; *RES* is the coal reserves, a measure of the energy resources endowment and the cost of energy in a region; *TEMP*1 is the average temperature in January and *TEMP*8 is the average temperature in August, two measures of the climatic conditions of a region; ε is the error term.

2.3. Differences in the Structure of Residential Energy Consumption

There are a wide variety of residential energy commodities, such as coal, coke, gasoline, diesel, natural gas, liquefied petroleum gas, coal gas, electricity, heat, *etc.* Regional residential energy consumptions not only differ in quantity, but also vary in consumption structure. In this paper, the proportion of electricity consumption and that of gasoline consumption in residential energy consumption are used as indicators to represent the diversity of the residential energy consumption structure in various regions of China (see Figure 2).

Figure 3 illustrates that there are comparatively large differences in residential energy consumption structure among various regions. In terms of the percentage of electricity consumption in residential energy consumption, the average proportion of the electricity consumption of 28 provinces in 2011 is 42%. The proportions of electricity consumption in Fujian (75%), Jiangsu (67%), Hainan (66%) and Guangxi (61%) are more than 60%, while the proportions of electricity consumptions in Inner Mongolia (16%), Xinjiang (18%), Heilongjiang (23%), Qinghai (25%), Shanxi (26%) and Tianjin (28%) are less than 30%. In terms of the proportion of gasoline consumption in residential energy consumption, the average level of 28 provinces in 2011 is 10%. The proportions of gasoline consumptions in Beijing (27%), Shandong (24%), Hainan (22%) and Jiangsu (20%) are above 20%, while the proportions of gasoline consumptions in Guizhou (1%), Inner Mongolia (1%), Gansu (1%) and Xinjiang (2%) are below 2%.

In order to analyze the factors that affect the regional differences in residential energy consumption structure, this paper selects and compares five provinces with similar consumption: Beijing, Anhui, Fujian, Guizhou and Shaanxi (see Table 1).

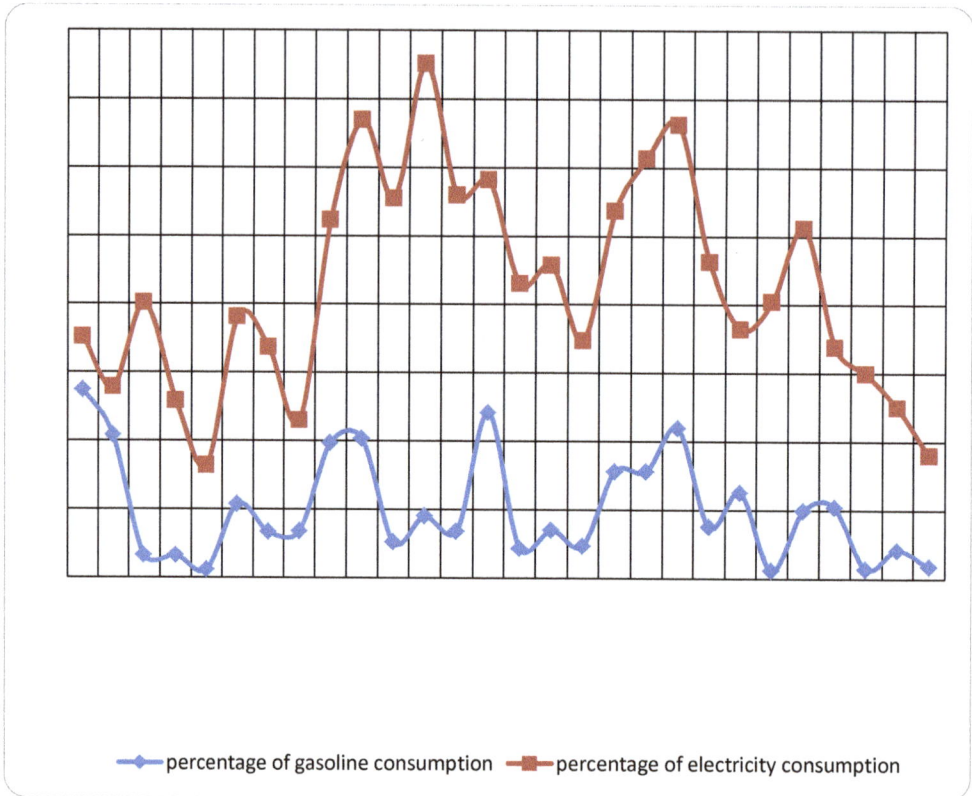

Figure 3. The percentages of electricity and gasoline consumption in different provinces (2011).

Table 1. Residential energy consumption in five provinces. tce, tons of standard coal equivalent.

	Beijing	Anhui	Fujian	Guizhou	Shaanxi
The total residential energy consumption (Ten thousand tce)	1305.8	1095.9	1142.1	1138.9	1218.2
Includes:					
Coal (ten thousand tons)	279.6	226.0	94.2	685.3	254.5
Gasoline (ten thousand tons)	243.5	39.2	70.6	8.7	86.5
Liquefied petroleum gas (ten thousand tons)	21.2	41.7	42.1	9.6	22.0
Natural gas (one hundred million cubic meters)	10.5	8.0	5.3	0.1	13.8
Heat (ten thousand tce)	99.3	71.0	0.0	0.0	92.1
Electricity (one hundred million kWh)	144.7	191.6	270.4	145.1	130.0
GDP per capita (yuan/person)	81,647	25,661	47,377	16,413	33,464
Average temperature in January (Celsius)	−4.5	0.3	8.0	−1.5	−2.8
Coal reserves per capita (tons/person)	18.9	134.0	11.6	169.1	287.8

Data Sources: "China Energy Statistical Yearbook 2012" and Provincial Statistical Yearbooks.

As listed in Table 1, the areas with similar overall levels of residential energy consumption differ greatly in the consumption of coal, gasoline, gas, heat and electricity and other end-use energy products, resulting in the structural differences of residential energy consumption. The reasons can be decomposed as follows:

(1) Economic development level. Take Beijing and Shaanxi for example: their consumptions of coal, gas, heat and electricity are more or less similar. However, their gasoline consumptions are quite different, lying in the different levels of economic development in both regions: Beijing's GDP per capita is approximately 2.5-times that of Shaanxi. Generally speaking, family cars become more and more popular with the increase of economic development level, and gasoline consumptions increase correspondingly. Thus, the difference in economic development level affects gasoline consumption and, thereby, influences the residential energy consumption structure of a region.

(2) Climatic condition. China spreads from Sanya city at 18° north latitude to the northern most county, Mohe, at 53° north latitude, covering tropical monsoon climate, subtropical monsoon climate, temperate monsoon climate, temperate continental climate and alpine climate. In summer, the temperature difference between northern and southern China is relatively small. In winter, however, the difference is obvious and exceeds 50 °C. Thus, climatic condition may bring about the differences in residential energy consumption. In order to solve the space heating problem in winter, the cities that are located to the north of China's Qinling Mountain-Huaihe River (e.g., northern Shaanxi, northern Henan, Shandong, Hebei, Beijing, Tianjin, Shanxi, Gansu, Qinghai, Ningxia, Inner Mongolia, Xinjiang, the majority of Heilongjiang, Jilin, Liaoning, etc.) have built a heating pipeline network covering the whole of the urban areas. Heating services are provided centrally in winter by government-designated companies. Climatic condition, undoubtedly, is a crucial factor that has an important impact on the residential energy consumption structure, which can be demonstrated by the examples of Beijing and Fujian. Fujian is situated in the southeast coast, while Beijing lies in the northeast of the North China Plain. The temperature difference between these two areas in winter is significant: the average temperatures in January are eight degrees Celsius and −4.5 degrees Celsius, respectively. This leads to the fact that the space heating consumption in Beijing is equivalent to 100 million tce, while in Fujian, it is nearly zero.

(3) Energy resources endowment. In the areas with abundant coal reserves, people have easy access to coal at a relatively low price for the purposes of heating, cooking, lighting, etc., which reduces the demand for other energy commodities. For instance, the residential energy consumptions in Fujian and Guizhou are very similar. However, the differences in their coal consumptions are significant, which can be largely attributed to the differences in the coal resources of these two areas.

In summary, the main factors that lead to the regional differences in residential energy consumption structure include economic development level, climate condition and energy resources endowment. In order to further verify the impact of these factors, equations with respect to the proportion of electricity consumption and that of gasoline consumption in the residential energy consumption are established respectively as follows:

$$ELERATIO = \beta_0 + \beta_1 GDPP + \beta_2 COALP + \beta_3 TEMP1 + \varepsilon \tag{2}$$

$$OILRATIO = \beta_0 + \beta_1 GDPP + \beta_2 COALP + \beta_3 TEMP1 + \varepsilon \tag{3}$$

where *ELERATIO* is the proportion of electricity consumption; *OILRATIO* is the proportion of gasoline consumption in the residential energy consumption; *COALP* is coal reserves per capita.

3. Results and Discussions

3.1. Different Amount of Residential Energy Consumption

The residential energy demand equation is validated by using sectional data in 2011, and the results are shown in Table 2.

Table 2. Coefficients of influencing factors of residential energy demand. *REC* is the residential energy consumption; *POP* denotes population; *GDPP* represents GDP per capita, a measure of the economic development level of a region; *URB* is the urbanization rate; *RES* is the coal reserves, a measure of the energy resources endowment and the cost of energy in a region; *TEMP1* is the average temperature in January and *TEMP8* is the average temperature in August.

	Model 1	Model 2	Model 3	Model 4	Model 5	Model 6	Model 7
Constant	−104.10	−475.61	−484.64	−225.03	−232.03	−592.59	−792.43
	(245.82)	(588.56)	(372.38)	(243.54)	(232.13)	(800.4)	(793.35)
POP	0.25	0.26	0.26	0.26	0.27	0.27	0.27
	(0.03) ***	(0.03) ***	(0.03) ***	(0.03) ***	(0.03) ***	(0.03) ***	(0.03) ***
GDPP	0.01	0.00		0.01	0.01	0.01	0.01
	(0.00) **	(0.01)		(0.00) **	(0.00) **	(0.00) *	(0.00) *
URB		13.56	13.93				
		(19.47)	(5.85) **				
RES				0.87			0.68
				(0.47) *			(0.47)
TEMP1					−19.73	−22.91	−20.93
					(8.31) **	(10.81) **	(10.64) *
TEMP8						15.72	21.19
						(33.35)	(32.78)
R^2	0.74	0.74	0.74	0.77	0.79	0.79	0.81
Adjusted R^2	0.71	0.71	0.72	0.74	0.76	0.75	0.76

Notes: Standard errors are in brackets. Significance at the 0.10, 0.05 and 0.01 levels are indicated by *, ** and ***, respectively.

According to the results above, population highly correlates with residential energy consumption. Meanwhile, the higher the level of economic development of a region, the larger amount of energy it consumes. Therefore, in the basic regression model (Model 1), two variables are considered, *i.e.*, population and GDP per capita. Regression results demonstrate that the coefficients of *POP* and *GDPP* are positive and significant, respectively, at the 1% and 5% significance levels with an adjusted R^2 of 0.71, which suggests that the basic model can well explain the regional differences in residential energy consumption.

Building on Model 1, Model 2 incorporates the urbanization rate as an explanatory factor. The results clarify that neither *GDPP* nor *URB* is significant. This is because multicollinearity exists between *GDPP* and *URB*, and these two variables are highly correlated, with the correlation coefficient being 0.95 (see Table 3). Both the urbanization rate and GDP per capita are measures of the economic development level for a region (like two sides of a coin, both the urbanization rate and GDP per capita can represent the economic development level of a region. In other words, neither the progress of urbanization brings about the increase in GDP per capita, nor is urbanization driven by the increase in GDP per capita. These two variables are different indicators of the economic development level of one region). Thus, they should not be incorporated into the model at the same time. In Model 3, population and urbanization rate are chosen as explanatory variables. The regression results show that *URB* is positive and significant at the 5% significance level, indicating that the urbanization rate has a positive effect on residential energy consumption. The higher the urbanization rate in a specific area, the greater is the energy consumed by local residents.

Table3. Correlation coefficient between GDP per capita and urbanization rate.

	GDPP	URB
GDPP	1	
URB	0.95	1

Again, based on Model 1, Model 4 involves coal reserves, denoted by RES, as the explanatory variable, and analyzes how the energy cost influences residential energy consumption. Since it is difficult to find a standard indicator to measure the energy cost, coal reserves are used here to represent the energy cost as an approximation. Regression results of Model 4 show that the coefficient of RES is positive and significant at the 10% significance level, which reveals that the areas with abundant energy resources have a high level of residential energy consumption per capita.

Models 5 and 6, respectively, incorporate the average temperature in January and the average temperatures of January and August in the basic Model 1. The coefficient of *TEMP1* passes the hypothesis testing at the 5% significance level. This suggests that the regions with cold weather in winter have great demand for space heating and require more energy, which is consistent with the proposed hypothesis. The coefficient of *TEMP8* is positive, but not significant. A possible explanation is that air-conditioning accounts for a large part of residential energy consumption in summer, and the number of air-conditioners used correlates with the income level of local residents. An obvious example is that in the southern region of the Yangtze River, despite the intolerable heat in summer, the majority of rural residents and even some urban residents, due to the lower income levels, still choose to use traditional cooling methods, like electric fan and cattail leaf fan, to save electricity consumption.

The regression results of Model 7, which integrates all variables, are basically the same as those of other models presented before. It can be seen that the coefficients of population, GDP per capita, coal resources and the average temperatures in January and August are all in line with the previous hypothesis. However, the coefficients of coal reserves and the average temperature in August do not pass hypothesis testing. The reason why the coefficient of the average temperature in August is not significant has been explained in Model 6, while the possible reason why the coefficient of coal reserves is not significant can be attributed to China's energy price formation mechanism. In the energy sector, prices are heavily influenced by government intervention. Electricity prices are controlled by the Pricing Management Department of the State Council of the People's Public of China or authorized administrative department, *i.e.*, energy prices are set by the government. As a result, the energy price cannot objectively reflect the scarcity of energy resources in various regions, and the pricing effect on residential energy consumption is altered.

3.2. Different Structure of Residential Energy Consumption

Regression results derived by using cross-sectional data in 2011 are presented as follows:

$$ELERATIO = 0.46 - 0.00008COALP + 0.01TEMP1 + 0.0000005GDPP$$

$$(0.04) *** (0.00004) * (0.002) *** \quad (0.000001) \quad R^2 = 0.65$$

(4)

$$OILRATIO = 0.01 + 0.000003GDPP - 0.00004COALP + 0.003TEMP1$$

$$(0.01) (0.000003) *** (0.00002) * (0.001) ** R^2 = 0.64$$

(5)

According to the regression results, the main factors that have an impact on the percentage of electricity consumption in residential energy consumption include the average temperature in January and coal reserves per capita, while the impact brought by GDP per capita is not significant. In a region with abundant coal reserves, the lower cost of coal encourages people to consume more coal resources. Thus, the increase in the proportion of coal consumption lowers the percentage of electricity consumption indirectly. In terms of temperatures, space heating consumption plays an important part in residential energy consumption in the northern regions where the temperatures are normally low, which indirectly lowers the proportion of electricity consumption. In contrast, southern regions witness a higher proportion of electricity consumption overall.

The main factors that influence the proportion of gasoline consumption in residential energy consumption are GDP per capita, coal reserves per capita and the average temperature in January. The impact on gasoline consumption caused by GDP per capita is obvious. When it comes to the temperature, its influence on gasoline consumption is indirect. Heating services in the region where the average temperature in January is low are provided by coal, which makes the proportion of coal consumption higher and, in turn, decreases the proportion of the consumptions of other energy commodities, like the gasoline consumption. Coal resources also have a similar impact mechanism on gasoline consumption. In a region with abundant coal resources, the proportion of coal consumption is undoubtedly higher, which leads to a lower level of gasoline consumption, as it does to the proportion of electricity consumption.

4. Conclusions

This paper presents an analysis of the differences in residential energy consumption in various regions of China and the factors that lead to such differences. We consider both the total amount and the structure of residential energy consumption. Regarding the total amount of residential energy consumption, the main influencing factors include population, economic development, energy resources endowment and climatic condition. Generally speaking, the residential energy consumption amount is larger in provinces with more population, such as Guangdong, Henan, Hebei, Sichuan, Shandong and Hunan. Meanwhile, economic development level, energy resources endowment and climatic condition influence the energy consumption per capita. The higher the level of economic development of a region, such as Beijing, Tianjin and Shanghai, the larger its energy per capita. The more abundant the energy resources per capita in an area, e.g., Inner Mongolia, Shanxi and Xinjiang, the more residential energy consumption per capita. Considering the climatic condition, the average temperature in January has a significant impact on residential energy consumption per capita, especially in the regions where the heating demand in winter is strong, such as Jilin, Liaoning, Heilongjiang, *etc.*, and energy consumption per capita is relatively large.

With regards to the residential energy consumption structure, the main influencing factors include economic development level, energy resources endowment and climatic condition. This paper uses the proportion of electricity consumption and that of gasoline consumption in residential energy consumption as indicators to represent the regional differences in residential energy consumption structure. The empirical analysis reveals that economic development level has a significantly positive impact on gasoline consumption. The higher the level of economic development of a region, the more family car ownership, the greater gasoline consumption and, consequently, the higher the proportion of gasoline consumption in residential energy consumption is. On the other hand, economic development level has no significant impact on the proportion of electricity consumption. The abundance of coal reserves negatively correlates with the proportion of electricity consumption and that of gasoline consumption. This is because the proportion of coal consumption is relatively higher in such regions, which indirectly lowers the proportion of electricity consumption and that of gasoline consumption. In addition, the average temperature In January, as an indicator of the climatic condition, has a significant impact on the residential energy consumption structure. The mechanism is that the colder a specific region, the higher the heating demand, which increases the percentage of heating consumption in residential energy consumption with a corresponding reduction in the proportion of electricity consumption and gasoline consumption.

The results of our study help understand Chinese residents' energy consumption demands in the future. Among major factors affecting residential energy consumption, the growth of China's total population is slow, and climate factors and resource endowments will not have significant changes in the short term, which indicates that the growth of residential energy consumption comes mainly from the increase in residential income and urbanization process. The current growth rate of income per capita of urban and rural residents is higher than the economic growth rate. In 2013,

China's urbanization rate had reached 53.7%, which will be on a rapid growth track according to the laws of urbanization proposed by Northam [27]. Thus, the increase in residential income and the acceleration of the urbanization process will inevitably be expected to bring about the rapid growth of residential energy consumption. In addition, during this process, people will gradually reduce the usage of coal for environmental protection purpose, which, in turn, indirectly increases the usage of clean energy and helps improve the quality of the environment.

Acknowledgments

This work was supported by the Major Research Plan of the National Natural Science Foundation of China (No. 91325302), the Fund for Creative Research Groups of the National Natural Science Foundation of China (No. 51121003), the National Natural Science Foundation of China (No. 41271543), the Fundamental Research Funds for the Central Universities (No. 2012WYB20) and the Specialized Research Fund for the Doctoral Program of Higher Education of China (No. 20130003110027).

Author Contributions

Weibin Lin contributed to the interpretation of the data and drafting the article; Bin Chen made contributions to the concept and design of the article; Shichao Luo collected and analyzed the data; Li Liang provided some useful advices and modified the draft.

Conflicts of Interest

The authors declare no conflict of interest.

References

1. Krigger, J.; Dorsi, C. *Residential Energy: Upper Saddle River*; Prentice Hall: Upper Saddle River, NJ, USA, 2009.
2. Nesbakken, R. Price sensitivity of residential energy consumption in Norway. *Energy Econ.* **1999**, *21*, 493–515.
3. Alberini, A.; Gans, W.; Velez-Lopez, D. Residential consumption of gas and electricity in the U.S.: The role of prices and income. *Energy Econ.* **2011**, *33*, 870–881.
4. Lenzen, M.; Wier, M.; Cohen, C.; Hayami, H.; Pachauri, S.; Schaeffer, R. A comparative multivariate analysis of residential energy requirements in Australia, Brazil, Denmark, India and Japan. *Energy* **2006**, *31*, 181–207.
5. Sarak, H. The degree-day method to estimate the residential heating natural gas consumption in Turkey: A case study. *Energy* **2003**, *28*, 929–939.
6. Isaac, M.; van Vuuren, D. Modeling global residential sector energy demand for heating and air conditioning in the context of climate change. *Energy Policy* **2009**, *37*, 507–521.

7. Zhu, D.; Tao, S.; Wang, R.; Shen, H.; Huang, Y.; Shen, G.; Wang, B.; Li, W.; Zhang, Y.; Chen, H.; *et al.* Temporal and spatial trends of residential energy consumption and air pollutant emissions in China. *Appl. Energy* **2013**, *106*, 17–24.

8. Zhang, Q. Residential energy consumption in China and its comparison with Japan, Canada, and USA. *Energy Build.* **2004**, *36*, 1217–1225.

9. Pachauri, S.; Jiang, L. The household energy transition in India and China. *Energy Policy* **2008**, *36*, 4022–4035.

10. Nakagami, H.; Murakoshi, C.; Iwafune, Y.; Jyukankyo Research Institute. International Comparison of Household Energy Consumption and Its Indicator. In Proceedings of the 2008 ACEEE Summer Study on Energy Efficiency in Buildings, Pacific Grove, CA, USA, 17–22 August 2008; Volume 8, pp. 214–224.

11. Brounen, D.; Kok, N.; Quigley, J. Residential energy use and conservation: Economics and demographics. *Eur. Econ. Rev.* **2012**, *56*, 931–945.

12. Tso, G.; Guan, J. A multilevel regression approach to understand effects of environment indicators and household features on residential energy consumption. *Energy* **2014**, *66*, 722–731.

13. Heinonen, J.; Junnila, S. Residential energy consumption patterns and the overall housing energy requirements of urban and rural households in Finland. *Energy Build.* **2014**, *76*, 295–303.

14. Chen, X.; Yuan, H. An empirical study on the factors affecting residential energy consumption behaviour in China. *Consum. Econ.* **2008**, *5*, 47–50. (In Chinese)

15. Nie, H.; Kemp, R. Index decomposition analysis of residential energy consumption in China: 2002–2010. *Appl. Energy* **2014**, *121*, 10–19.

16. Zhao, X.; Li, N.; Ma, C. Residential energy consumption in urban China: A decomposition analysis. *Energy Policy* **2011**, *41*, 644–653.

17. Qin, Y. Study on Chinese Residential Energy Consumption. Master's Thesis, Shanxi University of Finance & Economics, Taiyuan, China, 2013. (In Chinese)

18. Chen, J.; Wang, X.; Steemers, K. A statistical analysis of a residential energy consumption survey study in Hangzhou, China. *Energy Build.* **2013**, *66*, 193–202.

19. Fu, C.; Wang, W.; Tang, J. Exploring the sensitivity of residential energy consumption in China: Implications from a micro-demographic analysis. *Energy Res. Soc. Sci.* **2014**, *2*, 1–11.

20. Golley, J.; Meagher, D.; Xin, M. Chinese Household Consumption, Energy Requirements and Carbon Emissions. Available online: http://people.anu.edu.au/xin.meng/ Draft%20May%2012.pdf (accessed on 20 October 2014).

21. Xu, X. Analysis on Chinese Rural Residential Energy Consumption. Master's Thesis, Chinese Academy of Agricultural Sciences, Beijing, China, 2008. (In Chinese)

22. Lou, B. Study on Rural Households' Selection Behaviour in Residential Energy Consumption. Master's Thesis, Chinese Academy of Agricultural Sciences, Beijing, China, 2008. (In Chinese)

23. Zhang, N.; Xu, W.; Cao, P. Analysis of the factors that influenced rural households' residential energy consumption—Based on micro data of nine provinces. *Chin. J. Popul. Sci.* **2011**, *3*, 73–82. (In Chinese)

24. Li, G.; Nie, H.; Yang, Y. Regional disparities and influencing factors of rural energy consumption in China. *J. Shanxi Financ. Econ. Univ.* **2010**, *2*, 68–73.

25. Lun, F.; Canadell, J.; Xu, Z.; He, L.; Yuan, Z.; Zheng, D.; Li, W.; Liu, M. Residential energy consumption and associated carbon emission in forest rural area in China: A case study in Weichang County. *J. Mount. Sci.* **2014**, *11*, 792–804.

26. Suo, C.; Yang, Y.; Solvang, W. Analysis of influence factors of rural residence transformation on residential energy consumption. *Mod. Manag.* **2014**, *4*, 493–515.

27. Northam, R.M. *Urban Geography*; John Wiley: New York, NY, USA, 1979.

Emergy-Based Regional Socio-Economic Metabolism Analysis: An Application of Data Envelopment Analysis and Decomposition Analysis

Zilong Zhang, Xingpeng Chen and Peter Heck

Abstract: Integrated analysis on socio-economic metabolism could provide a basis for understanding and optimizing regional sustainability. The paper conducted socio-economic metabolism analysis by means of the emergy accounting method coupled with data envelopment analysis and decomposition analysis techniques to assess the sustainability of Qingyang city and its eight sub-region system, as well as to identify the major driving factors of performance change during 2000–2007, to serve as the basis for future policy scenarios. The results indicate that Qingyang greatly depended on non-renewable emergy flows and feedback (purchased) emergy flows, except the two sub-regions, named Huanxian and Huachi, which highly depended on renewable emergy flow. Zhenyuan, Huanxian and Qingcheng were identified as being relatively emergy efficient, and the other five sub-regions have potential to reduce natural resource inputs and waste output to achieve the goal of efficiency. The results of decomposition analysis show that the economic growth, as well as the increased emergy yield ratio and population not accompanied by a sufficient increase of resource utilization efficiency are the main drivers of the unsustainable economic model in Qingyang and call for polices to promote the efficiency of resource utilization and to optimize natural resource use.

Reprinted from *Sustainability*. Cite as: Zhang, Z.; Chen, X.; Heck, P. Emergy-Based Regional Socio-Economic Metabolism Analysis: An Application of Data Envelopment Analysis and Decomposition Analysis. *Sustainability* **2014**, *6*, 8618-8638.

1. Introduction

The process of economic development driven by industrialization can be seen as a transition from an agrarian to an industrial socio-metabolic regime [1]. The transition makes human society experience spectacular wealth accumulation, but it also causes large-scale ecological and social transformations [2]. Thus, the social-environmental relationship can be characterized as socio-economic metabolism [3,4]. The concept of socio-economic metabolism is applied to investigate the scale and composition of the socio-economic metabolic system [1,5–7] and to discuss its relationship with both economic growth [8], urbanization [5,9], industrialization [10,11] and environmental impacts [12]. The socio-economic metabolism perspective provides a useful framework for studying the interaction between human and natural systems [13] through quantifying the regional input-output amount and the structures of material and energy flows, the value of local resources and sustainability performance.

Nowadays, the term socio-economic metabolism has been widely applied in different fields, and related studies have been conducted at different levels, such as the household [14,15], industrial [10,16], urban [17–20] and regional level [21]. As for the regional level, related studies refer to quantifying material and the energy metabolic amount and the efficiency of different countries and regions, thus comparing the sustainability performance for different regional systems [22] by applying material flow analysis (MFA) [23], emergy analysis [20] and other integrated approaches (e.g., the multi-scale integrated analysis of socio-economic metabolism approach, MuSIASEM [21]).

As an instrument for aggregating various material flows into a few strategy indicators, MFA is applied to measure the weights of material inflows and outflows of socio-economic metabolism and can be broken into substance flow analyses (SFA), which deal with chemically-defined substances (such as phosphorus, copper, sulfur, *etc.*) in practical research [3]. However, the MFA-based socio-economic metabolism research has placed emphasis on the weight (quantities) of resource flows and ignored the varied qualities of material flows. Besides, the MFA-based metabolism studies provide strong and consistent evidence of the increasing consumption of resources in most economies, even in those economies that have focused their policies on dematerializing economic growth [24]. Emergy analysis, developed by Odum [25], is a technique of quantitative analysis that determines the values of resources, services and commodities in a common unit of solar energy, which allows all resources to be compared on a fair basis [26]. Therefore, emergy-based metabolism analysis, rather than the mass content of differing resources in the MFA, can overcome the limits of MFA that we mentioned above [3]. Moreover, emergy analysis provides an ecocentric bridge that connects economic and ecological systems and is a more holistic alternative to many existing methods for environmentally conscious decision making [27]. However, the emergy analysis has the problems of quantification, uncertainty and sensitivity, which also exist in all methods (such as MAF, LCA, exergy, *etc.*) that focus on a holistic view of industrial activity. In order to reduce the bias induced by the problems mentioned above, the paper integrated the emergy into data envelopment analysis, which is a nonparametric production frontier analysis approach and has been widely applied in micro- and macro-economic studies [24], to revise the results of emergy analysis and to provide decision makers with useful information regarding how to improve ecological efficiency. The paper also investigated the underlying determinant effects that influence the change of total emergy use by combining the methods of emergy and decomposition analysis. The present paper, taking Qiyang city and its eight sub-regions (Gansu Province, northwestern China) as the study case, investigates the features and findings of socio-economic metabolism.

2. Research Area

Qingyang city is located in the east part of Gansu province in northwestern China and ranges in latitude from $35°15'N$ to $37°10'N$ and in longitude from $106°20'E$ to $108°45'E$, with a total administrative area of 27,119 km^2, consisting of one district, including Xifeng (XF), and seven counties, including Qingcheng (QC), Ningxian (NX), Zhengning (ZN), Huanxian (HX), Heshui (HS), Huachi (HC) and Zhenyuan (ZY) (Figure 1). Qingyang is one of the cradles of traditional Chinese farming culture and also is rich in crude oil and raw coal, with 3.25 billion tons of crude oil, 12 billion tons of raw coal and 1358.8 billion cubic-meters of coal-bed methane.

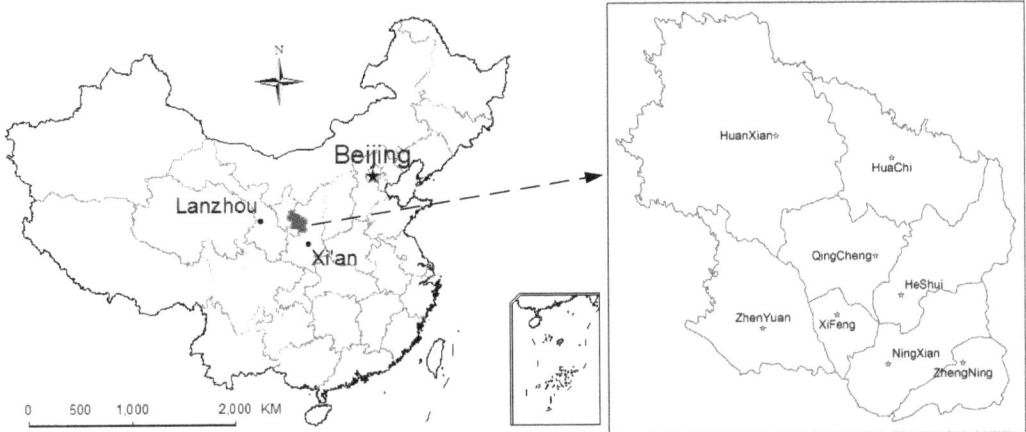

Figure 1. Qingyang city in China and the eight sub-regions.

In 2007, the GDP of Qingyang was $264 million, and GDP per capita was $1005, which was much lower than the average level of GDP per capita in China ($2640). Qingyang is in the fast transition period from traditional agriculture to rapid industrialization [28], particularly after the discovery of abundant energy resources and the implementation of China's Western Development Program in 2000. With the large amount of exploitation and processing of energy resources, environmental pressure has increased rapidly. Thus, in the transition period, searching for a sustainable development model is a major task faced by local officials.

Specifically, the eight sub-regions in Qingyang all have their own distinctive features in the aspects of geography, environment and economy. For instance, XF is the political center of Qingyang region and is largely dependent on the inputs of energy, materials and products from outside. QC, with a relatively longer history of oil exploitation, is the main oilfield. NX and ZN, in the southeastern part of Qingyang, have certain similar features, such as rich coal resources and relatively developed agriculture. HX, in northern Qingyang, has a very high rate (99.8%) of soil erosion and water loss, as well as an underdeveloped economy. HS and HC have a relatively sound ecological environment and mid-level economic development. Economic development in ZY largely depends on the processing of agriculture products, especially apricot. Thus, to achieve sustainable development of Qingyang as a whole, differentiated policies regarding sustainable development, targeting each sub-region are crucial, as the each sub-region has special characteristics.

3. Research Methods

3.1. Emergy and Emergy Based Indicators

Emergy analysis is an environmental accounting method, which considers the energy system for the thermodynamics of an open system, aims to evaluate the contributory value of different material flows to the ecological economic system [29,30]. Emergy is measured in solar embodied joules, abbreviated sej. Emergy analysis characterizes all materials, energy, capital and services in equivalents of solar energy, *i.e.*, how much emergy would be required to do a particular task if the solar radiation were the only input [27,31–33]. Thus, the key part of emergy analysis is to transform the various materials and capital in human activities to a unified unit, sej [25], in which the key parameter is the emergy transformity (Trf.). The transformity of solar radiation is assumed equal to one by definition (1.0 sej/J), while the transformities of all of the other materials, energy and services are calculated based on their convergence patterns through the biosphere hierarchy [34]. Regarding emergy flows, renewable, nonrenewable and feedback emergy are three streams at the input side of the system. Additionally, output product emergy and waste emergy are two steams at the output side (Figure 2). The major steps of emergy analysis include identifying the system boundary, collecting eco-economic data, establishing emergy flow accounting, calculating a set of indices and ratios and using them to conduct the socio-economic metabolism analysis. Theoretically, all of the material, information and capital flows through the target studied region should be diagramed and then calculated. However, that is usually difficult, due to the lack of databases, especially for a city located in an undeveloped area in China. Therefore, the paper carried out a black box study without probing into the socio-economic metabolic structure thoroughly.

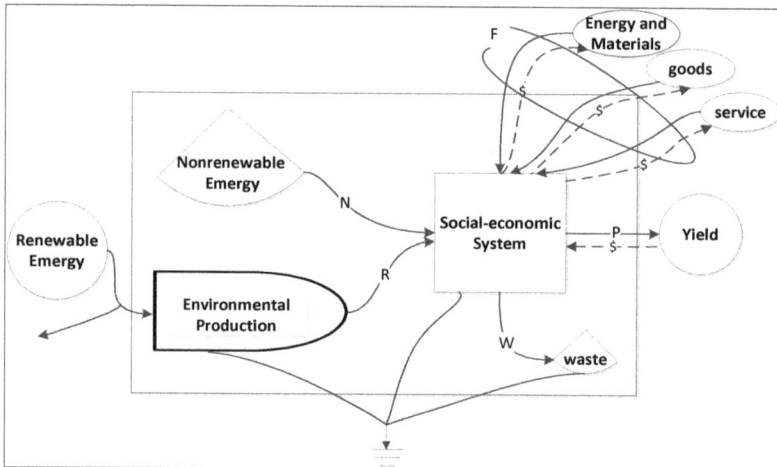

Figure 2. Emergy flows of a regional system.

Based on standard transformation, the emergy flow accounting was based on Qingyang and 8 sub-regions' raw data, which were from the QY Statistic Year Book (2001–2008), the XF, HX, HC, HS, ZN, ZY, NX, QC Statistic Year Books (2001–2008) [35–42] and investigations conducted on two planning programs: Qingyang Eco-City Development Program (investigation time period: 11–19 August 2008) and Qingyang Circular Economy Development Program (investigation time period: 20–29 April 2009); the corresponding indicators are given in Table 1. Additionally, the emergy transformity (Trf.) we used in this work relies on the transformities calculated by Odum and co-workers [33,43] (The averaged transformities of natural resources, products and processes used in this paper may lead to estimation bias to some extent, which was also shared by other approaches and is inevitable).

Table 1. The geographic and economic features of eight sub-regions and the whole (data of 2007).

		QY	XF	QC	ZN	NX	ZY	HX	HS	HC
Area (km²)		27,119	996	2693	1321	2654	3501	9236	2942	3776
Average altitude (m)		1,310	1421	1310	1460	1310	1500	1612	1436	1440
Average rainfall (m/a)		0.53	0.5271	0.498	0.63	0.56	0.39	0.646	0.56	0.46
Population (1000 person)		2626.9	341	331.8	240.9	535.8	523.3	349.1	168.1	131.7
Population growth rate (‰)		7.41	7.12	7.26	7.50	7.12	7.10	7.76	7.54	7.45
GDP (million $)		2640.98	675.96	697.00	114.86	208.18	188.98	151.37	99.49	442.58
GDP per capita ($)		1005	1605	2112	478	385	363	432	585	3404
Industrial Structures	Primary (%)	16.15	11.5	7.17	40.30	36.58	42.17	20.30	40.84	5.80
	Secondary (%)	59.22	51.9	81.66	12.90	24.50	20.67	46.60	24.32	84.80
	Tertiary (%)	24.63	36.6	11.17	46.80	38.91	37.16	33.10	34.84	9.30

Notes: The abbreviations QY, XF, QC, ZN, NX, ZY, HX, HS and HC in the table represent Qingyang city, Xifeng district, Qingcheng, Zhengning, Ningxian, Zhenyuan, Huanxian, Heshui and Huachi, respectively.

3.2. Data Envelopment Analysis

Data envelopment analysis (DEA), which is a non-parametric frontier approach to evaluate the relative efficiency of a set of homogeneous decision-making units (DMUs) featuring multiple inputs and outputs, has recently been widely applied to analyze energy, environmental and ecological efficiency [44,45]. In recent empirical studies in the macroeconomic literature, GDP is commonly used as the output, and capital, labor and natural resources (water, energy, land and other mineral resources) are used as the input. The DEA efficiency model considering pollutants can be divided into three types. The first model takes pollutants as investment costs and, thus, is used to as the input. The second model is the data transfer function method, which considers the pollutants as ordinary output after transferring the-smaller-the-better undesirable output into the-bigger-the-better desirable output. The third model is the distance function method first proposed by Färe *et al.* (1994) [46] and then further extended by Chung (1997) [47], Tone (2001) [48] and other scholars. Because it can redeem the defect of the first two methods, such as no reflection on the real production process, strong convexity constraints, *etc.* [45], the distance function method has been widely applied to analyze ecological (or environmental) efficiency and environmental performance.

To comprehensively capture the physical flow (including inputs and outputs) of social and economic activities and to measure the ecological efficiency precisely, the inputs should include all kinds of inputs coming from the natural system and waste discharged into the environment. Usually, these inputs and outputs are measured in different measurement units. The emergy-based accounting system can solve the unit problem. In this paper, we combined the emergy-based accounting system and the undesirable output DEA model (slack-based undesirable output model) proposed by Tone (2001) to evaluate the ecological efficiency (EE) of the socio-economic metabolic system in Qingyang. A more detailed explanation of the model and its underlying mathematical procedures can be found in [48,49].

The paper uses the emergy content of natural resource inputs, production and waste emissions instead of mass content to construct aggregate input and output factors. The input factors include the nonrenewable input emergy (N) and feedback emergy (F), and the output factors include both the output product emergy (P), which is taken as a desirable output indicator, and waste emergy (W), which is taken as an undesirable output indicator. The score of ecological efficiency obtained by the slack-based undesirable output model is bounded by zero and one. If the score is upper-bounded to one, this means that it represents the best performance. Moreover, the lower the score, the worse the level of ecological efficiency [24,50].

3.3. Decomposition Analysis

Decomposition analysis is widely applied to investigate the underlying determinant factors that influence the change of energy consumption, CO_2 emissions, material usage, *etc.* [51–54]. Decomposition analysis covers two kinds of specific methods. The first one is structural decomposition analysis (SDA), which handles the input-output model. The second one is index decomposition analysis (IDA), which uses sector- or regional- level data. Because of requiring less data, the IDA has been applied to environmental and resource issues more extensively than SDA [55]. The decomposition analysis carried out in this paper is IDA based on the advanced sustainability analysis (ASA) approach, which was developed by the Finland Futures Research Center [56–58].

Since the decomposition analysis is capable of assessing the efficiency in the use of a given input in affecting a final result, it can be considered a fundamental tool to ease the monitoring and evaluation of the sustainability of economies and productive sectors/processes [59]. The knowledge of the major factors that affect a process' performance is essential for the design of new policy instruments and the evaluation of the implemented measures over a desired pattern of sustainability [60,61]. Moreover, decomposition analysis is necessarily dealing with complex indicators to avoid the loss of information [62,63].

ASA is designed to investigate the relationship between changes in environmental, economic and social variables that are measured by any preferred indicator or index. An equation describing the relationship between the factors (e.g., intensive factor V/X_1 and extensive factor X_1) contributing to variable V can be expressed in its simplest form as follows:

$$V = \frac{V}{X_1} \times X_1 \tag{1}$$

The procedure can be applied to multiple actors, as well. The two-factor decomposition presented above can be continued by taking a result from the first decomposition as a starting point for further decompositions, and the new results can then be decomposed again. The equation that identifies the contributing variables can be formulated in a general form as follows:

$$V = \frac{V}{X_2} \times \frac{X_2}{X_3} \times ... \times \frac{X_{n-1}}{X_n} \times X_n \tag{2}$$

A more detailed explanation of the ASA approach and its underlying mathematical procedures can be found in [64] and the DECOIN (2008) [65] and SMILE (2011) [66] websites.

4. Results

4.1. Overview of the Emergy Results of Qingyang and the Eight Sub-Regions

To characterize the metabolic structure and efficiency of the socio-economic system in Qingyang and its eight sub-regions, we calculated the emergy indices (Table 2). We found that the socio-economic system in Qingyang depended heavily on nonrenewable emergy, which occupied 61.13% of the total emergy input in 2007. Renewable emergy was the second largest input (23.54), followed by the feedback emergy input (15.32). The high value of the environment loading ratio (ELR = 3.25) and the low value of the emergy investment ratio (EIR = 0.18) refer to economic growth being greatly devoted to local non-renewable resources, which leads to high environmental pressure. An ESI (2.51) between one and ten means that Qingyang developed a "producer"-oriented economy that highly relied on non-renewable resources (Table 3).

Table 2. Indicators for emergy analysis in Qingyang.

Category	Emergy Index	Meaning	Calculation
Emergy Flows	Renewable Input Emergy (R)	local renewable resource input	
	Nonrenewable Input Emergy (N)	local nonrenewable resource input	
	Feedback Emergy (F)	feedback emergy from outside regions	
	Total Emergy Input (U)	total emergy input	R + N + F
	Output Product Emergy (P)	output emergy of products	
	Waste Emergy (W)	waste emergy discharged to the environment	
Emergy Structure	The Ratio of R, N and F to U	The ratio reflects the contributions of renewable, nonrenewable and feedback emergy to the total emergy input of the regional system at the input side	R/U; N/U; F/U
Emergy-Efficiency	Net Emergy Yield Ratio (NEYR)	NEYR reflects the supporting capability of local resources to economic development; meanwhile, it accounts for whether the economic system has competitiveness in supplying primary energy and resource [33,68]	P/F
	Emergy Investment Ratio (EIR)	EIR shows the efficiency of the usage of feedback emergy compared with local renewable and nonrenewable emergy.	F/(N + R)
	Emergy Money Ratio (EMR)	EMR shows the level of economic development. In general, a developed economic system has lower EMR as fast money circulation, while a rural area has higher EMR as a large percentage of unpaid local emergy input [33].	U/GDP
	Emergy Per Capita (EPC)	EPC reflects the residents' living standard from the emergy perspective.	U/population
	Emergy Density (ED)	ED shows spatial concentration of emergy flow within the regional system. Usually, a developed economic system has higher ED [33,68].	U/Area
Environment Pressure	Emergy Waste Ratio (EWR)	the waste discharge level in the input-side perspective	W/U
	Waste Output Ratio (WOR)	the waste discharge level in the output-side perspective	W/P
	Environment Loading Ratio (ELR)	ELR shows the pressure of social-economic activities on the local ecosystem [67]	(F + N)/R
Integrated Indicator	Emergy Sustainability Index (ESI)	A sustainable system should have higher NEYR and lower ELR. Usually, when ESI < 1, it is a consumption system; when 1 < ESI < 10, it is an energetic system with enormous potentials for further development; while when ESI > 10, the system is economically lagging behind [69–71]	NEYR/ELR

XF, as the regional political center, mostly relied on feedback emergy, where the social-economic activity is supported by import emergy (96.76%). ZY also mainly relied on imported emergy (feedback emergy/total emergy input (F/U) is 54.99%). QC and ZN greatly depended on local non-renewable resources (U/nonrenewable input emergy (N) in the two sub-regions is greater than 50%). However, for HX and HC, the economic development highly relied on local renewable resources. The economic activities in NX and HS mainly depended on local renewable and non-renewable resources, respectively, but the dependence degree is less than HX and QC (Table 3).

The net emergy yield ratio (NEYR) is the ratio of the output emergy of the economic system to the feedback emergy from outside the region, which reflects the capability of local resources to support economic activities. Meanwhile, it shows the competitiveness of the economic system in supplying primary energy and resources [33]. The higher NEYR is, the higher the contribution of the regional economic system to the other regions is. In general, a value of NEYR of QY higher than five indicates that QY is an energy output region [67]. The NEYR of QC is the highest among the eight sub-regions, indicating the importance of QC in providing energy to the Qingyang economy. Additionally, this high value is due to the exploitation of non-renewable resources in QC. Conversely, the NEYR of XF is the lowest, which is caused by having the highest feedback emergy, as well as a higher EIR, emergy money ratio (EMR), emergy per capita (EPC) and emergy density (ED) compared to the average of QY, reflecting that XF has the strongest economic intensity and highest residential living conditions and ranks as the highest level of the economic system in QY city (Table 3).

Table 3. The emergy analysis results of socio-economic metabolism in Qiyang (2007). sej, solar embodied joules.

	QY	XF	QC	ZN	NX	ZY	HX	HS	HC
Emergy Flows									
R $(\times 10^{20}$ sej)	**41.8**	1.81	5.15	2.44	4.29	3.37	15.3	1.55	4.19
N $(\times 10^{20}$ sej)	**108.6**	0.34	25.1	4.85	3.76	0.25	1.27	2.71	0.44
F $(\times 10^{20}$ sej)	**27.2**	64.4	1.69	0.51	2.54	4.39	4.95	2.26	1.09
P $(\times 10^{21}$ sej)	**2.22**	1.41	6.07	0.72	3.22	1.76	5.63	0.79	0.63
W $(\times 10^{19}$ sej)	**39.23**	9.31	7.78	7.31	3.71	0.7	1.13	5.69	3.61
Emergy Structure									
R/ U (%)	**23.54**	2.73	16.12	31.33	40.49	42.19	71.08	23.79	73.24
N/U (%)	**61.13**	0.51	78.58	62.18	35.49	2.82	5.91	41.58	7.7
F/ U (%)	**15.32**	96.76	5.3	6.49	24.01	54.99	24.01	34.63	19.03
Emergy Efficiency									
NYER	**8.15**	0.22	35.9	14.24	12.66	4.01	11.37	3.52	5.75
EIR	**0.18**	29.87	0.06	0.07	0.31	1.22	0.3	0.53	0.24
EMR $(\times 10^{12}$ sej/$)	**6.732**	9.85	4.58	6.79	5.09	4.23	1.42	6.55	1.29
EPC $(\times 10^{15})$	**6.76**	19.5	9.62	3.24	2.16	1.53	6.16	5.66	17.7
ED $(\times 10^{11})$	**6.5**	66.8	11.9	5.91	4.37	2.28	2.33	3.23	6.18
Environmental Pressure									
EWR	**0.022**	0.014	0.02	0.09	0.035	0.009	0.005	0.087	0.063
WOR	**0.018**	0.07	0.01	0.1	0.01	0.003	0.002	0.07	0.06
ELR	**3.25**	35.68	5.2	2.19	1.47	1.37	0.41	3.2	0.37
ESI	**2.51**	0.006	6.9	6.5	8.61	2.92	27.94	1.1	15.74

Notes: The abbreviations QY, XF, QC, ZN, NX, ZY, HX, HS and HC in the table represent Qingyang city, Xifeng district, Qingcheng, Zhengning, Ningxian, Zhenyuan, Huanxian, Heshui and Huachi, respectively.

The environment loading ratio (ELR) is the ratio of the sum of feedback and nonrenewable emergy to renewable emergy, which reflects the pressure of the social-economic process on the local ecosystem. The ELR of XF is highest (ELR = 35.68), reflecting high environmental pressure as a result of social economic activity. The ELR of QC (5.2) is the second highest and above the Qingyang average, due to its larger consumption of non-renewable resources (such as crude oil exploitation). On the other hand, the ELRs of the other five sub-regions are below Qingyang average. These regions can be considered as the areas with lower environmental pressure in terms of resources extraction and use (Table 3).

The emergy investment ratio (EIR) is the ratio of feedback (purchased) inputs to local resources (F/N + renewable input emergy (R)). The EIRs of XF and ZY are greater than one, indicating that the two regions relied more on purchased inputs than locally-available resources. However, the EIRs of the other six regions are less than one, indicating the need for local resources in these regions (Table 3).

The emergy sustainability index (ESI) is the ratio of NEYR to ELR. When ESI < 1, it is a consumption system; when 1 < ESI < 10, it is an energetic system with the potential of further development; and when ESI > 10, the system is economically lagging behind [72]. In general, the ESI of QY is between one and 10, which reflects that QY is an energetic regional system and has the potential for future development. Regarding the eight sub-regions, the ESI of XF is less than one, which reveals that XF is a consumption system. The ESIs of HX and HC are higher than 10, which shows that both regions are economically behind and economic development needs to be accelerated and strengthened. The ESIs of NX, QC, ZY, HS and ZN are between one and 10, where the economic development is energetic and robust (Table 3).

4.2. Emergy-Based Data Envelopment Analysis

Based on the emergy accounting database, the paper applied the undesirable-output DEA model to estimate the production frontier and calculate the ecological efficiency (EE) score of eight sub-regions in Qingyang using the input-oriented GRS (generalized returns-to-scale) framework. It is arguable to assume that there exists a production frontier for the eight sub-regions, because of the significant differences in the characteristics of the economic structures of those sub-regions.

The average EE score of eight sub-regions increased from 0.669 in the year 2000 to 0.766 in 2002, then decreased to 0.701 in 2006 and, finally, achieved an EE score of 0.728 in 2007. The EE scores of ZY and HX equal one during 2000–2007, which means that the two regions were identified as being relatively efficient. QC could be identified as an efficient region after 2002. The EE scores of NX, HS and HC decreased from 1.000, 0.309 and 0.805 to 0.802, 0.252 and 0.766. As for ZN, the EE score increased from 0.697 to 0.805. The emergy efficiency of XF was lowest among the eight sub-regions during the whole period (the EE score was below 0.2 in most years). In 2007, only three regions (QC, ZN and HX) were identified as efficient regions (Table 4).

4.3. Decomposition Analysis of Total Emergy Use

In order to understand the main drivers of the changes of total emergy flow (U), a decomposition equation was developed according to Equations (1) and (2). The time change of total emergy use (U) during 2000–2007 was decomposed as follows:

$$U = \frac{U}{F} \times \frac{F}{R+N} \times \frac{R+N}{GDP} \times \frac{GDP}{POP} \times POP \qquad (1)$$

where F indicates feedback emergy flow, R is the value of all locally-available renewable emergy flows, N refers to the locally-available nonrenewable emergy flow, GDP is the value of gross domestic production and POP is the amount of population. According to Equation (3), changes of U are affected by five factors: U/F is the emergy yield ratio (EYR), which reflects the ability of a certain system to exploit available new emergy resources by investing local resources (local *versus* imported) [73]; F/(R + N) is the emergy investment ratio (EIR); (R + N)/GDP is the natural resource use per capital, which refers to the resource utilization rate; GDP/POP refers to the economic growth level; and POP refers to the population scale. The decomposition results of Qingyang and eight sub-regions are shown in Figures 3–11 with reference to the year 2000.

Table 4. Summary of emergy efficiency evaluation.

	2000	2001	2002	2003	2004	2005	2006	2007
XF	0.115	0.117	0.133	0.087	0.116	0.162	0.213	0.198
QC	0.423	0.951	1.000	1.000	1.000	1.000	1.000	1.000
ZN	0.697	1.000	0.827	0.810	0.802	0.808	0.807	0.805
NX	1.000	0.584	1.000	1.000	1.000	0.816	0.528	0.802
ZY	1.000	1.000	1.000	1.000	1.000	1.000	1.000	1.000
HX	1.000	1.000	1.000	1.000	1.000	1.000	1.000	1.000
HS	0.309	0.268	0.294	0.344	0.284	0.268	0.292	0.252
HC	0.805	0.764	0.873	0.762	0.738	0.759	0.770	0.766
Average	0.669	0.711	0.766	0.750	0.743	0.727	0.701	0.728

Notes: The abbreviations QY, XF, QC, ZN, NX, ZY, HX, HS and HC in the table represent Qingyang city, Xifeng district, Qingcheng, Zhengning, Ningxian, Zhenyuan, Huanxian, Heshui and Huachi respectively.

For Qingyang as a whole, the decomposition analysis (Figure 3) indicates that the factors, U/F, GDP/POP and POP, have increased the total emergy use and economic growth; GDP/POP was the main diver of the increasing emergy use. The factors (R + N)/GDP and F/(R + N) have decreased the emergy use. The negative contribution of the resource utilization rate, (R + N)/GDP, suggests policies to promote the efficiency of resource utilization.

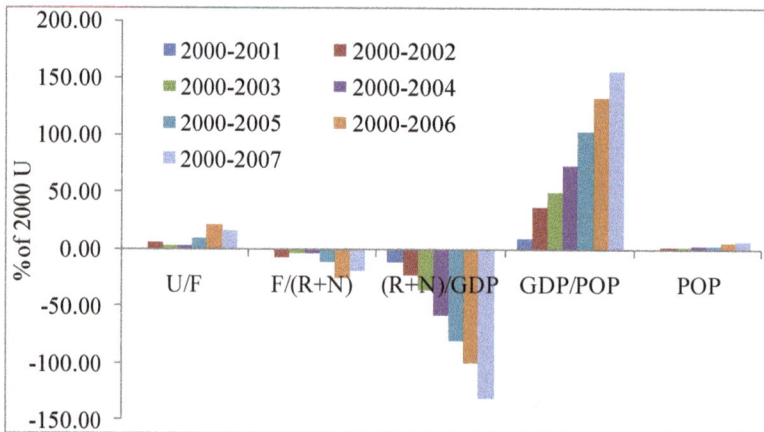

Figure 3. Advanced sustainability analysis (ASA) decomposition of the total emergy use (U) in Qingyang during 2000–2007; the contributions of five factors to the percentage of the 2000 total emergy use (U) level. POP, population.

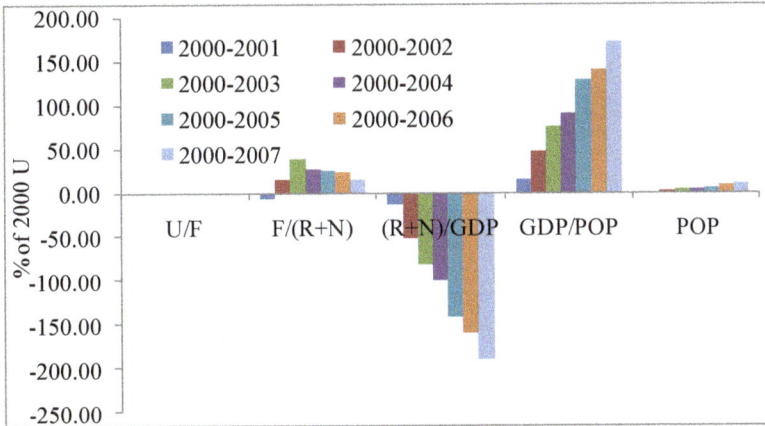

Figure 4. ASA decomposition of the total emergy use (U) in Xifeng during 2000–2007; the contributions of five factors to the percentage of the 2000 total emergy use (U) level.

In Xifeng (Figure 4), the considerable increase in the total emergy use can be seen in the decomposition results as an increase in the GDP/POP, F/(R + N) and POP components, and the economic growth was the main driving force. Since EIR = F/(R + N) indicates the efficiency of external investment in exploiting a unit of local resource, the EIR's contribution to the growth of emergy use means that the social economic system of Xifeng is fragile due to the fact that the local resource basis (R + N) cannot easily be increased and the non-renewable input emergy (N) is mainly contributing to the climate change. The factor (R + N)/GDP contributed to the decrease of emergy use considerably. The U/F factor slightly decreased in the investigated period.

In Qingcheng (Figure 5), the factors GDP/POP and U/F were the main contributors to the increase in total emergy use; while the factor F/(R + N) has decreased the emergy use. However, the factor (R + N)/GDP, unlike Qingyang and Xifeng, was fluctuating to a large extent, while finally decreasing.

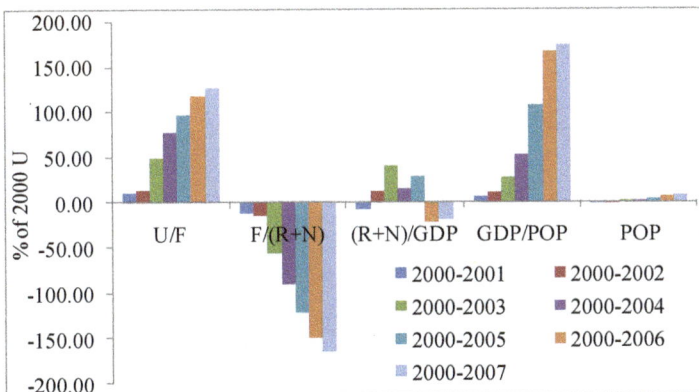

Figure 5. ASA decomposition of the total emergy use (U) in Qingcheng during 2000–2007, contributions of five factors to the percentage of the 2000 total emergy use (U) level.

For Zhengning (Figure 6), the decomposition result indicates that GDP/POP was the main driver of the emergy use increase, but (R + N)/GDP contributed to the decrease in emergy use. The U/F factor slightly decreased, except in the second period. Inversely, the F/(R + N) factor slightly increased, except in the third period.

In Ningxian (Figure 7), the factors F/(R + N) and GDP/POP were the main driving forces of the increase in total emergy use, and the contribution of F/(R + N) was more stable. The considerable and steady contribution of F/(R + N) to the U growth refers to the fact that the social economic system of Ningxian is more fragile than Xifeng, and an appropriate policy would be to optimize natural resource use (e.g., convert solar radiation into photovoltaic electricity to replace fossil fuels, recycling waste material resources to replace new raw imports, recycling water after appropriate treatment, *etc.*). The EIR (U/F) and resource utilization rate ((R + N)/GDP) have been the major factors decreasing emergy use. The effect of population scale is smaller than other factors.

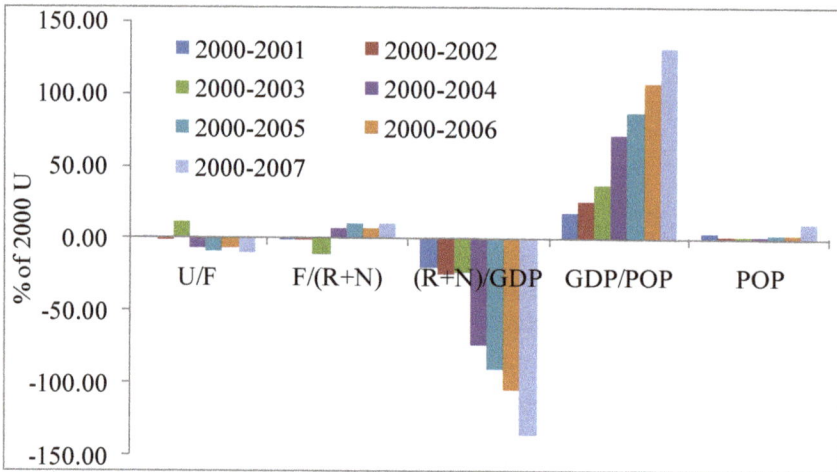

Figure 6. ASA decomposition of the total emergy use (U) in Zhengning during 2000–2007; the contributions of five factors to the percentage of the 2000 total emergy use (U) level.

In Zhenyuan (Figure 8), the fast economic growth (GDP/POP) was the most important factor contributing to the emergy use increase, due to its energetic economic development (Table 3). The factor F/(R + N) also has increased emergy use, which suggests policies to optimize natural resource use, but the effect is less than the economic growth. The factor (R + N)/GDP has decreased the emergy use, except in the third period, and the U/F also had negative effects on the emergy use increase, except in the first period.

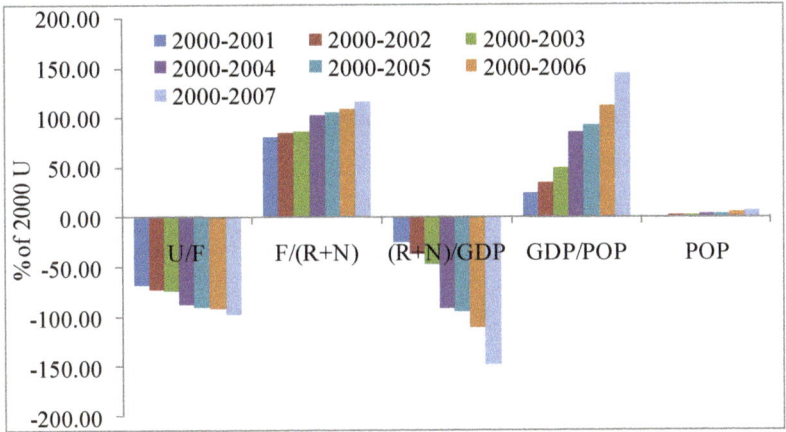

Figure 7. ASA decomposition of the total emergy use (U) in Ningxian during 2000–2007; the contributions of five factors to the percentage of the 2000 total emergy use (U) level.

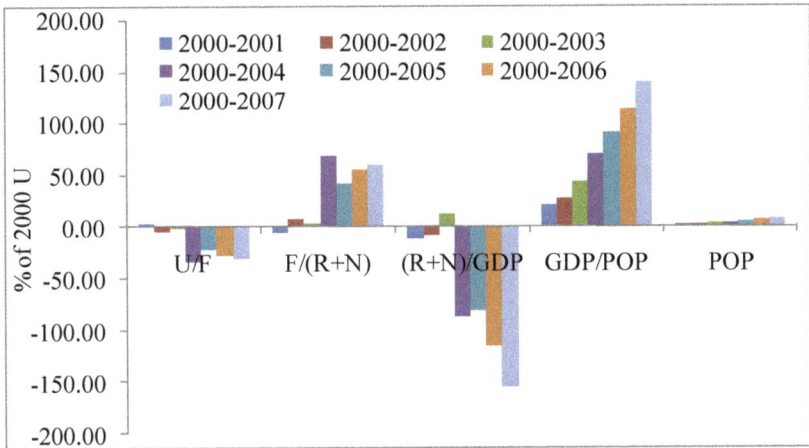

Figure 8. ASA decomposition of the total emergy use (U) in Zhenyuan during 2000–2007; contributions of five factors to the percentage of the 2000 total emergy use (U) level.

The economic growth in Huanxian (Figure 9) has increased the emergy use. However, the improvement of resource utilization efficiency has decreased the emergy use. The behavior of U/F and F/(R + N) was a bit more complex. For the factor U/F, from 2000 to 2003, the emergy yield ratio (EYR) increased, and after 2003, the parameter shows instead a decline, except in the last year. The factor F/(R + N) shows an inverse behavior.

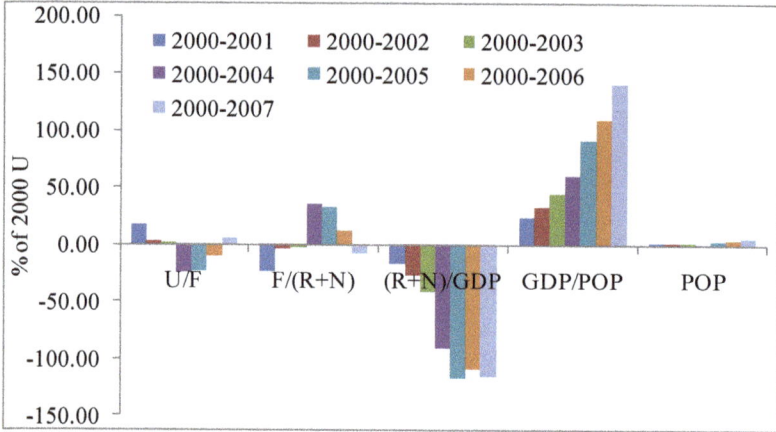

Figure 9. ASA decomposition of the total emergy use (U) in Huanxian during 2000–2007; contributions of five factors to the percentage of the 2000 total emergy use (U) level.

In Heshui (Figure 10), economic growth was the main driving factor of the emergy use increase, and the increase of F/(R + N) contributed to the increase of the total emergy use, although to a lesser extent than GDP/POP. However, the factor (R + N)/GDP showed a steady decline (less natural emergy input per unit of GDP), which strongly contributed to the decrease of the emergy use in the investigated period. The decreasing of U/F contributed to the decease of the total emergy use, but to a lesser extent than (R + N)/GDP.

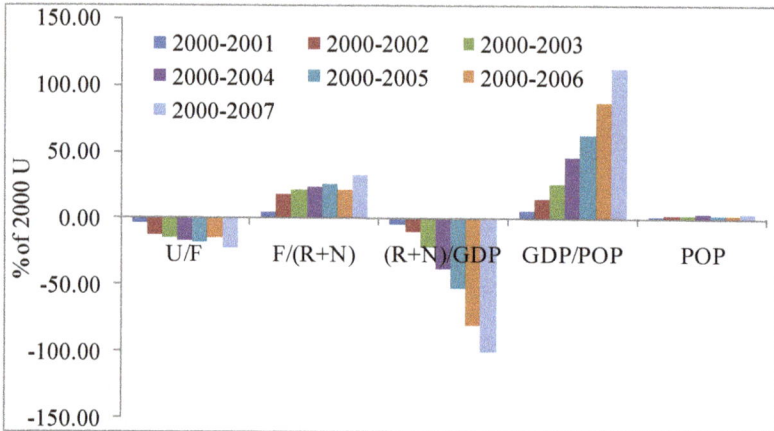

Figure 10. ASA decomposition of the total emergy use (U) in Heshui during 2000–2007; the contributions of five factors to the percentage of the 2000 total emergy use (U) level.

In Huachi (Figure 11), the factors GDP/POP and (R + N)/GDP, respectively, contributed to increasing and decreasing total emergy use. The factors U/F and F/(R + N) slightly changed in the investigated period. The factor U/F increased before 2003, but after 2003, the parameter showed instead a decline. Inversely, the factor F/(R + N) decreased before 2003, but after 2003, the parameter showed instead an ascent.

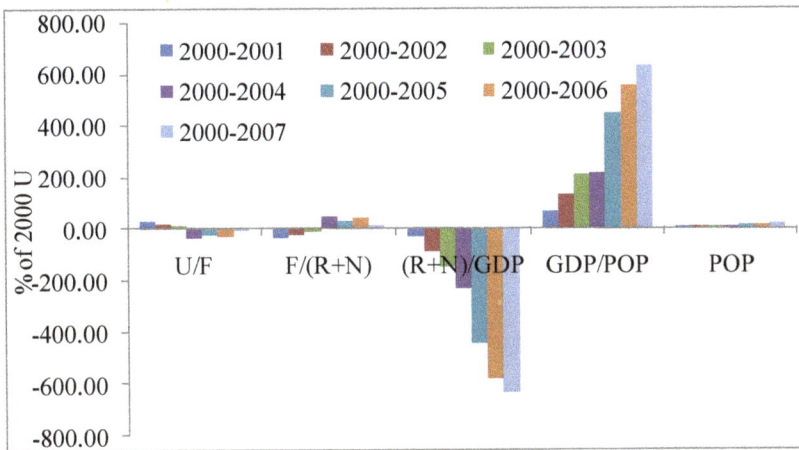

Figure 11. ASA decomposition of the total emergy use (U) in Huachi during 2000–2007; the contributions of five factors to the percentage of the 2000 total emergy use (U) level.

5. Summary and Conclusions

The main purpose of this study is to investigate metabolic structure and efficiency and, further to identify driving factors, so that we can employ an emergy-based analysis of the socio-economic metabolism of Qingyang city over 2000–2007 by combining data envelopment analysis (DEA) and decomposition analysis. By integrating emergy accounting into the DEA framework, the paper quantified the ecological efficiency of the socio-economic metabolic system, which can be applied to express the potential efficiency improvement and to make comparisons across different regions. The use of a multi-method analysis helps to avoid misleading interpretations based only on emergy-based indicators.

The investigated regional socio-economic metabolism greatly relied on non-renewable emergy flows and feedback (purchased) emergy flows, except Huanxian and Huachi, which highly depended on renewable emergy flow. Analysis of emerge indicators for sub-regions in Qingyang shows that Qingcheng, with the highest NEYR (35.9), is the main energy supplier in Qingyang, followed by Zhengning (14.24), Ningxian (12.66) and Huanxian (11.37). However, the ELR value of Qingcheng (5.2), which was above the Qingyang average (3.25), indicates that the current economic activities are not sustainable, as they relied mainly on local non-renewable emergy inputs. The highest EIR (29.87) and ELR (35.68) of Xifeng refer to the fact that the region relied more on purchased inputs than locally-available resources, and the region's current economic model has led to great environmental pressure. The ELRs of the other five regions indicate that these regions achieved a balance between the emergy use of renewable and non-renewable resources. The ESI results indicate that Xifeng developed a "consumer"-oriented economy; Ningxian, Qingcheng,

Zhenyuan, Heshui and Zhengning developed "producer"-oriented economies; and Huanxian and Huachi were economically behind, and their economic development needs to be accelerated and strengthened.

The emergy-based data envelopment analysis showed that the eight sub-regions, on average, had an EE level of 0.728 in 2007, which implies that Qingyang as a whole has the potential to reduce natural resource inputs and waste output. The EE scores (1.00) of Zhenyuan, Huanxian and Qingcheng indicate that the three regions can be considered as relatively emergy-efficient areas. The EE scores of Xifeng and Heshui are the lowest in Qingyang, and the scores in 2007 were 0.198 and 0.252, which suggest that the two regions could reduce natural resource consumption and waste emission to a great extent without scarifying any desirable output. The ecological efficiency of Zhengning, Ningxian and Huachi is higher than the average level of the whole of Qingyang. During the investigated period, the economic growth was the main driving factor of the increase in the total emergy use in Qingyang and its sub-regions. The contribution of the population scale factor is very small. The factor (R + N)/GDP contributed to the decrease of emergy use considerably (except Qingcheng), which suggests policies to promote the efficiency of resource utilization and, thus, to decrease emergy use further. The positive contribution of the environmental investment ratio to the growth of total emergy use in Xifeng, Zhengning, Ningxian, Zhenyuan, Huanxian, Heshui and Huachi demonstrates that the social economic system of the above seven regions is fragile, especially for Ningxian, Zhenyuan and Heshui.

The integrated approach used in this study is suggested as a tool to design future scenarios of resource use and ecological efficiency in the near future. The results of socio-economic metabolism analysis in Qingyang suggest policies to promote the efficiency of resource utilization, especially for Xifeng and Qingcheng. The fragile social economic systems of Xifeng, Ningxian, Zhenyuan, Heshui and Huachi suggest that series appropriate policies should be adopted to optimize natural resource use (e.g., convert solar radiation into photovoltaic electricity to replace fossil fuels, recycling waste material resources to replace new raw imports, recycling water after appropriate treatment, etc.).

It is worth noting that we cannot probe the structures of socio-economic metabolism that are regarded as crucial for policy makers to coordinate the system and improve regional eco-efficiency, because of insufficient statistical data in this economically lagging area. In addition, we employed the uniformed transformity of natural resources, products and processes, which may lead to estimation bias of the regional features to a certain extent. Nevertheless, we believe our findings provide solid and meaningful results that can provide useful implications for policy makers.

Acknowledgments

The authors would like to thank Wenting Jiao and the other two anonymous reviewers for their helpful comments on this paper and would like to acknowledge the financial support from the Natural Science Foundation of China (41301652, 41101126 and 41261112), the Specialized Research Fund for the Doctoral Program of Higher Education (20120211120026) and the Fundamental Research Funds for the Central Universities (lzujbky-2013-132). Special thanks go to the Green Talent Program of BMBF Germany and the Alexander von Humboldt Foundation of Germany.

Author Contributions

Zilong Zhang and Xingpeng Chen conceived and designed the study; Zilong Zhang contributed to data collection, data processing and draft paper as well as paper revised; Peter Heck contributed to data analysis and paper revised.

Conflicts of Interest

The authors declare no conflict of interest.

References

1. Krausmann, F.; Fischer-Kowalski, M.; Schandl, H.; Eisenmenger, N. The Global Sociometabolic Transition. *J. Ind. Ecol.* **2008**, *12*, 637–656.
2. Fischer-Kowalski, M.; Krausmann, F.; Pallua, I. A sociometabolic reading of the Anthropocene: Modes of subsistence, population size and human impact on Earth. *Anthropocene Rev.* **2014**, *1*, 8–33.
3. Huang, S.-L.; Lee, C.-L.; Chen, C.-W. Socioeconomic metabolism in Taiwan: Emergy synthesis *versus* material flow analysis. *Resou. Conser. Recycl.* **2006**, *48*, 166–196.
4. Fischer-Kowalski, M.; Haberl, H. Tons, joules, and money: Modes of production and their sustainability problems. *Soc. Nat. Resour.* **1997**, *10*, 61–85.
5. Huang, S.L.; Chen, C.W. Urbanization and Socioeconomic Metabolism in Taipei. *J. Ind. Ecol.* **2009**, *13*, 75–93.
6. Zhang, Y.; Liu, H.; Li, Y.T.; Yang, Z.F.; Li, S.S.; Yang, N.J. Ecological network analysis of China's societal metabolism. *J. Environ. Manag.* **2012**, *93*, 254–263.
7. Fischer-Kowalski, M. Society's Metabolism: The Intellectual History of Materials Flow Analysis, Part I, 1860–1970. *J. Ind. Ecol.* **1998**, *2*, 61–78.
8. Krausmann, F.; Gingrich, S.; Eisenmenger, N.; Erb, K.H.; Haberl, H.; Fischer-Kowalski, M. Growth in global materials use, GDP and population during the 20th century. *Ecol. Econ.* **2009**, *68*, 2696–2705.
9. Kennedy, C.; Pincetl, S.; Bunje, P. The study of urban metabolism and its applications to urban planning and design. *Environ. Pollut.* **2011**, *159*, 1965–1973.
10. Anderberg, S. Industrial metabolism and the linkages between economics, ethics and the environment. *Ecol. Econ.* **1998**, *24*, 311–320.
11. Janssen, M.A.; van den Bergh, J.; van Beukering, P.J.H.; Hoekstra, R. Changing industrial metabolism methods for analysis. *Popul. Env.* **2001**, *23*, 139–156.
12. Fischer-Kowalski, M.; Amann, C. Beyond IPAT and Kuznets curves: Globalization as a vital factor in analysing the environmental impact of socio-economic metabolism. *Popul. Env.* **2001**, *23*, 7–47.
13. Lee, C.L.; Huang, S.L.; Chan, S.L. Synthesis and spatial dynamics of socio-economic metabolism and land use change of Taipei Metropolitan Region. *Ecol. Model.* **2009**, *220*, 2940–2959.

14. Liu, J.R.; Wang, R.S.; Yang, J.X. Metabolism and driving forces of Chinese urban household comsumption. *Popul. Env.* **2005**, *26*, 325–341.

15. Yang, D.W.; Gao, L.J.; Xiao, L.S.; Wang, R. Cross-boundary environmental effects of urban household metabolism based on an urban spatial conceptual framework: A comparative case of Xiamen. *J. Clean. Prod.* **2012**, *27*, 1–10.

16. Tian, J.P.; Shi, H.; Chen, Y.; Chen, L.J. Assessment of industrial metabolisms of sulfur in a Chinese fine chemical industrial park. *J. Clean. Prod.* **2012**, *32*, 262–272.

17. Abel, W. The metabolism of the city. *Sci. Am.* **1965**, *213*, 178–193.

18. Newman, P.W.G. Sustainability and cities: Extending the metabolism model. *Landsc. Urban. Plan.* **1999**, *44*, 219–226.

19. Kennedy, C.; Cuddihy, J.; Engel-Yan, J. The changing metabolism of cities. *J. Ind. Ecol.* **2007**, *11*, 43–59.

20. Zhang, Y.; Yang, Z.F.; Yu, X.Y. Evaluation of urban metabolism based on emergy synthesis: A case study for Beijing (China). *Ecol. Model.* **2009**, *220*, 1690–1696.

21. Siciliano, G.; Crociata, A.; Turvani, M. A Multi-level Integrated Analysis of Socio-Economic Systems Metabolism: An Application to the Italian Regional Level. *Environ. Policy Gov.* **2012**, *22*, 350–368.

22. Geng, Y.; Liu, Y.; Liu, D.; Zhao, H.X.; Xue, B. Regional societal and ecosystem metabolism analysis in China: A multi-scale integrated analysis of societal metabolism (MSIASM) approach. *Energy* **2011**, *36*, 4799–4808.

23. Bringezu, S.; Schütz, H.; Moll, S. Rationale for and interpretation of economy-wide materials flow analysis and derived indicators. *J. Ind. Ecol.* **2003**, *7*, 43–64.

24. Hoang, V.N. Analysis of resource efficiency: A production frontier approach. *J. Environ. Manag.* **2014**, *137*, 128–136.

25. Brown, M.T.; Ulgiati, S. Understanding the global economic crisis: A biophysical perspective. *Ecol. Model.* **2011**, *223*, 4–13.

26. Zhang, L.M.; Xue, B.; Geng, Y.; Ren, W.X.; Lu, C.P. Emergy-Based City's Sustainability and Decoupling Assessment: Indicators, Features and Findings. *Sustainability* **2014**, *6*, 952–966.

27. Hau, J.L.; Bakshi, B.R. Promise and problems of emergy analysis. *Ecol. Model.* **2004**, *178*, 215–225.

28. Zhang, Z.; Chen, X.; Jiao, W.; Lu, C. Dynamic evolution of the coupled environmental-economic system of Qingyang, Gansu Province: Based on emergy theory and econometric method. *Acta Scien. Circum.* **2010**, *30*, 2125–2135. (In Chinese)

29. Odum, H.T.; Peterson, N. Simulation and evaluation with energy systems blocks. *Ecol. Model.* **1996**, *93*, 155–173.

30. Huang, S.L.; Lee, C.L.; Chen, C.W. Socioeconomic metabolism in Taiwan: Emergy synthesis *versus* material flow analysis. *Resou. Conser. Recycl.* **2006**, *48*, 166–196.

31. Brown, M.T.; Herendeen, R.A. Embodied Energy Analysis and Emergy Analysis: A Comparative View. *Ecol. Econ.* **1996**, *19*, 219–235.

32. Odum, H.T. *Environmental Accounting: Emergy And Environmental Decision Making*; Wiley: New York, NY, USA, 1996.

33. Ulgiati, S.; Odum, H.T.; Bastianoni, S. Emergy Use, Environmental Loading and Sustainability: An Emergy Analysis of Italy. *Ecol. Model.* **1994**, *73*, 215–268.

34. Ulgiati, S.; Ascione, M.; Zucaro, A.; Campanella, L. Emergy-based complexity measures in natural and social systems. *Ecol. Indic.* **2011**, *11*, 1185–1190.

35. Heshui Bureau of Statistics. *Heshui County Statistics Year Book (2001–2008)*; China Statistics Press: Beijing, China, 2001–2008.

36. Huanxian Bureau of Statistics. *Huan County Statistics Year Book (2001–2008)*; China Statistics Press: Beijing, China, 2001–2008.

37. Ningxian Bureau of Statistics. *Ningxian Statistics Year Book (2001–2008)*; China Statistics Press: Beijing, China, 2001–2008.

38. Qingcheng Bureau of Statistics. *Qingcheng County Statistics Year Book (2001–2008)*; China Statistics Press: Beijing, China, 2001–2008.

39. Qingyang Bureau of Statistics. *Qingyang Statistic Year Book (2001–2008)*; China Statistics Press: Beijing, China, 2001–2008.

40. Xifeng Bureau of Statistics. *Xifeng District Statistics Year Book (2001–2008)*; China Statistics Press: Beijing, China, 2001–2008.

41. Zhengning Bureau of Statistics. *Zhengning Statistics Year Book (2001–2008)*; China Statistics Press: Beijing, China, 2001–2008.

42. Zhenyuan Bureau of Statistics. *Zhenyuan County Statistics Year Book (2001–2008)*; Chins Statistics Press: Beijing, China, 2001–2008.

43. Lan, S.; Qin, P.; Lu, H. *Emergy Synthesis of Ecological-Economic System*; Chemical Press: Beijing, China, 2002. (In Chinese)

44. Liu, J.S.; Lu, L.Y.Y.; Lu, W.-M.; Lin, B.J.Y. A survey of DEA applications. *Omega* **2013**, *41*, 893–902.

45. Song, M.; An, Q.; Zhang, W.; Wang, Z.; Wu, J. Environmental efficiency evaluation based on data envelopment analysis: A review. *Renew. Sust. Energy Rev.* **2012**, *16*, 4465–4469.

46. Färe, R.; Grosskopf, S.; Norris, M.; Zhang, Z. Productivity growth, technical progress, and efficiency change in industrialized countries. *Am. Econ. Rev.* **1994**, *84*, 66–83.

47. Chung, Y.H.; Färe, R.; Grosskopf, S. Productivity and Undesirable Outputs: A Directional Distance Function Approach. *J. Environ. Manag.* **1997**, *51*, 229–240.

48. Tone, K. A slacks-based measure of efficiency in data envelopment analysis. *Eur. J. Oper. Res.* **2001**, *130*, 498–509.

49. Tone, K. Dealing with Undesirable Outputs in DEA: A Slacks-Based Measure (SBM) Approach. In Proceedings of the North American Productivity Workshop III, Toronto, ON, Canada, 22–25 June 2004; pp: 44–45.

50. Camarero, M.; Castillo, J.; Picazo-Tadeo, A.J.; Tamarit, C. Eco-Efficiency and Convergence in OECD Countries. *Environ. Resour. Econ.* **2013**, *55*, 87–106.

51. Ang, B.W.; Zhang, F.Q. A survey of index decomposition analysis in energy and environmental studies. *Energy* **2000**, *25*, 1149–1176.

52. Weinzettel, J.; Kovanda, J. Structural Decomposition Analysis of Raw Material Consumption. *J. Ind. Ecol.* **2011**, *15*, 893–907.

53. Xu, X.Y.; Ang, B.W. Index decomposition analysis applied to CO_2 emission studies. *Ecol. Econ.* **2013**, *93*, 313–329.

54. Zhang, F.Q.; Ang, B.W. Methodological issues in cross-country/region decomposition of energy and environment indicators. *Energy Econ.* **2001**, *23*, 179–190.

55. Ang, B.W.; Xu, X.Y. Tracking industrial energy efficiency trends using index decomposition analysis. *Energy Econ.* **2013**, *40*, 1014–1021.

56. Kaivo-oja, J.; Luukkanen, J.; Malaska, P. Advanced sustainability analysis. In *Our Fragile World. Challenges and Opportunities for Sustainable Development*; EOLSS Publishers: Oxford, UK, 2001; Volume 2, pp. 1529–1552.

57. Vehmas, J.; Malaska, P.; Luukkanen, J.; Kaivooja, J.; Hietanen, O.; Vinnari, M.; Iivonen, J. *Europe in Global Battle of Sustainability: Rebound Strikes Back? Advanced Sustainability Analysis*; Turun kauppakorkeakoulu: Turku, Finland, 2003.

58. Kaivo-oja, J.; Luukkanen, J.; Malaska, P. Sustainability evaluation frameworks and alternative analytical scenarios of national economies. *Popul. Environ.* **2001**, *23*, 193–215.

59. Hoffren, J.; Luukkanen, J.; Kaivo-Oja, J. Statistical decomposition modelling on the basis of material flow accounting. *J. Ind. Ecol.* **2000**, *4*, 4.

60. Jungnitz, A. Decomposition Analysis of Greenhouse Gas Emissions and Energy and Material Inputs in Germany. Available online: http://www.petre.org.uk/pdf/sept08/pETRE_WP1a_GWS_Jungnitz.pdf (accessed on 24 November 2014).

61. Ghisellini, P.; Zucaro, A.; Viglia, S.; Ulgiati, S. Monitoring and evaluating the sustainability of Italian agricultural system. An emergy decomposition analysis. *Ecol. Model.* **2014**, *271*, 132–148.

62. Ridolfi, R.; Andreis, D.; Panzieri, M.; Ceccherini, F. The application of environmental certification to the Province of Siena. *J. Environ. Manag.* **2008**, *86*, 390–395.

63. Raugei, M. Emergy indicators applied to human economic systems—A word of caution. *Ecol. Model.* **2011**, *222*, 3821–3822.

64. Vehmas, J.; Luukkanen, J.; Kaivo-oja, J.; Panula-Ontto, J.; Allievi, F. Key Trends of Climate Change in The Asean Countries. Available online: https://www.utu.fi/fi/yksikot/ffrc/julkaisut/e-tutu/Documents/eBook_2012-5.pdf (accessed on 24 November 2014).

65. Decoin. Development and Comparison of Sustainability Indicators. Available online: http://www.decoin.eu/ (accessed on 24 November 2014).

66. Smile. Synergies in Multi-Scale Inter-Linkages of Eco-Systems. Available online: http://www.smile-fp7.eu/ (accessed on 24 November 2014).

67. Brown, M.T.; Ulgiati, S. Emergy-based Indices and Ratios to Evaluate Sustainability: Monitoring Economies and Technology toward Environmentally Sound Innovation. *Ecol. Eng.* **1997**, *9*, 51–69.

68. Geng, Y.; Zhang, P.; Ulgiati, S.; Sarkis, J. Emergy Analysis of an Industrial Park: The Case of Dalian, China. *Sci. Total. Environ.* **2010**, *408*, 5173–5283.

69. Siche, J.R.; Agostinho, F.; Ortega, E.; Romeiro, A. Sustainability of Nations by Indices: Comparative Study between Environmental Sustainability Index, Ecological Footprint and the Emergy Performance Indices. *Ecol. Econ.* **2008**, *66*, 628–637.

70. Brown, M.T.; Ulgiati, S. Energy Quality, emergy, and Transformity: H.T. Odum's Contributions to Quantifying and Understanding Systems. *Ecol. Model.* **2004**, *178*, 201–213.

71. Ulgiati, S.; Brown, M.T. Quantifying the Environmental Support for Dilution and Abatement of Process Emissions: The Case of Electricity Production. *J. Clean. Prod.* **2002**, *10*, 335–348.

72. Ulgiati, S.; Brown, M.T. Monitoring Patterns of Sustainability in Natural and Man-Made Ecosystems. *Ecol. Model.* **1998**, *108*, 23–36.

73. Zucaro, A.; Ripa, M.; Mellino, S.; Ascione, M.; Ulgiati, S. Urban resource use and environmental performance indicators. An application of decomposition analysis. *Ecol. Indic.* **2014**, *47*, 16–25.

Factors Affecting Migration Intentions in Ecological Restoration Areas and Their Implications for the Sustainability of Ecological Migration Policy in Arid Northwest China

Yongjin Li, David López-Carr and Wenjiang Chen

Abstract: Ecological migration policy has been proposed and implemented as a means for depopulating ecological restoration areas in the arid Northwest China. Migration intention is critical to the effectiveness of ecological migration policy. However, studies on migration intention in relation to ecological migration policy in China remain scant. Thus this paper aims to investigate the rural residents' migration intentions and their affecting factors under ecological migration policy in Minqin County, an ecological restoration area, located at the lower terminus of Shiyang River Basin in arid Northwest China. The data for this study come from a randomly sampled household questionnaire survey. Results from logistic regression modelling indicate that most residents do not intend to migrate, despite rigid eco-environmental conditions and governance polices threatening livelihood sustainability. In addition to demographic and socio-economic factors, the eco-environmental factors are also significantly correlated with the possibility of a resident intending to migrate. The implications of the significant independent variables for the sustainability of ecological migration policy are discussed. The paper concludes that ecological migration policies may ultimately be more sustainable when taking into account household interests within complex migration intention contexts, such as household livelihoods dynamics and environmental change.

Reprinted from *Sustainability*. Cite as: Li, Y.; López-Carr, D.; Chen, W. Factors Affecting Migration Intentions in Ecological Restoration Areas and Their Implications for the Sustainability of Ecological Migration Policy in Arid Northwest China. *Sustainability* **2014**, *6*, 8639-8660.

1. Introduction

Managing population size in ecological restoration areas through resettlement policy is one of the strategic measures for both environmental and development aims in China [1,2]. This type of resettlement is always called ecological migration when related to migration policy aimed at rehabilitating the degenerated eco-environment [3]. In China, ecological migration policy has seemingly been designed to achieve rural development and eco-environmental rehabilitation simultaneously through one policy intervention [4]. Many authors approve ecological migration as a preferred approach for protecting arid Northwest China's fragile eco-environment, even though they also recognize its potential problems, such as the mismatch between traditional production style in the sending area and new lifestyles in the receiving area [5,6]. However, other authors argue that the primary environmental rationale behind ecological migration is largely inadequate and that there is insufficient justification to point toward ecological migration as the only possible solution [7,8]. Despite ecological migration policy extant in China since the 1990s, potential migrants' voices are ineffectively heard in the process.

Arid Northwest China accounts for over 20% of China's total land area, which includes Xinjiang Uygur Autonomous Region, the western part of Inner Mongolia Autonomous Region, Hexi Corridor in Gansu Province, and Qilian Mountains and Qaidam Basin in Qinhai Province, with a main landscape of desert, high mountains and great basins forming its characteristic topography [9]. This area is situated in the deep hinterland of Eurasia with scarce precipitation and greatly varied air temperature. In these extremely arid great basins, such as Tarim Basin, Junggar Basin and Hexi Corridor, the local peoples have a long history of utilizing inland river water to irrigate crops and develop sandy and alpine pastures for animal husbandry [9]. Originating in the high mountain snow melt, the water flows eventually to the Gobi Desert.

In the past 50 years, about four million hectares of man-made oases have newly been developed in all of the river basins, and about 622 reservoirs with a storage capacity of 6.6 billion cubic meters have been built [10]. In addition to the fast urbanization in the middle parts of the river basins, the large scale utilization of surface water in the middle reaches and over-exploitation of ground-water in the lower reaches have not only facilitated the steady development of social economy but also resulted in serious eco-environmental problems in the arid area, especially in the lower reaches of these river basins. Problems include shortened runoff courses of most rivers, shrinking or dried up terminal lakes, declining quantity and quality of surface water in the lower reaches, more serious soil salinization and desertification, and seriously degraded or destroyed vegetation [10]. In recent years, strong dust-storms have occurred frequently across the region. They are thought to be a result of continuing deterioration of the environment relating to current practices in water use and agriculture [11].

With the aim to systematically restore the degenerated eco-environment in the inland river basins, especially in the end terminuses, some integrative river basin governance planning or policies have been proposed and implemented, such as that in Shiyang River Basin located in the east of the Hexi Corridor in Gansu Province. The Shiyang River Basin is one of the earliest to have been developed and is one of the most overexploited inland basins in northwest China [11]. The shortage of surface water and overexploitation of ground-water have caused serious eco-environmental and social problems, such as desertification and environmentally forced out-migration in Minqin County, the lower terminus of Shiyang River Basin.

In November 2007, the Central government approved an ecological restoration plan named "Key Governance Planning for Shiyang River Basin" (Chinese Pinyin: *Shiyanhe Liuyu Zhongdian Zhili Guihua*) with the total investment from 2006–2020 to reach 4.7 billion RMB Yuan (about 0.64 billion US$ in 2007), of which over 1 billion Yuan will be invested in the territory of Minqin. Apart from the costly engineering projects, the management policies involved in the planning can be categorized into three types. First, environmental policies, which mainly include shutting motor-pumped wells, decreasing cultivated land area, and restricting pumped ground-water by controlling electricity supply and IC-card rationing equipment installed on the motor-pumped well's mouth. These measures aim to reduce groundwater and surface water consumption, and allocate more water to desert vegetation restoration in the end terminus. Second, economic policies that include constructing greenhouses for vegetables and warm barns for livestock husbandry, shifting from crop farming to fruit trees or forage grass, and encouraging peasants to do non-farm work through labor-skills training. And the third, ecological migration policy, which plans for out-migration of 10,500 residents from the Minqin County's marginal land neighboring the desert. To implement this plan, the government has drawn up specific stimulation approaches including subsidizing 6000 Yuan per capita on the condition of the household head signing an agreement to abandon the household's local water and land rights.

A relatively long history of desertification-induced out-migration in Northwest China, including a large number of voluntary out-migrants from Minqin County in the past three decades [12], precedes the execution of ecological migration policy [7]. When the ecological migration policy was initially passed in tandem with other kinds of policies as part of the river basin governance planning in recent years, residents remaining behind who suffered exacerbated desertification might be less capable of out-migrating than earlier voluntary migrants, because migration is often expensive, and those most vulnerable to environmental change are usually poor [13]. Still the questions remain: Do the residents have the propensity to migrate? What factors influence their migration intentions? Answering these questions is helpful to successfully implement the ecological migration policy in the ecological restoration area, but there is scant study of these critical questions.

According to the theory of reasoned action proposed by Ajzen and Fishbein [14], migration intention remains the dominant determinant of migration behavior [15]. This paper will utilize Minqin County as a case to study the factors affecting the local residents' migration intentions and discuss their implications on the sustainability of ecological migration policy in arid Northwest China. The findings will implicate China's ecological migration policy, so that population, environment, and development are more harmoniously related in the arid Northwest China area. Furthermore, findings could contribute to the future research and resettlement policy applications in other similarly marginal environments globally. In the following sections, we first describe the study area and data sources. Then, we propose an analytical model and a framework of factors affecting migration intentions in an ecological restoration area. Theories and hypotheses about the relationships between the proposed drivers and migration intention are then reviewed briefly. Finally, results of logistic regression analysis are used as the basis for in-depth discussion of the implications of the significant factors for ecological migration policy in Arid Northwest China.

2. Data and Methods

2.1. Study Area

Minqin County is located in Northwest China (Figure 1), a hotspot of severe water shortage and desertification [16–19]. Minqin County is located at the lower terminus of the Shiyang River Basin, one of the three inland river basins in the Hexi Corridor of Gansu Province, Northwest China. The geographical location lies between 102°52′~E103°50′E, 38°22′N~39°6′N. Minqin County is surrounded by the Badain Jaran Desert and the Tengger Desert from the west, north, and east.

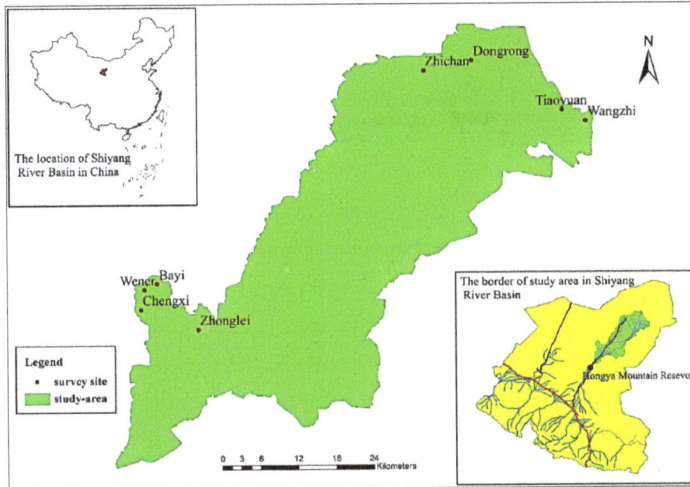

Figure 1. Map of study area.

Since the Hongya Mountain Reservoir at the south edge of Minqin County was constructed in 1958, natural surface water flowing to Minqin County has disappeared. All of the surface water allocated to Minqin County is controlled by the Hongya Mountain Reservoir. Along with urbanization and industrial development growing rapidly in the middle portion of the Shiyang River Basin, water storage in the reservoir decreased from an annual average of 545 million cubic meters in the 1950s to 136 million cubic meters in the 1990s [20]. As surface water volume decreased, a large amount of groundwater was pumped to fill the gap, allowing the cultivated land area and agricultural structure to remain largely unchanged. In 2000, the ratio of consumed groundwater volume to surface water volume reached 7:1 [21]. As the groundwater table continuously decreased, a large amount of vegetation deteriorated, and groundwater quality and soil quality declined.

Because Minqin County is located at the lower terminus of Shiyang River Basin, the oasis remains fertile and productive. Before 2002, although the marginal area was threatened by desertification and decreasing groundwater quality, residents had relatively larger farm areas and could earn more income from agriculture than rural residents in the other part of the Shiyang River Basin. After 2002, and especially since 2007, since the "Key Governance Planning for Shiyang River Basin" has been implemented, the quantity of basic livelihood assets, such as arable land quota and water availability for agriculture, has decreased greatly in the study area, even though the trend of eco-environmental deterioration has been effectively curbed, and the groundwater table has continued to rise [22].

2.2. Source of Data

The data applying in this study come from two resources. One is a random household structured questionnaire survey conducted by the authors in eight villages in Minqin County from December 2010 to January 2011. It elicited data regarding the respondents' migration intentions and variables affecting these intentions. The information collected includes household demographic characteristics (composition, age, education and migration), household livelihood assets/strategies, and the respondents' subjective evaluation of the local eco-environmental status and trends according to the sustainable livelihood framework [23]. Another is China's fifth and sixth population census data at the county level in 2000 and 2010, which was used to analyze the macro-demographic characteristics and dynamics in Minqin County.

In the questionnaire survey, the households were randomly selected, and the interviewees within the selected households were all adults, nearly always the household head or spouse. The sampling frame for the household survey comprises about 9500 rural households, which are homes to about 40,000 residents, in the 55 villages within nine northwesterly desert-neighbored townships of Minqin County. The multistage Probability Proportionate to Size (PPS) sampling method was utilized to create the household sample. There are four stages in the sampling process. In the first stage, four townships were sampled out of the nine marginal townships in proportion to their population size. At the second stage, eight marginal villages were selected from all villages of the four sampled townships proportionate to their population size. The eight sampled villages were Wen'er, Bayi, Chengxi from Daba Township; Wangzhi, Tiaoyuan from Donghu Township; Zhonglei from Sanlei Township; and Dongrong, Zhichan from Xiqu Township (Figure 1). At the third stage, the investigators randomly selected 40 households in each of the eight villages. We used a systematic sampling method based on the detailed household roster for each of the marginal villages provided by the secretary of the village branch of the Chinese Communist Party (CCP). At the intra-household sampling stage, if the household head or spouse was at home, that person became the interviewee. Otherwise, the interviewee would be a household member over the age of 18 whose birthday was closest to December 1. Face-to-face questionnaire administration was used on site, whereby an interviewer presents the questions orally and completes the questionnaire on the spot. The expected sample size was 320 and the final valid sample size was 308. Geographical homogeneity among the sampled villages allowed this relatively small sample size to satisfy the acceptable sampling error of between 5% and 6% at the 95% level of confidence [24].

2.3. Analytical Model

The binary logistic regression model is always used to explore the factors affecting adoption of some specific agricultural technologies for rural sustainable development in arid Northwest China such as [25–27], and to examine the determinants of migration intentions among developing countries such as [15,28,29]. In this study, we also used a binary logistic regression model to evaluate rural residents' intention to migrate and the predicting factors. In the questionnaire, we asked the respondent whether he/she has the intention to migrate out of his/her hometown. The options are Yes and No. Migration intention as dependent variable (DV) is measured by dichotomy 1 (Yes) and 0 (No). The Logistic Regression Model used in this paper is as follows:

$$\text{logit } P = \ln(\frac{P}{1-P}) = \alpha + \beta_1 x_1 + \beta_2 x_2 + \text{K} + \beta_p x_p \tag{1}$$

In the above equation, P indicates the possibility of having propensity for migration. $x_1, x_2, ..., x_p$ indicate various factors affecting migration intention. α is the constant indicating intercept in the model, and $\beta_1, \beta_2, ..., \beta_p$ indicate the coefficients of various factors in the model. All the computations in this paper are processed by IBM SPSS Statistics 19.0.

2.4. Predictors

Unlike previous studies which mainly took demographic characteristics and social-economic factors as predominant predictors of migration intentions [28–30], this paper added the predictors of political-economic and eco-environmental factors into the analytical models (Figure 2). This idea is inspired by the sustainable livelihood framework [23]. The rationale for this consideration is that the migration intentions studied in this paper are promoted mainly by the governmental policies portfolio, which will impact the migration intentions directly by ecological migration policy and indirectly by feedback loops among governmental policies, rural household livelihoods assets/strategies, and local environmental change.

The name and definitions of the predicting factors are listed in Figure 2. The type of measure and the descriptive statistics for all these variables in the empirical models are given in Tables 1 and 2, respectively. Explanatory variables and their justifications are discussed below.

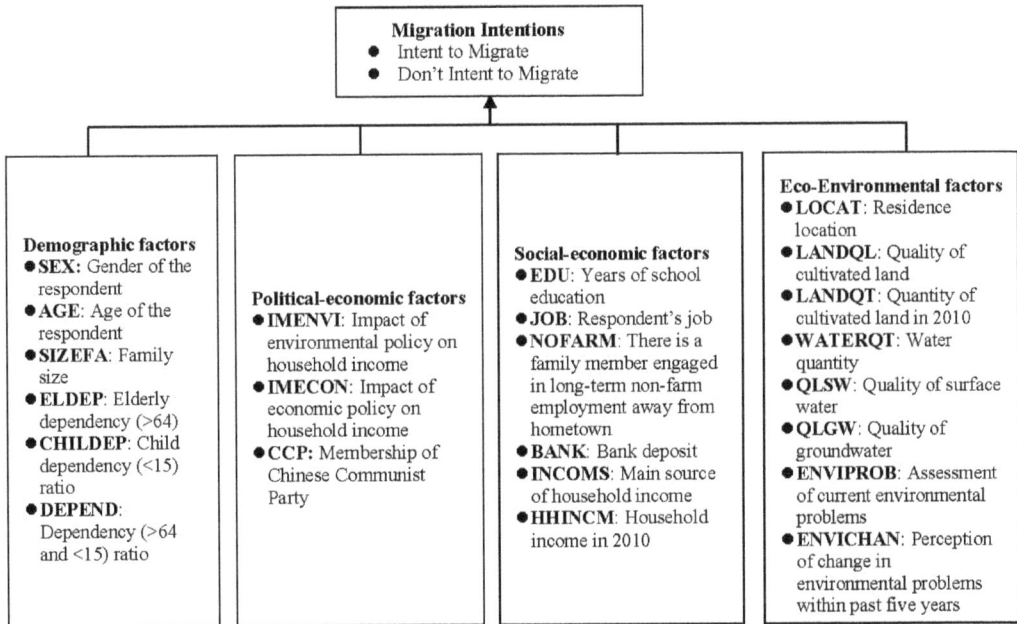

Figure 2. Factors affecting migration intentions in the ecological restoration area.

Table 1. Definition of variables used in the analysis.

Variables	Type of Measure	H_0 [a]
INTENT	Dichotomous (1 if yes, 0 if no)	
SEX	Dummy (0, female; 1, male)	+
AGE	Years	-
SIZEFA	Numbers	?
ELDEP	Percent	-
CHILDEP	Percent	-
DEPEND	Percent	-
IMENVI	Dummy (0 if increase, 1 if decrease)	-
IMECON	Dummy (0 if increase, 1 if decrease)	-
CCP	Dummy (1 if yes, 0 if no)	+
EDU	Years	+
JOB	Dummy: full-time farm work (reference), part time non-farm employment (-PNF), full time non-farm employment (-FNF)	+
NOFARM	Dummy (1 if yes, 0 if no)	+
BANK	Ordinal: 0 = "having no bank deposit", 1 = "Below 10,000 Yuan", 2 = "10,000~20,000 Yuan", 3 = "20,000~30,000 Yuan", 4 = "30,000~40,000 Yuan", 5 = "40,000~50,000 Yuan", 6 = "Above 50,000 Yuan"	+
INCOMS	Dummy: farm employment (reference), non-farm employment (-NFE), remittances (-REM), government subsidies (-GOS)	+, ~, -
HHINCM	1000 Yuan (a Chinese unit of currency)	+
LOCAT	Dummy: near dam/near desert = 0/1	+
LANDQL	Ordinal: 1 = "very bad", 2 = "bad", 3 = "normal", 4 = "good", 5 = "very good"	-
LANDQT	Mu (a Chinese unit of area, 1 Mu = 1/15 ha)	-
WATERQT	Dummy: sufficient both for domestic use and agriculture production (reference); sufficient for domestic use, but insufficient for agricultural production (-SDIA); sufficient for agricultural production, but insufficient for domestic use (-SAID); insufficient for both domestic use and agricultural production (-IBDA)	+
QLSW	Dummy: suitable for human daily life (reference); suitable for livestock drink (-SFLS); only suitable for irrigation (-SFIR); not suitable for irrigation (-NFIR). Ordinal: [3,2,1,0]	+
QLGW	Dummy: suitable for human daily life (reference); suitable for livestock drink (-SFLS); only suitable for irrigation (-SFIR); not suitable for irrigation (-NFIR). Ordinal: [3,2,1,0]	+
ENVIPROB	Ordinal: [0,32] = [no problem, very serious problems]= $\sum_{i=1}^{8} ENVIPROB_{ij}$, j = [0,1,2,3,4] = [no problem, a little, not serious, serious, very serious], i = [1,2,...,8] = [sand storms (-SS), land desertification (-LD), land salinization (-LS), groundwater mineralization (-GM), shortage of water resources (SWR), discarding used plastic film in the field (-DUPF), vegetation deterioration (-VD), and converting forest to farmland (-CFF)]	+
ENVICHAN [b]	In Model 1: Ordinal: [−16,16] = [sharply deteriorated, greatly ameliorated] = $\sum_{i=1}^{8} ENVICHAN_{ij}$, j = [−2,−1,0,1,2] = [sharply deteriorated, some deteriorated, no change, some ameliorated, greatly ameliorated], the meaning of i is same as that in ENVIPROB;	-
	In Model 2: Dummy (if sharply deteriorated within past five years = 1, others = 0)	+

[a] H_0 = Hypothesized relationship with Migration Intentions, "+" denotes positive, "-" denotes negative, and "?" denotes indeterminate; [b] Model 1 and 2 are described in Section 2.5.

Table 2. Descriptive Statistics.

Type	Variables	N	Min	Max	Mean	SD
	INTENT	308	0	1	0.26	0.44
Demographic	SEX	308	0	1	0.72	0.45
	AGE	306	18	81	50.41	10.84
	SIZEFA	308	1	9	4.17	1.38
	ELDEP	308	0	100.00	11.48	22.32
	CHILDEP	308	0	60.00	8.76	14.65
	DEPEND	308	0	100.00	21.70	25.72
Political-economic	IMENV	306	0	1	0.90	0.30
	IMECO	304	0	1	0.03	0.17
	CCP	308	0	1	0.11	0.31
Social-economic	EDU	307	0	12	6.98	3.91
	JOB-PNF	304	0	1	0.22	0.42
	JOB-FNF	304	0	1	0.03	0.16
	NOFAM	307	0	1	0.32	0.47
	BANK	308	0	6	0.4	1.08
	INCOMS-NFE	305	0	1	0.06	0.23
	INCOMS-REM	305	0	1	0.01	0.10
	INCOMS-GOS	305	0	1	0.01	0.08
	HHINCM	298	0.6	255	26.83	22.23
Eco-environmental	LOCAT	308	0	1	0.48	0.5
	LANDQL	301	1	5	3.47	0.96
	LANDQT	298	0	50	10.28	4.61
	WATERQT-SDIA	308	0	1	0.79	0.41
	WATERQT-SAID	308	0	1	0.02	0.13
	WATERQT-IBDA	308	0	1	0.15	0.36
	QLSW	306	1	3	1.81	0.69
	QLGW	305	0	3	2.22	0.91
	ENVIPROB	285	2	32	20.87	4.26
	ENVICHAN	271	−16	10	−3.66	5
	Valid N (listwise)	229				

2.4.1. Demographic Factors

SEX, AGE, SIZEFA, ELDEP, CHILDEP and DEPEND measure gender of the respondent, age of the respondent, family size, elderly dependency (>64) ratio, child dependency (<15) ratio, and dependency (>64 and <15) ratio, respectively.

According to Grieco and Boyd [31], gender has a core influence on the statuses of males and females, their roles, and stages in the life-cycle. These help determine people's position in society and, therefore, the opportunities women and men have to consider in moving to the pre- migration stage. Many previous studies suggest that gender roles impact men's and women's migration intentions and behavior differently [28,29]. In this study, males were expected to have more propensities to migrate than females, because the social norms and attitudes tend to be less friendly toward women's active pursuit of economic activities outside the home. This, in turn, discourages or prevents many women from realizing their migration plans [28].

In many instances, older respondents may not migrate because their attachment to their community tends to be stronger than that of younger respondents [32]. Therefore, AGE was expected to be negative for migration intention. Size of family also determines how the household will manage in times of climate-related events. The larger the family size, the more vulnerable it may be in times of decreasing natural livelihood capital. The needs of a larger household will be difficult to provide for compared to a smaller one where just a few people have to be attended to. On the other hand, larger households might be able to more easily diversify their income by sending one of their members elsewhere for cash labor without losing essential household labor [32]. Therefore, the expected sign of SIZEFA is indeterminate.

De Jong [29] shows that the presence of children or elderly dependents increases intention to migrate for men because of increased financial family resource needs, but reduces intention to move for women because of dependent care responsibility. In this study, the dependent variable is rural residents' migration intention that is prescribed by ecological migration policy as virtually permanent family out-migration. Because of the uncertainty about livelihood approaches after out-migration, especially for higher dependency ratio households that are short of laborers, the ELDEP, CHILDEP and DEPEND were supposed to be negative to migration intention.

2.4.2. Political Economic Factors

The political economic factors mainly include IMENVI, IMECON and CCP, which measure respondents' subjective evaluation of the impact on household income of, respectively, environmental policy, economic policy, and membership of Chinese Communist Party.

Along with ecological migration policy, environmental and economic policies are essential elements in the river basin governance planning. If these policies decrease the household income after implementation, the migration intention will be strengthened. Therefore, IMENVI and IMECON were expected to be negative to migration intention. China is a one-party country. If the respondent had a membership of Chinese Communist Party (CCP) that governs the country, as the result of obedience to government, the migration intentions were expected to be stronger than those of non-members.

2.4.3. Social-Economic Factors

EDU, JOB, NOFARM, BANK, INCOMS, HHINCM indicate years of schooling, respondent's job, a family member engaged in long-term non-farm employment away from respondent's hometown, bank deposit, main source of household income, and household income in 2010.

If the respondent had more education or had a job other than farmer, he/she will have more opportunity to seek a livelihood in a new place [29,30]. Therefore, EDU and JOB were expected to be positive to migration intention. If a household member engaged in long-term non-farm employment away from his/her hometown, there will be a greater social network which is advantageous to out-migration [28,30]. Thus, NOFARM was expected to be positive to migration intention too.

If a household had more bank deposits or annual income, it would have more financial capital for out-migration. Therefore, BANK and HHINCM were also expected to be positive to migration intention.

If the main resource of household income comes from non-farm employment (-NFE), the family will depend less on land resource and will have a greater propensity to migrate. If the household income comes mainly from remittances (-REM), the propensity to migrate is uncertain. One possible case is that the family will migrate as it depends less on land resource; another is that the family will not migrate to avoid the possible increase in the consumption portion of the remittance in a new place [29]. If the main resource of household income comes from government subsidies (-GOS), the family will have less propensity to migrate, in case the government subsidies are withdrawn after they leave their place of origin. Therefore, INCOMS-NFE was expected to be positive to migration intention, INCOMS-GOS was expected to be negative to migration intention, and the expected indication of INCOMS-REM is indeterminate.

2.4.4. Eco-Environmental Factors

LOCAT, LANDQL, LANDQT, WATERQT, QLSW, QLGW, ENVIPROB, ENVICHAN denote, respectively, residence location, quality of cultivated land, quantity of cultivated land in 2010, water quantity, quality of surface water, quality of groundwater, respondent's assessment of current environmental problems and perception of change in environmental problems within the past five years.

Historically, the voluntary emigrants from Minqin County resided in the northern towns near the desert [33]. Therefore, households near the desert were expected to have more propensities to migrate than those near the dam. Land and water resources are the critical livelihood assets of rural residents in arid areas. The better the quality of the land or water resources, the less willing families are to migrate [34]. The quantity of the land or water resources will impact the migration intentions in the same way. According to the definition of the eco-environmental factors (Table 1), the indications of LANDQL and LANDQT were expected to be negative, and the indication of WATERQT (-SDIA, SAID, or -IBDA), QLSW (-SFLS or -SFIR), and QLGW (-SFLS or -SFIR) were expected to be positive.

Environmental problems have long been the impetus of out-migration [35,36]. If the surrounding environment has more serious problems, the resident family will be more likely to migrate. Thus, the indications of ENVIPROB were expected to be positive. Government rehabilitation of the degraded environment will affect migration intention also. If the respondents perceived greater environmental amelioration, their families will be less inclined to migrate. So the indication of ENVICHAN was expected to be negative.

2.5. Measurement and Analysis

The measurements of all the predictors are listed in Table 1. In the questionnaire, two Likert Scales [37] (pp. 197–199) were utilized to measure the respondent's assessment of eight environmental problems' current statuses and perception of the change in these environmental problems within the past five years. These environmental problems are the following: sandstorm, land desertification, land salinization, groundwater mineralization, shortage of water resources, discarding used plastic film in the field, vegetation deterioration, and converting forest to farmland. The first scale is 0, 1, 2, 3, 4, indicating no problem, a little, not serious, serious and very serious; and the second scale is −2, −1, 0, 1, 2, indicating sharply deteriorated, somewhat deteriorated, no change, somewhat ameliorated, and greatly ameliorated. The Reliability Statistics, Cronbach's Alpha, for the two scales are 0.656 and 0.754 (computed by IBM SPSS Statistics 19.0, based on 285 and 271 valid cases, accounting for 92.5% and 88% of the total, separately), which means the two scales have acceptable internal consistency reliability. In the analysis, we measure the respondent's assessment of the current status of eight environmental problems, and the perception of the change of these environmental problems up to two independent variables, which are "Assessment of current environmental problems" (ENVIPROB) and "Perception of change in environmental problems within past five years" (ENVICHAN), with the interval scale of (0,32) and (−16,16) indicating (no problem, very serious) and (sharply deteriorated, greatly ameliorated).

We implemented two models to explore factors affecting intention to migrate. In Model 1, the predictors of ENVIPROB, ENVICHAN, QLSW and QLGW are used as interval scale variables. If they are actually or nearly significant in Model 1, they will be substituted with dummy variables separated by their components (as described in Table 1) in Model 2 for further analysis. Whether or not a variable is included in the models is determined by its forward stepwise p-value. In each step, a factor can be entered into the model when its p-value is less than 0.45; it must be excluded when its p-value is more than 0.5. The significant variables from the model with higher Model Chi-Square will be used in the final model for final interpretation.

3. Results

3.1. Macro-Demographic Characteristics of Minqin County

By comparing the population size and structure of Minqin in 2000 (Figure 3a) and 2010 (Figure 3b), we observe an astonishing ageing and depopulation process in the study area. The Fifth National Population Census shows the resident population of Minqin in 2000 was 302,085, while the Sixth National Population Census of China shows that Minqin county had 241,251 residents in 2010. This represents over 60,000 fewer residents, a decrease of approximately 20% of Minqin County's population during the first decade of the 21st century. The population age structure also changed significantly. In 2000, the percentage of age 0–14 was 28.9%, while in 2010, it decreased to 16.4%. The percentage of 65+ increased from 5.7% in 2000 to 9.6% in 2010.

3.2. Migration Intentions and Affecting Factors

Only 80 respondents, approximately one quarter of the 308, have intention to migrate. In contrast, 228 respondents do not intend to migrate (Table 3). Younger respondents had a higher ratio of intention to migrate, while respondents over the age of 65 are more willing to migrate than those aged 50–64 (Table 3).

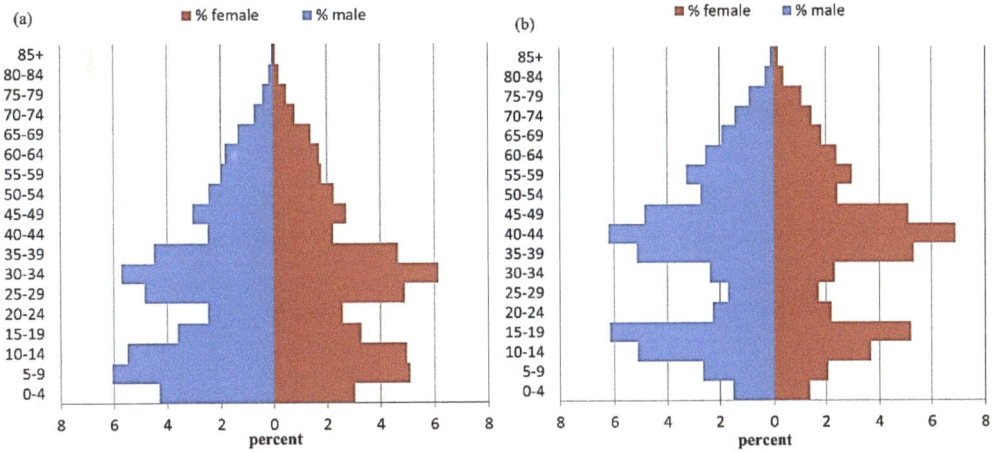

Figure 3. (**a**) Population Pyramid for Minqin 2000; (**b**) Population Pyramid for Minqin 2010. Source: China's 5th and 6th Population Census Data.

Table 3. Migration intentions by age.

| | | Migration Intentions | | Total | Ratio of Yes to Total |
		No	Yes		
	18–29	2	3	5	0.60
	30–39	25	12	37	0.32
Valid	40–49	82	43	125	0.34
	50–64	90	14	104	0.13
	65+	28	7	35	0.20
Missing		1	1	2	0.50
Total		228	80	308	0.26

As Model 1 demonstrates, ENVICHAN (perception of change in environmental problems within the past five years) has a significant relationship with "migration intention". The QLGW (quality of groundwater) is nearly significant. In model 2, we substitute eight dummy variables for ENVICHAN (perception of change in environmental problems within past five years), and three dummy variables for QLGW (quality of groundwater). The Coefficients for the two logistic regression models of the Log Odds of intention to migrate are listed in Table 4. From the Model Chi-Square and corresponding p-value (Table 4), Model 2 is more suitable than Model 1 for explaining the relationships between various independent factors and migration intention. We selected 11 variables with a significance level less than 0.10 and one nearly significant but theoretically important variable (LOCAT) from Model 2 as the independent variables entered in the final model (Table 5).

Table 4. Coefficients for various logistic regression models of the Log Odds of intending to migrate.

Variables	Model 1 (Forward Stepwise, "Conditional") CRITERIA = PIN(0.45) POUT(0.5)			Model 2 (Forward Stepwise, "Conditional") CRITERIA = PIN(0.45) POUT(0.5)		
	b	Exp(b)	p	b	Exp(b)	p
INTERCEPT	4.139	62.722	0.016	1.879	6.544	0.194
AGE	−0.077 ***	0.926	0.000	−0.075 ***	0.928	0.000
SIZEFA				0.189	1.208	0.163
ELDEP	0.025 **	1.025	0.002	0.038 **	1.038	0.008
DEPEND				−0.011	0.989	0.326
IMECON	−0.976	0.377	0.428	−1.058	0.347	0.361
CCP	0.647	1.91	0.215			
EDU	−0.049	0.952	0.297	−0.039	0.962	0.417
BANK	−0.229	0.795	0.213	−0.210	0.810	0.265
INCOMS-NFE				1.665 *	5.287	0.016
HHINCM	0.030 *	1.03	0.026	0.039 **	1.039	0.002
LOCAT	−0.797	0.451	0.265	−0.854	0.426	0.108
LANDQL				−0.139	0.870	0.462
LANDQT	−0.079 *	0.924	0.058	−0.110 *	0.896	0.016
WATERQT-IBDA	0.848 *	2.334	0.067	1.081 *	2.948	0.025
QLGW	−0.602	0.547	0.136	--	--	--
-SFIR	--	--	--	1.317 *	3.731	0.018
ENVICHAN	−0.09 *	0.914	0.012			
-SS	--	--	--	−0.672	0.511	0.262
-LD	--	--	--	1.208 *	3.348	0.063
-LS	--	--	--	2.062 ***	7.858	0.000
-GM	--	--	--	−0.640	0.527	0.222
-DUPF	--	--	--	−1.182 *	0.307	0.088
-VD	--	--	--	1.296 *	3.656	0.059
Nagelkerke R Square	0.202			0.332		
Model chi-square	34.409			68.097		
Model p	0.001			0.000		
Degrees of freedom	12			20		

$* p < 0.10; ** p < 0.01; *** p < 0.001.$

From the final model, AGE (respondents' age), ELDEP (household elderly (>64) dependency ratio), INCOMS-NFE (household income primarily from non-farm employment), HHINCM (household gross income in 2010 (K Yuan)), LANDQT (quantity of cultivated land in 2010), WATERQT-IBDA (water quantity is insufficient for both domestic use and agricultural production), QLGW-SFIR (groundwater quality is suitable only for irrigation), ENVICHAN-LS and ENVICHAN-VD (perceiving the problems of "land salinization" and "vegetation deterioration" as sharply deteriorating within the past five years) significantly predict whether or not a resident has an intention to migrate (Model $p < 0.001$). The Exp (b), odds ratios, of the final model (Table 5) suggests that the odds of migration intent will increase by 2.5% if the household elderly (>64) dependency ratio increases one percentage point, and by 3.5% if annual household income increases by 1000 Yuan. The odds of having the intention to migrate are four times greater for residents with household income derived primarily from non-farm employment than for residents who earn wages primarily from farm work. The odds of opining that "water quantity is insufficient for both domestic use and agricultural production" relating to a propensity for migration are two times greater than the odds of believing "water quantity is sufficient for domestic use and agricultural production". If an informant opines that groundwater quality is suitable only for irrigation, the odds of having the intention to migrate become nearly three times greater than for those who feel groundwater quality is suitable for daily life. And if an informant perceives the problem of "vegetation deterioration" or "land salinization" as having sharply deteriorated during the prior five years, the odds of propensity for migration become four to five times greater than the odds of those who do not perceive these problems. Conversely, the odds of having an intention to migrate decreases by 7% if the respondent's age increases one year, and decreases 9% if the quantity of household cultivated land in 2010 increases one Mu (Mu is a China's area unit. 1 Mu = 1/15 ha).

Table 5. Final Logistic Regression Model for the Log Odds of intending to migrate.

Type	Variables	B	Exp(B)	Sig.
	INTERCEPT	1.297	3.660	0.166
Demographic factors	AGE ***	−0.070	0.932	0.000
	ELDEP **	0.024	1.025	0.002
Socio- economic factors	INCOMS-NFE *	1.421	4.142	0.033
	HHINCM ***	0.035	1.035	0.001
Eco-environmental factors	LOCAT	−0.575	0.562	0.209
	LANDQT*	−0.098	0.907	0.015
	WATERQT-IBDA *	0.842	2.321	0.057
	QLGW-SFIR *	1.066	2.904	0.026
	ENVICHAN-LD	0.527	1.694	0.272
	ENVICHAN-LS ***	1.569	4.804	0.001
	ENVICHAN-DUPF	−1.025	0.359	0.116
	ENVICHAN-VD *	1.308	3.698	0.049

Hosmer and Lemeshow Test: Chi-square, 7.048; d.f., 8; sig., 0.531. Model chi-square = 62.643 ***; d.f., 12. Nagelkerke R Square 0.294. * $p < 0.10$; ** $p < 0.01$; *** $p < 0.001$.

4. Discussion

4.1. The Sustainability of Ecological Migration Policy in Arid Northwest China

The history of China's ecological migration could be traced back to the resettlement of residents from poor areas with a harsh substantial environment and fragile ecology in Western China's Provinces of Ningxia and Gansu by the provincial government in 1980s and 1990s [6]. The meaning of "ecological migration" in Chinese literature [1,2,38,39] is different from "environmental migration" in English literature [40–42]. Although both ecological migration and environmental migration belong to the forced migration category, the main driving forces of the former are government and environment, while, for the latter, governmental force is absent.

There is a relatively longer history of environmental migration in the study area [33]. From the analysis of macro-demographic characteristics of Minqin County, we can see that there is a tremendous depopulation and ageing process in Minqin County from 2000–2010. Although the Key Governance Planning for Shiyang River Basin has been implemented since 2007, the implementation of ecological migration policy by the government began in 2009. That is to say, the 60,000 out-migrants could not result from ecological migration policy. The survey results of this study revealed that most remaining residents in the study area have no propensity to migrate. From the demographic trends in the past, we forecast with a high degree of confidence that the population number in Minqin may reach dynamic equilibrium by the adjustment of local environmental conditions and economic opportunities.

Foggin [7] has argued that the ecological migration policy remains an untested social experiment at an enormous scale—with potentially devastating long-term (generational) social, cultural, and possibly environmental consequences, some of them irreversible. To enhance resilience of the coupled social-ecological system in flexible rather than rigid ways, new governance approaches will need to consider the role of migration: support the needs of migrants, and also of those who remain behind [35]. As suggested by Warner [35], the government should establish new modes of governance to improve society's ability to manage environmentally induced migration, rather than persuading residents to migrate by a one-time migration subsidy.

Although population pressure' often deemed a major cause of land degradation in arid and semi-arid lands (ASALs) [43,44], recurrent voluntary environmental out-migration could not only weaken the tension between population and environment, but could export environmental impacts elsewhere while also increasing social vulnerability [45]. We do not oppose the strategy of ordered resettlement to reduce population pressure directly, but we argue that rural households' concerns about long-term livelihood sustainability determine their migration intention and behavior. The government might usefully pay more attention to create more profitable economic opportunities and more attractive living environments in other places to decrease population density in degraded arid lands, rather than treat ecological migration as an engineering approach that focuses on a specific size and a limited period.

4.2. The Implications of Household Income Amount and Structure for Ecological Migration Policy

A higher household income means the respondent's family has more money to move. In this study, the respondent will have a higher probability of migration intention when his household has more annual income. This is the same as the result found by De Jong, Root, Gardner, Fawcett and Abad [15] who, nonetheless, also found that the money to move has little or no direct impact on actual migration behavior. That is, improvement of household income can foster migration intention, but does not necessarily guarantee that migration will follow.

This study also reveals that if the main source of household income is non-farm employment, the respondent would have a significantly higher possibility of intention to migrate. This finding differs from the result of De Jong [29], which indicates that, in the case of Thailand, non-farm industry of a household has no significant impact on current intentions to move. Another study of migrations and behavior in a rural Philippine province by the same author [15] shows that for actual migration behavior, prior migration experience becomes a dominant explanatory factor. If a rural household's main income is from non-farm employment, some family members are most probably migrant labors in an urban area. This kind of migrant labor experience would foster more actual migration behavior according to De Jong, Root, Gardner, Fawcett and Abad [15].

Based on the results of this paper and other studies, we suggest that the government could do much to increase rural household income as well as opportunities to get more non-farm income, two significant predictors of migration intention. The increased proportion of non-farm earning in whole household income would not only foster migration intention, but also facilitate actual migration action [15].

4.3. The Implications of Arable Land and Irrigation Water Resources Quota for Ecological Migration Policy

Arable land and water resources remain the critical restrictive factors among natural resources necessary for agriculture, especially in arid areas. The relatively long history of voluntary out-migration in the study area, especially among the northern villages bordering desert, is driven mainly by declining arable land and irrigation water resources. The implementation of the river governance program in Shiyang River Basin has amplified the shortage of arable land and irrigation water resources quota. The remarkable diminution of arable land and water resources for agricultural production has resulted in immediate reduction of the natural capitals for the rural households' traditional livelihoods. To cope with these tensions, migration may be one of the adaptive livelihood strategies [15,35]. This viewpoint is backed by the out-migration history in the study area.

As the inland river governance planning has dual aims, to rehabilitate the eco-environment and to improve the victims' livelihood, the measures to decrease the arable land and irrigation water resources quota must be accompanied by measures to diversify livelihood and to develop water/land-saving industries. Otherwise, the migration intentions could not be transformed into actual migration actions, and the environmental conflicts might be transformed into social conflicts. The results of this study indicate that the respondents have more odds of intending to migrate when they have less arable land quota or when they deem water resources insufficient both for domestic use and agricultural production. A compulsory ecological migration policy is not recommended; with such a policy, the forced out-migration would be a failure of the social-ecological system to adapt as stated by Warner [35].

4.4. The Implications of Water Quality and Eco-Environmental Conditions for Ecological Migration Policy

The results show that the possibility of a respondent having migration intention will be higher when he/she believes the quality of groundwater is worse, and that the quality of surface water does not have a significant effect on migration intention. These results are consistent with the assertion that groundwater is the major source of irrigation water; the region is limited in surface water resources [46]. In the past five decades, the excessive exploitation of groundwater and the decreasing surface water supplements have caused continuous decline in the groundwater table. Because infiltrated irrigation water is the main source of groundwater recharge, large amounts of saline matter were transferred from topsoil to groundwater in the process of infiltration of the inspissated irrigation water after evaporation. As a result, the quality of groundwater declined gradually and became less suitable for irrigation [21], forcing some people in the victimized areas to out-migrate [33].

Evidently, the eco-environment remains a critical predictor for migration intentions in our case study of marginal communities in the Minqin County. The results show that the perception of change in environmental problems within the past five years impacts migration intentions significantly. This is consistent with the results of increasing studies about the effects of environmental change on population migration [35,47,48]. The main points of these studies could be summarized by stating that positive environmental characteristics decreased out-migration and negative environmental characteristics increased out-migration [49]. Also, some people will be trapped in areas that expose them to serious risk. Even in the context of quite significant environmental change posing serious threats to the sustainability of livelihoods [36], environmental change may further erode household resources in such a way that migration becomes less, rather than more, likely. All of these statements have realistic counterparts in Minqin County, such as the relatively long history of voluntary out-migration and the overwhelming majority of respondents without migration intention in our survey.

Specifically, this study found that changes in the environmental problems of land salinization and vegetation deterioration have significant effects on migration intentions. Land salinization is adversely affecting grain production. Vegetation deterioration is the cardinal symptom of environmental problem in the study area. The antidote to these two problems is increasing surface water supply to Minqin County. This is one of the main policies in the Key Governance Planning for Shiyang River Basin. However, the aim of the policy is to rehabilitate the deteriorated wild environment bordering deserts, rather than to conserve the arable land for rural residents. In fact, the arable land quota has decreased dramatically as ordered by the governance planning.

Another interesting finding in this study is that the assessment of current environmental problems doesn't have significant effects on migration intentions. This is contrary to the effects of the perception of change in environmental problems within the past five years as discussed above. Using the decision framework for environmental induced migration proposed by Renaud, *et al.* [50] as an analogy, the assessment of current environmental problems is slow onset changes. The perception of change in environmental problems within the past five years is rapid onset changes. A slow onset change may lead to voluntary migration because the environmental effects are more difficult to detect and disentangle from other drivers, particularly economic [36]. A rapid onset change is likely to immediately displace people or communities who have to flee in order to save their lives [50]. As stated by Renaud, Dun, Warner and Bogardi [50], rapid onset hazards are not necessarily of natural origin; their trigger can be caused by social or economic factors such as the arable land quota change in the study area.

4.5. The Implications of Demographic Characteristics for Ecological Migration Policy

The effects of the demographic variables are largely consistent with the previous studies but also reveal some differences. As the surveyed migration intentions in this study are formed within the context of ecological migration policy, migration among respondents would largely take the form of permanent family migration, which is different from the overwhelming pattern of temporary individual migration in other areas lacking rigid resettlement policies, such as the case of Hubei province in China [28]. Therefore, we did not consider migration intention of men and of women as two independent samples as in some influential studies (e.g., [29]). Instead, we used gender of the respondent as a predictor in the analytical models. We found that it has no significant correlation with migration intentions. This result is consistent with Yang [28].

Among the demographic factors, respondent's age as a significant predictor with negative indication is consistent with the previous studies [28,29]. However, the household elderly (>64) dependency ratio as another significant predictors of migration intention with positive indication remains seemingly counterintuitive. From the perspective of the new economics of labor migration (NELM) [51], a household with higher elderly dependency ratio may be more vulnerable to, and more relatively deprived by, a harsh environment with a shortage of labor forces. Therefore, this kind of household can be expected to have a stronger incentive to migrate than one with lower elderly dependency ratio. The results of this study also show that the household child dependency (<15) ratio does not have significant effects on the respondent's migration intention. This could be attributed to no significant difference in the number of children among rural households in China since the implementation of birth control policy from the late 1970s.

5. Conclusions

By employing Minqin County in the terminus of Shiyang River Basin as a case, the predictors of residents' intentions to out-migrate under ecological migration in arid Northwest China are investigated. As the study area has a relatively long history of eco-environmental degradation and population out-migration, the ecological migration policy faces a population with less ability and intention to migrate. The survey results show that most of the residents in the marginal communities of Minqin County do not intend to migrate; indeed only a small fraction desires to migrate. This is consistent with the larger migration literature. Most people remain in origin areas even in areas of high out-migration. Those who do migrate are usually unwitting migrants: they would prefer to remain in their origin area if they felt they could afford—financially and emotionally—to do so [52,53].

The policy implications for government and the public is that, in addition to demographic and socio-economic factors, the eco-environmental factors of water quantity, groundwater quantity, land quantity and change trends of these problems are also significantly correlated with the possibility of a resident intending to migrate. Additionally, the study provides some evidence that inland river basin governance policies impact rural household livelihood assets and environment quality, both significant predictors of migration intention.

Inland river basin governance policies had mixed and somewhat complex impacts on household livelihoods and environmental integrity, and then on migration intentions. As policies are implemented, water availability and quality, soil quality, and vegetation cover may increase or recover gradually, thereby impeding migration intention. If household annual gross income and the proportion of income derived from non-farm employment are increased by the economic policies, as anticipated by the government, migration intention may increase. However, currently most of the residents' household incomes are derived from household agriculture which impedes the intention to migrate. We argue that ecological migration policy may be ultimately unsuccessful if implemented in a compulsory manner, or even encouraged by the local government out of motivation for financial subsidies from the central government as warned by Wang [54]. The complex interplay among policies, household livelihoods, environmental change, and migration intention deserves further investigation. The Ecological Migration Policy will ultimately be more sustainable when taking into account household interests within complex migration intention contexts, such as household livelihoods dynamics and environmental change.

Acknowledgments

This work was supported by the National Social Science Foundation of China under Grant number 11BSH059. It is also one part of Yongjin Li's postdoctoral research report that was done in the School of Life Sciences, Lanzhou University. We thank the Human-Environment Dynamics Lab and Department of Geography at UCSB for providing personnel and facilities support and an encouraging environment for the writing of this paper. We thank Fengmin Li in the School of Life Sciences, Lanzhou University, for his kindly help and advices for this study. We are also grateful for the comments and criticisms of an earlier version of this manuscript by the journal's anonymous reviewers.

Author Contributions

Yongjin Li designed the research, conducted the survey and completed the paper; David López-Carr contributed to the framework of affecting factors and data analysis; Wenjiang Chen contributed to the questionnaire design.

Conflicts of Interest

The authors declare no conflict of interest.

References

1. China National Report on the Implementation of the United Nations Convention to Combat Desertification. Available online: http://hmhfz.forestry.gov.cn/uploadfile/main/2010-9/file/2010-9-1-081984c8ffbd42b2b70a10fdb9d4efa8.pdf (accessed on 20 September 2013).
2. Wang, Z.; Song, K.; Hu, L. China's largest scale ecological migration in the three-river headwater region. *Ambio* **2010**, *39*, 443–446.
3. Zhang, L.; Liu, J. Key issues of ecological migration in northern deserted areas of China. *Chin. J. Ecol.* **2009**, *28*, 1394–1398. (In Chinese)

4. Dickinson, D.; Webber, M. Environmental resettlement and development, on the steppes of inner mongolia, PRC. *J. Dev. Stud.* **2007**, *43*, 537–561.

5. Song, J. The Origin and Related Policies of China's Ecological Migration. Available online: http://www.mzb.com.cn/zgmzb/html/2005-10/14/content_26853.htm (accessed on 18 March 2014). (In Chinese)

6. Liang, F. Study of China's eco-migration. *J. China Three Gorges Univ. Hum. Soc. Sci.* **2011**, *33*, 11–15. (In Chinese)

7. Foggin, J.M. Rethinking "ecological migration" and the value of cultural continuity: A response to wang, song, and hu. *Ambio* **2011**, *40*, 100–101.

8. Foggin, J.M. Depopulating the tibetan grasslands: National policies and perspectives for the future of tibetan herders in qinghai province, China. *Mt. Res. Dev.* **2008**, *28*, 26–31.

9. Ren, M.E. *Essentials of China's Physical Geography*; The Commercial Press: Beijing, China, 1999; p. 430. (In Chinese)

10. Wang, G.; Cheng, G.; Xu, Z. The utilization of water resource and its influence on eco-environment in the northwest arid area of China. *J. Nat. Resour.* **1999**, *14*, 109–116. (In Chinese)

11. Ma, J.; Wang, X.; Edmunds, W.M. The characteristics of ground-water resources and their changes under the impacts of human activity in the arid northwest China—A case study of the shiyang river basin. *J. Arid Environ.* **2005**, *61*, 277–295.

12. Qing, X.; Li, D.; Pan, Y. The study on influencing factors and forcastable model of ecological migration—A case study on the terminal area of the minqin basin. *Northwest. Popul.* **2007**, *28*, 41–44. (In Chinese)

13. Black, R.; Bennett, S.R.; Thomas, S.M.; Beddington, J.R. Climate change: Migration as adaptation. *Nature* **2011**, *478*, 447–449.

14. Ajzen, I.; Fishbein, M. *Understanding Attitudes and Predicting Social Behavior*; Prentice-Hall: Upper Saddle River, NJ, USA, 1980.

15. De Jong, G.F.; Root, B.D.; Gardner, R.W.; Fawcett, J.T.; Abad, R.G. Migration intentions and behavior: Decision making in a rural philippine province. *Popul. Environ.* **1985**, *8*, 41–62.

16. Sun, D.F.; Dawson, R.; Li, B.G. Agricultural causes of desertification risk in minqin, China. *J. Environ. Manag.* **2006**, *79*, 348–356.

17. Zhang, X.; Wang, X.; Yan, P. Re-evaluating the impacts of human activity and environmental change on desertification in the minqin oasis, China. *Environ. Geol.* **2008**, *55*, 705–715.

18. Xie, Y.; Chen, F.; Qi, J. Past desertification processes of minqin oasis in arid China. *Int. J. Sustain. Dev. World Ecol.* **2009**, *16*, 260–269.

19. Zhang, K.C.; Qu, J.J.; Liu, Q.H. Environmental degradation in the minqin oasis in northwest China during recent 50 years. *J. Environ. Syst.* **2004**, *31*, 357–365.

20. Ding, H.; Wang, G.; Huang, X. Runoff reduction into hongyashan reservoir and analysis on water resources crisis of minqin oasis. *J. Desert Res.* **2003**, *23*, 84–89. (In Chinese)

21. Li, D.; Ma, J.; Nan, Z. Characteristic of groundwater drawdown and its sustainable development countermeasure in minqin basin. *J. Desert Res.* **2004**, *24*, 734–739. (In Chinese)

22. Zhao, Y. Two Obligatory Targets of "Key Governance Planning for Shiyang River Basin" Have Been Accomplished Eight Years Ahead of Schedule. Available online: http://szb.gsjb. com/jjrb/html/2012-10/09/content_89632.htm (accessed on 20 August 2013). (In Chinese)

23. DFID Sustainable Livelihoods Guidance Sheets. Available online: http://www.eldis.org/vfile/ upload/1/document/0901/section2.pdf (accessed on 2 November 2013).

24. De Vaus, D.A. *Surveys in Social Research*, 5th ed.; Allen & Unwin: Crows Nest, Australia, 2002.

25. He, X.; Cao, H.; Li, F. Econometric analysis of the determinants of adoption of rainwater harvesting and supplementary irrigation technology (rhsit) in the semiarid loess plateau of China. *Agric. Water Manag.* **2007**, *89*, 243–250.

26. Zhang, W.; Li, F.; Xiong, Y.; Xia, Q. Econometric analysis of the determinants of adoption of raising sheep in folds by farmers in the semiarid loess plateau of China. *Ecol. Econ.* **2012**, *74*, 145–152.

27. He, X.; Cao, H.; Li, F. Factors influencing the adoption of pasture crop rotation in the semiarid area of China's loess plateau. *J. Sustain. Agric.* **2008**, *32*, 161–180.

28. Yang, X. Determinants of migration intentions in hubei province, China: Individual *versus* family migration. *Environ. Plan. A* **2000**, *32*, 769–788.

29. De Jong, G.F. Expectations, gender, and norms in migration decision-making. *Popul. Stud. J. Demogr.* **2000**, *54*, 307–319.

30. Sandefur, G.D.; Scott, W.J. A dynamic analysis of migration: An assessment of the effects of age, family and career variables. *Demography* **1981**, *18*, 355–368.

31. Grieco, E.M.; Boyd, M. *Women and Migration: Incorporating Gender into International Migration Theory*; Working Paper WPS 98–139; Florida State University: Tallahassee, FL, USA, 1998.

32. Abu, M.; Codjoe, S.; Sward, J. Climate change and internal migration intentions in the forest-savannah transition zone of ghana. *Popul. Environ.* **2014**, *35*, 341–364.

33. Bai, J.; Jin, X.; Yang, D. The characteristics and casues of out-migration in ecological fragile area: Case of minqin county, gansu province. *J. Nanjing Coll. Popul. Program. Manag.* **2012**, *28*, 9–13. (In Chinese)

34. Gray, C.L. Soil quality and human migration in kenya and uganda. *Globle Environ. Chang. Hum. Policy Dimens.* **2011**, *21*, 421–430.

35. Warner, K. Global environmental change and migration: Governance challenges. *Globle Environ. Chang. Hum. Policy Dimens.* **2010**, *20*, 402–413.

36. Geddes, A.; Adger, W.N.; Arnell, N.W.; Black, R.; Thomas, D.S.G. Migration, environmental change, and the challenges of governance. *Environ. Plan. C* **2012**, *30*, 951–967.

37. Neuman, W.L. *Social Research Methods: Quantitative and Qualitative Approaches*; Allyn and Bacon: Boston, MA, USA, 2003.

38. Zhou, J.; Shi, G.Q.; Sun, Z.G.; Li, J.Y. Ecological migration of tarim river basin: Conflict and harmony between human and river. In Proceedings of the 4th International Yellow River Forum on Ecological Civilization and River Ethics, Zhengzhou, China, 20–23 October 2009; Yellow River Conservancy Press: Zhengzhou, China, 2010; pp. 90–97.

39. Dong, C.; Liu, X.M.; Klein, K.K. Land degradation and population relocation in northern China. *Asia Pac. Viewp.* **2012**, *53*, 163–177.

40. Adamo, S.B. Environmental migration and cities in the context of global environmental change. *Curr. Opin. Environ. Sustain.* **2010**, *2*, 161–165.

41. Dun, O.; Gemenne, F. Defining "environmental migration". *Forced Migr. Rev.* **2008**, *31*, 10–11.

42. Swain, A. Environmental migration and conflict dynamics: Focus on developing regions. *Third World Q.* **1996**, *17*, 959–974.

43. Qi, F.; Wei, L.; Jianhua, S.; Yonghong, S.; Yewu, Z.; Zongqiang, C.; Haiyang, X. Environmental effects of water resource development and use in the tarim river basin of northwestern China. *Environ. Geol.* **2005**, *48*, 202–210.

44. Li, X.Y.; Xiao, D.N.; He, X.Y.; Chen, W.; Song, D.M. Evaluation of landscape changes and ecological degradation by gis in arid regions: A case study of the terminal oasis of the shiyang river, northwest China. *Environ. Geol.* **2007**, *52*, 947–956.

45. Gemenne, F. What's in a name: Social vulnerabilities and the refugee controversy in the wake of hurricane katrina. In *Environment, Forced Migration and Social Vulnerability*; Springer: New York, NY, USA, 2010; pp. 29–40.

46. Sun, Y.; Kang, S.Z.; Li, F.S.; Zhang, L. Comparison of interpolation methods for depth to groundwater and its temporal and spatial variations in the minqin oasis of northwest China. *Environ. Modell. Softw.* **2009**, *24*, 1163–1170.

47. Massey, D.; Axinn, W.; Ghimire, D. Environmental change and out-migration: Evidence from nepal. *Popul. Environ.* **2010**, *32*, 109–136.

48. Black, R.; Adger, W.N.; Arnell, N.W.; Dercon, S.; Geddes, A.; Thomas, D. The effect of environmental change on human migration. *Globle Environ. Chang. Hum. Policy Dimens.* **2011**, *21*, S3–S11.

49. Gray, C.L. Environment, land, and rural out-migration in the southern ecuadorian andes. *World Dev.* **2009**, *37*, 457–468.

50. Renaud, F.G.; Dun, O.; Warner, K.; Bogardi, J. A decision framework for environmentally induced migration. *Int. Migr.* **2011**, *49*, e5–e29.

51. Stark, O.; Bloom, D.E. The new economics of labor migration. *Am. Econ. Rev.* **1985**, *75*, 173–178.

52. Hunter, L.M. The association between environmental risk and internal migration flows. *Popul. Environ.* **1998**, *19*, 247–277.

53. Carr, D. Population and deforestation: Why rural migration matters. *Progr. Hum. Geogr.* **2009**, *33*, 355–378.

54. Wang, X. Ecological migration: A complex story. Review of yuan yuan's book titled "ecological migration policies and local government practice". *Open Times* **2011**, *2*, 154–158. (In Chinese)

Understanding Relationships among Agro-Ecosystem Services Based on Emergy Analysis in Luancheng County, North China

Fengjiao Ma, A. Egrinya Eneji and Jintong Liu

Abstract: Exploring the relationship between different services has become the focus of ecosystem services research in recent years. The agro-ecosystem, which accounts for one-third of the global land area, provides lots of services but also disservices, depending on resources provided by other systems. In this paper, we explored the agro-ecosystem from four aspects: a summary of different indicators in the agro-ecosystem, input and output changes with time, relationships between different ecosystem services and disservices, and resource contribution to major services, using Luancheng County of North China as the study area. We then used emergy analysis to unify all the indicators. The conclusions were that the agro-ecosystem maintained provisioning and regulating services but with increasing volatility under continued growth in production inputs and disservice outputs. There was a positive correlation between most of the different services and disservices. Rainfall and groundwater resources were the most used input resources in the agro-ecosystem and all other major ecosystem services depended directly on them.

Reprinted from *Sustainability*. Cite as: Ma, F.; Eneji, A.E.; Liu, J. Understanding Relationships among Agro-Ecosystem Services Based on Emergy Analysis in Luancheng County, North China. *Sustainability* **2014**, *6*, 8700-8719.

1. Introduction

Ecosystem services, defined as the benefits humans derive from the ecosystem, has become the focus of ecosystem research in recent years [1–5]. The increasing focus on ecosystem services research started from the recognition and monetary valuation of the benefit flows from ecosystems to society, such as mapping supply and demand and assessing the current and future status of ecosystem services [6]. The research has progressed in recent times to the mechanism of providing ecosystem services and management based on different ecosystem services, using different methods [7–12]. While managing multiple ecosystem services simultaneously is important, it is also extremely challenging. Humanity has invested substantial effort into engineering the ecosystem to produce desired services such as food, timber, and fodder but often at the expense of other services of the ecosystem such as flood control, *etc.* Exploring the relationship between different services has become the focus of research in recent years, with the aim of improving our ability to sustainably manage the ecosystem to provide multiple services [13,14].

The agro-ecosystem accounts for one-third of the global land area with high productivity and is dependent on other systems. Extrapolating global trends from 1960 onward, Tilman *et al.* [15] predicted that by 2050, cropland will increase by 23% and pasture land by 16%. Hence, agriculture accounts for a massive and growing share of the Earth's surface. It was proposed that the agro-ecosystem could provide other services, such as provision of clean air, retaining and recycling of nutrients, mitigation against climate change, *etc.*, in addition to food alone when evaluated based on the scope of ecosystem services [16–18]. Although agro-ecosystem services have been assigned relatively low value [19] when compared with other terrestrial and aquatic ecosystems, they offer the best chance of increasing global ecosystem services provision [20]. Porter *et al.* [20] estimated the ecosystem services of a combined food and energy agro-ecosystem that simultaneously produces food, fodder, and bioenergy. Sandhu *et al.* [21] investigated and quantified the value of ecosystem services of the organic and conventional arable system. Both of them tried to improve the agro-ecosystem services. New estimation frameworks or methods have been widely explored to gain a more accurate estimate of services from the agro-ecosystem. Schulte *et al.* [22] provided a framework for managing soil-based ecosystem services for the sustainable intensification of agriculture. Robinson *et al.* [23] and Dominati *et al.* [24] also developed their soil frameworks to evaluate ecosystem services. Dominati *et al.* [25] used a soil change-based methodology to quantify and value the services from agro-ecosystems. In addition, Ibarra *et al.* [26] valued the ecosystem services of urban wetlands using an agro-ecosystem approach.

While an agro-ecosystem provides important provisioning services, it also creates disservices and consumes resources from other systems [20,27]. The consumption of water, emissions of greenhouse gases, and discharging of underutilized fertilizer adversely affect human beings. Despite real differences, few studies distinguish among ecosystem services, ecosystem disservices, and resource consumption in their evaluation of an ecosystem. For example, the agro-ecosystem needs water for irrigation, which should be classified as resource consumption, but is often considered as an ecosystem disservice [28–30]. In this study, we first developed a conceptual framework and explored the structures and changes in agro-ecosystem input resources, output services, and disservices, and then analyzed the relationships among different services. Finally, we explored the relationship between input resources and major services, taking Luancheng County (North China) as a case study.

2. Conceptual Framework

Ecosystem services are defined as the benefits people obtain from the ecosystem in the Millennium Ecosystem Assessment (MEA), which is generally consistent with current usage in the literature, although some scientists noted that this definition mixes "ends" and "means" [31–33]. Considering this ambiguity, Boyd and Banzhaf opined that "the final ecosystem services are components of nature, directly enjoyed, consumed, or used to yield human well-being" for developing the national-scale environmental welfare accounting and performance assessment [34]. Fisher and Turner [33,35] considered ecosystem services to be the aspects of the ecosystem utilized (actively or passively) for human well-being. However, ecosystem services are defined differently depending on the research goals [35]. Here, we adopted the definition in MEA, since the focus of our research was not on distinguishing whether the benefits belong to final services or not. "Ecosystem disservices" is another controversial term with different definitions. Zhang *et al.* [27] characterized ecosystem disservices as a reduction in productivity or increase in production costs. Lyytimäki *et al.* [36] considered disservices to be ecosystem functions disturbed by human activities. We define ecosystem disservices as adverse effects on human beings and the ecosystem, considering the resources cost separately in our framework.

Separately measuring the components of ecosystem inputs and outputs adds clarity to ecosystem evaluation and can enhance the recognition of the relationship between the ecosystem and human beings. The ecosystem inputs refer to the resource consumption by the ecosystem while the ecosystem outputs include the ecosystem services and disservices. Since this study focused on the agro-ecosystem, the components of resource inputs include renewable climate resources, the nonrenewable underground water resource, and purchased renewable and nonrenewable resources from human economical feedback. Underground water is separated from renewable resources because the groundwater resource is yearly over-exploited and the recharge rate is much less than the exploitation rate. Groundwater could not therefore be classified under renewable resources. We divided the ecosystem services outputs into provisioning services, regulating services, and supporting services. The ecosystem disservices outputs included the loss of inorganic fertilizer (classified under provisioning disservices) and loss of soil (supporting disservices).

The entire framework for the study is shown in Figure 1. The agro-ecosystem, being our research focus, is located in the middle of the framework. Three inputs—renewable climate resources, the nonrenewable groundwater resource, and economic feedback resources—are arranged to the left side. The ecosystem services and disservices are shown on the right side of the framework.

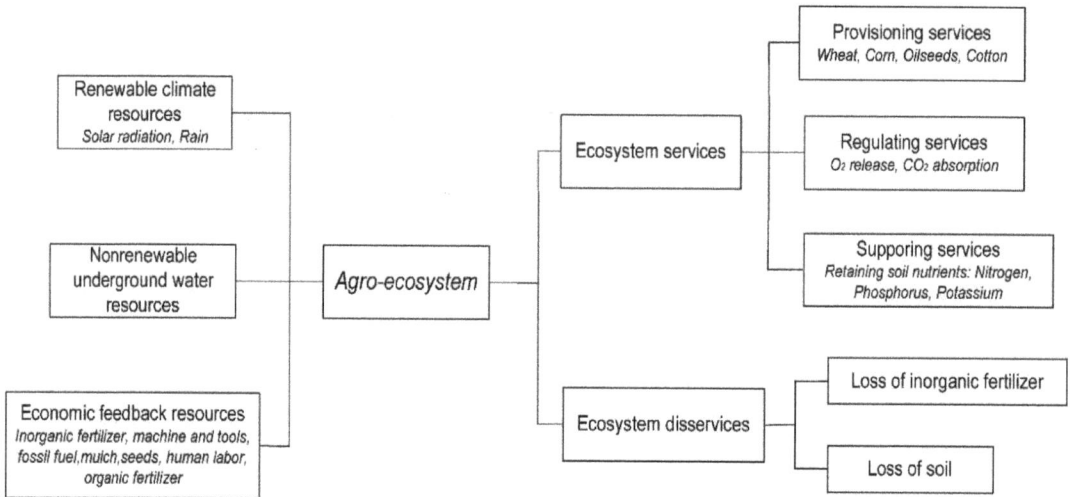

Figure 1. Conceptual framework for evaluating the agro-ecosystem.

3. Materials and Methods

3.1. Study Area

The North China Plain (NCP) is one of the most productive and intensively cultivated agricultural regions in China. About 50% of the nation's wheat and 33% of its maize are produced in this region. Our study area, Luancheng County, is located in NCP and is a typical high-production agro-ecosystem. It is a traditionally agricultural county. Agriculture accounts for 23.8% of GDP, greater than Hebei Province's ratio of 15%, although Hebei Province is a major grain-producing province in China. The yield of wheat in the area was 7210.5 kg/ha and that of corn was 8730.0 kg/ha in 2005. The yearly government subsidy to farmers is about 4200 Yuan/ha (nearly 700 $/ha), with strict regulations to protect farmland from illegal occupation to ensure national food security.

The area is characterized by a warm temperate continental monsoon climate with an annual mean temperature of 12.7 °C, with the highest temperature (26.4 °C) in July and the lowest (3.9 °C) in January. Mean solar radiation is 724 kJ/(cm^2·a) and the annual sunshine hour is 2521.8 h. Annual precipitation is about 536 mm, 2/3 of which is concentrated in summer. The geomorphology is piedmont alluvial plain and topography is flat with meadow cinnamon soil type. The groundwater resource is abundant with salinity of 0.5–1.0 g/L and the water table is shallow. However, the water table has continued to decline in successive years due to severe overexploitation for irrigation. The contradiction between water scarcity and irrigation of the agro-ecosystem has increasingly intensified in this region.

3.2. Methods

3.2.1. Emergy Analysis Theory

Ecosystem services valuation methods include the economic valuation system and ecological valuation system. The economic valuation method has difficulties when services are not marketed. Emergy analysis is an ecological valuation method based on thermodynamic principles, which translates different inputs and outputs of an ecosystem into the same solar emjoule (sej) unit using solar energy as the base energy [37]. The emergy theory estimates the ecocentric value rather than the human-centric value [38,39]. This is in direct contrast to the economic view [39].

Although a controversial methodology, the emergy analysis offers a number of advantages, as it provides: (1) a way to bridge economic and ecological systems; (2) an objective means by which to quantify and value non-market inputs into a system; (3) a common unit that allows for a comparison of all resources; and (4) a more holistic alternative to many existing methods of decision-making. But critics of the emergy analysis generally complain that the method: (1) lacks formal links with related concepts in other disciplines; (2) lacks adequate details on the underlying methods; (3) is computationally and data intensive and (4) is based on sweeping generalizations that remain unproven [39–42]. Nowadays, emergy analysis is used to assess the sustainability of regional development, agricultural practices, and preservation and restoration of the natural environment, although controversy remains as to its use for ecosystem services valuation [40].

Based on the emergy theory, value does not only rely on human preferences and willingness to pay, but instead stems from the work of the biosphere to develop and stabilize an ecosystem structure, growth, organization, and diversity [43]. Jørgensen and Nielsen [44] stated that a complete diagnosis, focused on the ecosystem services, could be developed through the use of complementary indicators such as emergy and eco-exergy. Pulselli et al. [45] considered ecosystem services as a counterpart of emergy flows to ecosystem. Although the valuation methodology has no consensus, some studies [43,45–50] had tried to link the ecosystem services and emergy analysis, which is an evaluation from a donor-side approach [51,52].

3.2.2. Data Collection

The inputs and outputs structures in an agro-ecosystem are illustrated succinctly in Figure 2.

Data sources: Long-term (1984–2008) annual mean climatic data for solar radiation and rainfall were taken from the National Ecosystem Research Network of China and Luancheng County weather station. Socioeconomic data were obtained from statistical yearbooks of the local government. Soil data were taken from the National Ecosystem Research Network of China, the second national soil survey data, and relevant literature. Other parameters like products' economic coefficient, the price value of O_2 release, CO_2 fixation, *etc.* were obtained from the literature (see Table A1).

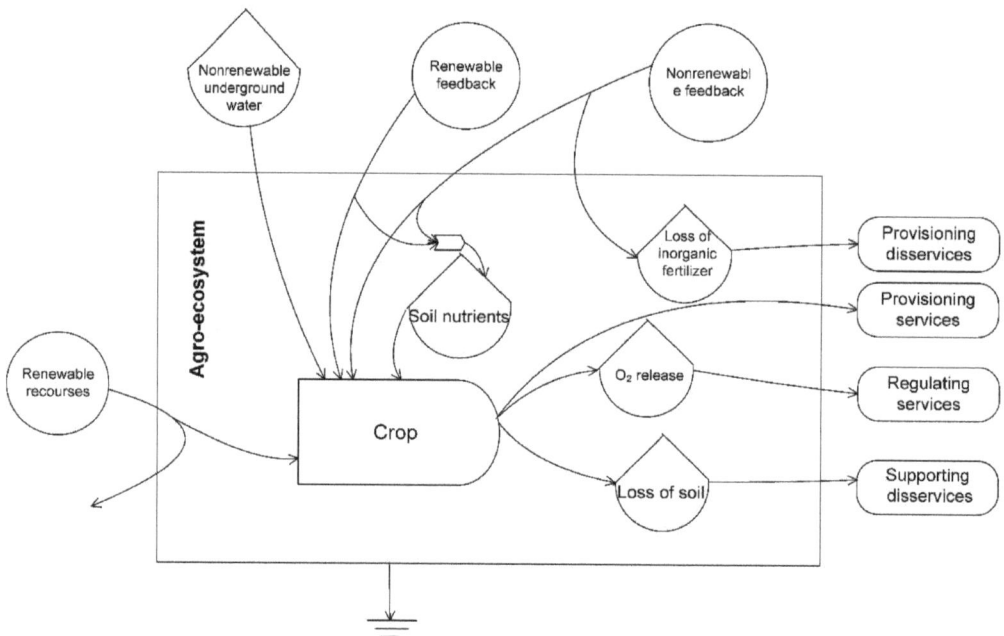

Figure 2. Emergy flows in the agro-ecosystem.

3.2.3. Data Analysis

All the data (indicators) were standardized using the emergy analysis method. This was done by transforming the different inputs and outputs into energy or mass data and then multiplying by the appropriate transformities (cited from the literature or calculated from our data) to obtain the emergy (see Table A1). Relationships between different services and the consumption resources were analyzed by bivariate correlation and step-wise linear regression using the SPSS and EViews software.

4. Results and Discussion

4.1. Summary of Different Indicators in the Agro-Ecosystem of Luancheng County

The calculated emergy values of basic structures (inputs and outputs) are shown in Table 1. The uncertainties, which referred to parameter uncertainties (u^2) in this paper, were measured by the Variance method [53,54]. As can be seen from Table 1, the largest inputs to the ecosystem were underground water, inorganic fertilizer, machines and tools, and fossil fuel. The use of machines and tools increased considerably over the years due to intensive mechanization of farm operations. Underground water showed a very large variability; with zero inputs in some years, rainfall was high enough to sustain crops. The largest inputs were those of the economic feedback resources (see Figure 1), except underground water, which counts as a nonrenewable resource. Both economic feedback resources and underground water are nonrenewable resources that have exerted an enormous pressure on the sustainability of the agro-ecosystem.

Table 1. Classical statistics of the calculated emergy.

	Unit	Number	Minimum	Maximum	Mean	Standard Deviation	Uncertainty
Input resources							
Solar energy absorbed	10^{18} sej	25	1.13	1.74	1.47	0.14	0.02
Rain (geopotential)	10^{17} sej	25	3.23	11.21	5.49	1.66	2.76
Rain (chemical)	10^{18} sej	25	5.65	19.59	9.60	2.91	8.47
Underground water energy	10^{19} sej	25	0	9.44	3.91	2.94	8.64
Seeds	10^{18} sej	25	4.58	9.84	6.37	1.74	3.03
Human labor	10^{18} sej	25	3.39	9.90	6.68	2.50	6.25
Organic fertilizer	10^{17} sej	25	0.37	1.41	1.07	0.39	0.15
Inorganic fertilizer	10^{19} sej	25	3.54	7.85	5.68	1.12	1.25
Pesticides	10^{17} sej	25	0	6.85	3.70	2.35	5.52
Mulch	10^{16} sej	25	0	3.61	1.28	1.42	2.02
Machines and tools	10^{20} sej	25	0.40	1.68	1.09	0.53	0.28
Fossil fuel	10^{19} sej	25	2.77	8.05	4.68	1.72	2.96
Output ecosystem services and disservices							
Wheat	10^{20} sej	25	1.09	1.88	1.44	0.22	0.05
Corn	10^{20} sej	25	0.15	2.17	1.69	0.41	0.17
Oil seeds	10^{19} sej	25	0.16	1.05	0.44	3.75	14.1
Cotton	10^{19} sej	25	0	9.92	1.29	2.11	4.45
O_2 release	10^{20} sej	25	0.90	2.14	1.70	0.29	0.08
CO_2 fixation	10^{20} sej	24	−3.4	3.46	0.41	1.73	2.99
Nitrogen	10^{19} sej	12	1.77	3.63	2.92	0.49	0.24
Phosphorus	10^{19} sej	12	0.61	4.21	1.82	1.03	1.06
Potassium	10^{19} sej	11	0.89	1.44	1.24	0.14	0.02
Loss of inorganic fertilizer	10^{19} sej	25	0.96	2.10	1.52	0.30	0.09
Loss of soil	10^{18} sej	25	1.86	3.86	2.90	0.48	0.23

The inputs resources of Chinese agriculture generally were not the same as those in Luancheng County [55]. For example, water is south China was taken as a renewable resource, considering that rainfall and surface water are relatively abundant there. Also, the proportion of renewable resources was much lower than the nonrenewable feedback resources. In addition, the biggest input of nonrenewable resources in Chinese agriculture was inorganic fertilizer, rather than the machines and tools in Luancheng County. There are two reasons for this phenomenon: (1) the mechanization of agriculture in Luancheng County has been more complete than in other regions because about 66% of China's land area is mountainous, which is a great limitation to the use of large agricultural machines; and (2) the irrigation water in Luancheng County depends on underground water, which needs more machines to pump the water from underground.

The largest output indicators were wheat, corn, O_2 release, and CO_2 fixation (see Table 1). Ecosystem provisioning services outputs ,which are the main benefits from the agro-ecosystem under study, are indicated mainly by yields of wheat and corn (the two major crops in the area; they also gave the largest outputs). The O_2 release and CO_2 fixation, which are regulating services, have always been ignored, although their values were as large as those of wheat and corn. For the ecosystem disservices outputs, losses of inorganic fertilizer and soil were much less than the main ecosystem services (Table 1). However, they tended to increase significantly over time and their effects may slowly become obvious in the future.

It has been confirmed that provisioning services were the largest output from an agro-ecosystem, whether based on the emergy analysis as in this work or on monetary valuation [28–30]. However, soil loss, which classifies as a supporting disservice, was quite different in this work from Chinese agriculture generally. Soil loss was classed as a nonrenewable natural resource and was "consumed" in so large an amount that it was only exceeded by inorganic fertilizer consumption [55]. It is true that lands that have undergone soil erosion account for almost one-third of total Chinese arable land, contrary to our observations in Luancheng County. However, the loss of soil was opposite to conserving soil (supporting services), so we defined it as supporting disservices rather than consumption in the agro-ecosystem.

4.2. Changes in Different Structures of Inputs and Outputs

The overall profile of the agro-ecosystem can be seen in Figure 3. The input resources were expressed in negative values since they were derived from other systems and consumed by the agro-ecosystem. The ecosystem services had positive values while disservices had negative values. Overall, the negative values increased consistently while the positive values increased firstly and then decreased. Specifically, the economic production system had the largest negative values (accounting for >70% of the total negative values) with a rising trend. The groundwater system followed with a fluctuating growth. The provisioning services showed mostly positive values, followed by the regulating services. Both of the main positive services increased at the initial period, only to decrease progressively thereafter. The raw data for supporting services were collected from 1998 to 2008 and the supporting services showed a stable trend during this period (Figure 3). The ecosystem services took a relatively small proportion.

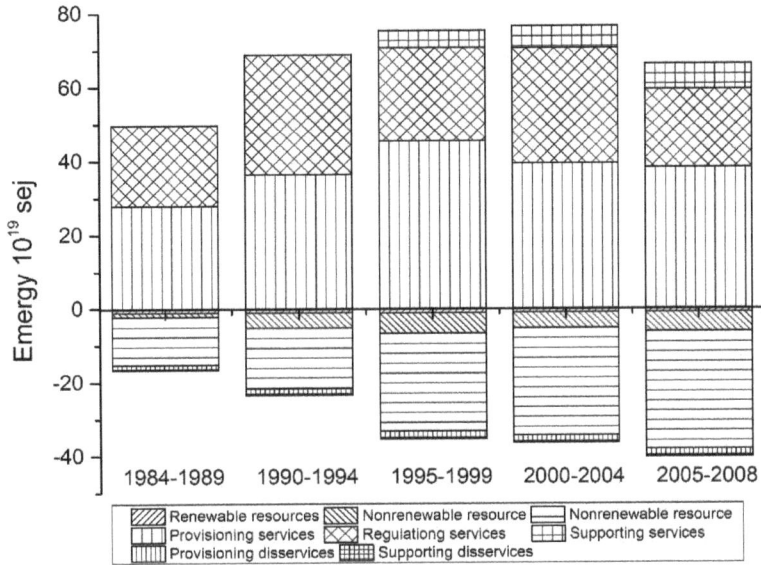

Figure 3. Changes in different indicators of the agro-ecosystem of Luancheng County, China.

Previous research on agro-ecosystem services mainly focused on the benefits derived therefrom but the ecosystem disservices have become gradually recognized in recent years [27–30]. However, there is no consensus on these disservices. For example, some authors considered groundwater consumption as an ecosystem disservice [30], while some defined it as a regulating service [28,29]. Here, we consider it as a resource input from the groundwater system. If our analysis was performed based on the definitions of others, the results would be quite different. We defined ecosystem disservices as useless or harmful outcomes for humans. For example, loss of inorganic fertilizer is useless for humans but harmful to other ecosystems, such as eutrophication of water bodies or pollution of underground water. The potential disservice might be much larger than the emergy value of the loss of inorganic fertilizer. If our conceptual framework is supported by life cycle assessment, the value of ecosystem disservices might even be larger than that shown here.

4.3. Correlation between Different Services and Disservices

We explored the relationships between different indicators of services and disservices, considering there were some synergistic or trade-off relationship between different services. The results are shown in Table 2. The relationships between indicators of ecosystem services and disservices were mainly positive. Cotton production (provisioning services) and potassium and some of the phosphorus supplying services (supporting services) had negative correlations with other services. Cotton production negatively correlated with other services because the arable land was limited and wheat and corn competed for much of the cotton acreage. This also resulted in some possibility of a trade-off relationship between provisioning services, but not synergistic relationships. The reason for the negative relationship between potassium nutrients and other services was the imbalanced (or irrational) fertilization. Fertilizer application in the study area mainly focused on nitrogen fertilizer. The phosphate and, especially, potassium content of soil remained progressively low since they are applied as mere supplements to support the substantial increase in production. On the other hand, the nitrogen content showed a positive correlation, confirming that nitrogen fertilizer was applied excessively.

The relationships between other provisioning services and regulating services were synergistic, meaning that increasing the provisioning services brought about increases in regulating services without trade-off effects. However, the ecosystem disservices were also positively related with major services, suggesting that more and more burdens were imposed on the agro-ecosystem with its growing services.

Most previous studies have shown that growth in provisioning services resulted in a decline in regulating services [56]. For example, increasing the grain field was at the expense of increasing the level of soil erosion in mountain areas [57]. However, our research area had no significant reduction in regulating services because the agro-ecosystem had no serious soil erosion, given the plain terrain. The second reason for the negative correlation was the land use change—the conversion of forest or grassland to arable land, leading to a decline in regulating services and an increase in provisioning services. However, the agro-ecosystem of Luancheng County has been the same for a long time (having not been recently transformed from other ecosystems). In addition, the oxygen supply service was related to productive ability so that it kept increasing with the growing provisioning services. Thus, the relationship between provisioning services and regulating services is dependent on the area or location.

Table 2. Correlation (matrix) among different services and disservices of the agro-ecosystem in Luancheng County, China.

	Corn	Oilseeds	Cotton	O₂ release	CO₂ fixation	Nitrogen	Phosphorus	Potassium	Loss of chemical fertilizer	Loss of soil
Wheat	0.532	0.502	−0.451	0.924 **	0.21	0.374	−0.422	−0.568	0.194	−0.070
Corn		0.451	−0.850 **	0.797 **	−0.073	0.824 **	0.04	−0.261	0.664 **	0.713 **
Oilseeds			−0.615 *	0.631 *	0.093	0.255	−0.303	−0.126	0.249	0.124
Cotton				−0.706 *	−0.068	−0.765 **	−0.105	0.159	−0.779 **	−0.653 *
O₂ release					0.155	0.596 *	−0.308	−0.485	0.416	0.345
CO₂ fixation						0.128	0.011	−0.049	0.067	0.462
Nitrogen							0.171	−0.132	0.509	0.679 *
Phosphorus								0.454	0.117	0.369
Potassium									−0.144	0.205
Loss of chemical fertilizer										0.676 *

Note: ** $p < 0.01$, * $p < 0.05$.

4.4. Regression Analysis of Resource Inputs and Major Services Outputs

Provisioning services are the only focus of an agro-ecosystem regardless of other long-term services and disservices. Farmers apply very large inputs (consumptions) in order to get more yields from the fields. We took wheat and corn yields as examples and performed a regression analysis to explore the contributions of different resources to the yields, considering that wheat and corn account for 85% of all provisioning services in Luancheng County. The resulting regression equation for wheat was (see Table 3):

$$Y_w = 7.304 \times 10^{19} + 1.009G + 6.488 \times R + 0.731 \times F \qquad (1)$$

where Y_w was wheat yield, G was groundwater resources, R was rainfall resources, and F was fossil fuel. We used a stepwise regression algorithm to obtain this regression equation. The Ramsey's regression specification error test (Ramsey RESET) was used to determine the appropriateness of the linear relationship using EViews software. The probability of F-statistic was 0.1601, being larger than 0.1. Since the functional form was valid at the confidence level of 0.1, we established that the most effective indicators were underground water, rainfall, and fossil fuel. The fossil fuel was used in the regression equation because it was needed for pumping groundwater for irrigation.

Table 3. Results of multiple stepwise regression analysis.

Services	Variables	Coefficients	Standard Error	t-value	Sig. (p-value)
Wheat	Constant	7.30×10^{19}	1.49×10^{19}	4.89	0
	Groundwater	1.009	0.139	7.273	0
	Rainfall	6.488	1.218	5.328	0
	Fossil fuel	-0.731	0.18	-4.064	0.001
	Adjusted R^2		0.675		
	Standard Error of the Estimate		1.27×10^{19}		
Corn	Constant	5.38×10^{19}	3.30×10^{19}	1.629	0.118
	Fertilizer	1.693	0.635	2.667	0.014
	Groundwater	0.485	0.243	2	0.058
	Adjusted R^2		0.449		
	Standard Error of the Estimate		3.05×10^{19}		
O_2 release	Constant	4.98×10^{19}	1.55×10^{19}	3.207	0.004
	Groundwater	1.155	0.121	9.524	
	Rainfall	7.372	1.161	6.348	
	Adjusted R^2		0.787		
	Standard Error of the Estimate		1.33×10^{19}		

We then performed path analysis to decompose correlations into different components, considering that explanatory variables could interact with each other. The results of path analysis are shown in Table 4. Both the Pearson correction coefficient, which is equal to the sum of the direct and indirect path coefficients, and the direct path coefficient of underground water were the largest. Although fossil fuel had the biggest indirect path coefficient, its Pearson correction coefficient was negative. This is because the input of fossil fuel was greater for years with less rainfall, in which the yields of wheat were relatively low. The independents were rain water, groundwater, and fuel for extracting water (fossil). This suggests that wheat production depends heavily on water, especially underground water.

154

Table 4. Summary of correlation coefficients for wheat.

Explanatory Variables	Pearson Correction Coefficient	Direct Path Coefficient	Indirect Path Coefficients			
			Underground Water	Rainfall	Fossil Fuel	Total
Underground water	0.521	1.328		−0.582476	−0.224238	−0.806714
Rainfall	0.016	0.892	−0.867184		−0.010116	−0.8773
Fossil fuel	−0.017	−0.562	0.529872	0.016056		0.545928

The regression equation for corn was (see Table 3):

$$Y_c = 5.379 \times 10^{19} + 1.693 \times C + 0.485 \times G \quad (2)$$

where Y_c was corn yield, C was the fertilizer resource, and G was the groundwater resource. A Ramsey RESET test showed that the probability of F-statistic was 0.0468, being larger than 0.01, hence the functional form was correct at the confidence level of 0.01. The water supply was not included in the regression because the corn grew during the rainy season (June to September) and the rainfall is usually sufficient for growth. The results of path analysis are shown in Table 5. The supply of fertilizer had a larger direct path coefficient and the underground water had larger indirect path coefficients with corn yield. The Pearson correction coefficient of fertilizer was little larger than for the underground water, showing that both fertilizer and underground water were important inputs for corn.

Table 5. Summary of correlation coefficients for corn.

Explanatory Variables	Pearson Correction Coefficient	Direct Path Coefficient	Indirect Path Coefficients		
			Fertilizer	Underground Water	Total
Fertilizer	0.635	0.464		0.170868	0.170868
Underground water	0.576	0.348	0.227824		0.227824

It could be seen that different provisioning services were sensitive to different resources from the regressions of the major crop yield. Therefore, we should rationally allocate the different resources according to crop requirements.

In addition to provisioning services, the regulating services were also important ecosystem services in which oxygen release accounted for >80%. So we performed a regression analysis of oxygen release service and the resulting equation was (see Table 3):

$$Y_o = 4.980 \times 10^{19} + 1.155 \times G + 7.372 \times R \quad (3)$$

where Y_o was oxygen release, G was the groundwater resource, and R was the rainfall resource. A Ramsey RESET test showed that the probability of F-statistic was 0.1919, which was larger than 0.1. Therefore, the functional form was correct at the confidence level of 0.1. The contributory factors were only groundwater and rainfall, both being water resources. The results of path analysis (Table 6) showed that both the direct and indirect path coefficients of underground water were larger than those of rainfall, hence the Pearson correction coefficient of underground water was also larger. Thus, it could be noted that the regulating service, just like the provisioning services of wheat, was water-dependent and that water resources, especially underground water, directly determine how much service could be given.

Table 6. Summary of correlation coefficients for oxygen release.

Explanatory Variables	Pearson Correction Coefficient	Direct Path Coefficient	Indirect Path Coefficients		
			Underground Water	Rainfall	Total
Underground water	0.669	1.184		−0.515217	−0.515217
Rainfall	0.017	0.789	−0.773152		−0.773152

The influential factors of ecosystem services mainly center on land use change [58], climate change [59], and other natural factors [60], but the major driving factor differed according to region. The lack of water resources was the biggest challenge in North China. Our results also showed that the water resources directly determine the agro-ecosystem services and that groundwater supported this ecosystem. Thus, using the groundwater scientifically to optimize ecosystem services remains the key challenge in Luancheng County.

5. Conclusions

The high-yielding agro-ecosystem of Luancheng County in North China maintained provisioning services and regulating services but with increasing volatility with continued growth in farm (consumption) inputs and disservices outputs. Most of the relationships between different services and disservices of the agro-ecosystem were positive. However, cotton fields under provisioning services and soil potassium under supporting services had negative correlations with the others. The rainfall and groundwater resources were the most contributory input resources in the agro-ecosystem of Luancheng County and all other major ecosystem services depended on them directly.

Acknowledgment

The study was supported by the Project of the Main Direction Program of Knowledge Innovation of CAS (KSCX2-EW-J-5). Anthony Egrinya Eneji thanks the CAS Senior Visiting Fellowship for the opportunity. We are grateful for the support of Lipu Han and Yueyan Liu.

Author Contributions

Jintong Liu and Fengjiao Ma designed this research; Fengjiao Ma performed the calculations and analyzed the data; and A. Egrinya Eneji and Fengjiao Ma wrote the paper. All authors have read and approved the final manuscript.

Appendix

Table A1. Transformities, sources and formulation of raw data used for emergy analysis.

	Unit	Transformity (sej/unit)	Formulation of Raw Data	Sources of Raw Data
Solar radiation	J	1 [a]	Solar radiation = arable area × solar radiation intensity	e
Rain (geopotential)	J	8888 [a]	Rain geopotential = arable area × elevation × average rainfall × density × gravitational acceleration	e
Rain (chemical)	J	15,444 [a]	Rain chemical energy = arable area × average rainfall × Gibbs energy × density	e
Underground water	J	This study [(1)]		e
Seeds	J	7.86×10^4 [b]	Seed = amount of seed per unit area × arable area	f
Human labor	J	8.10×10^4 [c]	Human labor = the amount of human labor per unit area × arable area	f
Organic fertilizer	g	2.70×10^6 [a]	Organic fertilizer = amount of organic fertilizer per unit area × arable area	fg
Inorganic fertilizer	g	3.80×10^9 [d]	Raw data	h
Pesticide	g	1.60×10^9 [a]	Raw data	h
Mulch	g	3.80×10^8 [a]	Raw data	h
Machine and tools	J	7.50×10^7 [d]	Raw data	h
Fossil fuel	J	1.59×10^5 [a]	Raw data	h
Wheat	J	6.80×10^4 [a]	Wheat energy = wheat yield × wheat calorific value	gh
Corn	J	8.52×10^4 [d]	Corn energy = corn yield × corn calorific value	gh
Oilseeds	J	8.60×10^4 [a]	Oilseeds energy = oilseeds yield × oilseeds calorific value	gh
Cotton	J	8.60×10^5 [a]	Cotton energy = cotton yield × cotton calorific value	gh
O_2 release	$	4.94×10^{12} [c]	NPP = produce yield × (1-moisture content of each products)/each products economic coefficient. Price of O_2 Release = NPP × (32/30) × the price of O_2 production	chi
CO_2 fixation	J	6.25×10^4 [a]	Energy of CO_2 fixation = accumulation of organic matter in soil × calorific value of organic matter	eglmno
Nitrogen	g	4.60×10^9 [c]	Nitrogen = arable area × topsoil thickness × density × percentage content of nitrogen	eglmno
Phosphorus	g	1.78×10^{10} [c]	Phosphorus = arable area × topsoil thickness × density × percentage content of Phosphorus	eglm

Table A1. *Cont.*

	Unit	Transformity (sej/unit)	Formulation of Raw Data	Sources of Raw Data
Potassium	g	1.74×10^{9} [c]	Potassium = arable area × topsoil thickness × density × percentage content of Potassium	eglmo
Loss of inorganic fertilizer	g	3.80×10^{9} [d]	Loss of inorganic fertilizer mass = mass of inorganic fertilizer × inorganic fertilizer loss rate	hq
Loss of soil		This study [2]		

Note: [1] Emergy of groundwater = energy of groundwater × transformity of groundwater. The amount of irrigation water was calculated using Yuan's approach [61]; this method relies on meteorological data and crop yield. Transformity of groundwater = energy of groundwater × transformity of rainfall × update time [62]; the relationship between groundwater update time (Y) and groundwater table (m) was Y = 0.13x + 6.73 in North China Plain, according to Wei [63]; [2] Emergy of topsoil loss = loss of topsoil organic matter × transformity of organic matter + loss of topsoil nitrogen × transformity of nitrogen+ loss of topsoil phosphorus × transformity of phosphorus + loss of topsoil potassium × transformity of potassium; [3] References for transformity: [a] [62]; [b] [64]; [c] [65]; [d] [66]; [4] References for raw data: e [67]; f [68]; g [69]; h Hebei Bureau of Statistics [70]; i Chinese Agricultural Yearbook [71]; j [72]; k State Forestry Administration [73]; l [74]; m [75]; n [76]; o [77]; p [78]; q [79]; r [80].

Conflicts of Interest

The authors declare no conflict of interest.

References

1. Benayas, J.M.R.; Newton, A.C.; Diaz, A.; Bullock, J.M. Enhancement of Biodiversity and Ecosystem Services by Ecological Restoration: A Meta-Analysis. *Science* **2009**, *325*, 1121–1124.
2. Daily, G.C.; Matson, P.A. Ecosystem services: From theory to implementation. *Proc. Natl. Acad. Sci. USA* **2008**, *105*, 9455–9456.
3. Kinzig, A.; Perrings, C.; Chapin, F.; Polasky, S.; Smith, V.; Tilman, D.; Turner, B. Paying for ecosystem services—Promise and peril. *Science* **2011**, *334*, 603–604.
4. Schroter, D.; Cramer, W.; Leemans, R.; Prentice, I.C.; Araujo, M.B.; Arnell, N.W.; Bondeau, A.; Bugmann, H.; Carter, T.R.; Gracia, C.A.; *et al.* Ecosystem service supply and vulnerability to global change in Europe. *Science* **2005**, *310*, 1333–1337.
5. Tallis, H.; Kareiva, P.; Marvier, M.; Chang, A. An ecosystem services framework to support both practical conservation and economic development. *Proc. Natl. Acad. Sci. USA* **2008**, *105*, 9457–9464.
6. Lu, Y.; Liu, S.; Fu, B. Ecosystem service: From virtual reality to ground truth. *Environ. Sci. Technol.* **2012**, *46*, 2492–2493.
7. Bohlen, P.J.; Lynch, S.; Shabman, L.; Clark, M.; Shukla, S.; Swain, H. Paying for environmental services from agricultural lands: An example from the northern Everglades. *Front. Ecol. Environ.* **2009**, *7*, 46–55.

158

8.   Bugalho, M.N.; Caldeira, M.C.; Pereira, J.S.; Aronson, J.; Pausas, J.G. Mediterranean cork oak savannas require human use to sustain biodiversity and ecosystem services. *Front. Ecol. Environ.* **2011**, *9*, 278–286.

9.   Chazdon, R.L. Beyond deforestation: Restoring forests and ecosystem services on degraded lands. *Science* **2008**, *320*, 1458–1460.

10. Koch, E.W.; Barbier, E.B.; Silliman, B.R.; Reed, D.J.; Perillo, G.M.E.; Hacker, S.D.; Granek, E.F.; Primavera, J.H.; Muthiga, N.; Polasky, S.; *et al.* Non-linearity in ecosystem services: Temporal and spatial variability in coastal protection. *Front. Ecol. Environ.* **2009**, *7*, 29–37.

11. Maestre, F.T.; Quero, J.L.; Gotelli, N.J.; Escudero, A.; Ochoa, V.; Delgado-Baquerizo, M.; Garcia-Gomez, M.; Bowker, M.A.; Soliveres, S.; Escolar, C.; *et al.* Plant species richness and ecosystem multifunctionality in global drylands. *Science* **2012**, *335*, 214–218.

12. Plummer, M.L. Assessing benefit transfer for the valuation of ecosystem services. *Front. Ecol. Environ.* **2009**, *7*, 38–45.

13. Bennett, E.M.; Peterson, G.D.; Gordon, L.J. Understanding relationships among multiple ecosystem services. *Ecol. Lett.* **2009**, *12*, 1394–1404.

14. Power, A.G. Ecosystem services and agriculture: tradeoffs and synergies. *Philos. Trans. R. Soc. B Biol. Sci.* **2010**, *365*, 2959–2971.

15. Tilman, D.; Fargione, J.; Wolff, B.; D'Antonio, C.; Dobson, A.; Howarth, R.; Schindler, D.; Schlesinger, W.H.; Simberloff, D.; Swackhamer, D. Forecasting agriculturally driven global environmental change. *Science* **2001**, *292*, 281–284.

16. Gordon, L.J.; Finlayson, C.M.; Falkenmark, M. Managing water in agriculture for food production and other ecosystem services. *Agric. Water Manag.* **2010**, *97*, 512–519.

17. Qiu, J.; Turner, M.G. Spatial interactions among ecosystem services in an urbanizing agricultural watershed. *Proc. Natl. Acad. Sci. USA* **2013**, *110*, 12149–12154.

18. Swinton, S.M.; Lupi, F.; Robertson, G.P.; Hamilton, S.K. Ecosystem services and agriculture: Cultivating agricultural ecosystems for diverse benefits. *Ecol. Econ.* **2007**, *64*, 245–252.

19. Costanza, R.; dArge, R.; deGroot, R.; Farber, S.; Grasso, M.; Hannon, B.; Limburg, K.; Naeem, S.; Oneill, R.V.; Paruelo, J.; *et al.* The value of the world's ecosystem services and natural capital. *Nature* **1997**, *387*, 253–260.

20. Porter, J.; Costanza, R.; Sandhu, H.; Sigsgaard, L.; Wratten, S. The value of producing food, energy, and ecosystem services within an agro-ecosystem. *Ambio J. Hum. Environ.* **2009**, *38*, 186–193.

21. Sandhu, H.S.; Wratten, S.D.; Cullen, R.; Case, B. The future of farming: The value of ecosystem services in conventional and organic arable land. An experimental approach. *Ecol. Econ.* **2008**, *64*, 835–848.

22. Schulte, R.P.O.; Creamer, R.E.; Donnellan, T.; Farrelly, N.; Fealy, R.; O'Donoghue, C.; O'hUallachain, D. Functional land management: A framework for managing soil-based ecosystem services for the sustainable intensification of agriculture. *Environ. Sci. Policy* **2014**, *38*, 45–58.

23. Robinson, D.A.; Hockley, N.; Cooper, D.M.; Emmett, B.A.; Keith, A.M.; Lebron, I.; Reynolds, B.; Tipping, E.; Tye, A.M.; Watts, C.W.; *et al.* Natural capital and ecosystem services, developing an appropriate soils framework as a basis for valuation. *Soil Biol. Biochem.* **2013**, *57*, 1023–1033.

24. Dominati, E.; Patterson, M.; Mackay, A. A framework for classifying and quantifying the natural capital and ecosystem services of soils. *Ecol. Econ.* **2010**, *69*, 1858–1868.

25. Dominati, E.; Mackay, A.; Green, S.; Patterson, M. A soil change-based methodology for the quantification and valuation of ecosystem services from agro-ecosystems: A case study of pastoral agriculture in New Zealand. *Ecol. Econ.* **2014**, *100*, 119–129.

26. Ibarra, A.A.; Zambrano, L.; Valiente, E.L.; Ramos-Bueno, A. Enhancing the potential value of environmental services in urban wetlands: An agro-ecosystem approach. *Cities* **2013**, *31*, 438–443.

27. Zhang, W.; Ricketts, T.H.; Kremen, C.; Carney, K.; Swinton, S.M. Ecosystem services and dis-services to agriculture. *Ecol. Econ.* **2007**, *64*, 253–260.

28. Chang, J.; Wu, X.; Liu, A.; Wang, Y.; Xu, B.; Yang, W.; Meyerson, L.A.; Gu, B.; Peng, C.; Ge, Y. Assessment of net ecosystem services of plastic greenhouse vegetable cultivation in China. *Ecol. Econ.* **2011**, *70*, 740–748.

29. Chang, J.; Wu, X.; Wang, Y.; Meyerson, L.A.; Gu, B.; Min, Y.; Xue, H.; Peng, C.; Ge, Y. Does growing vegetables in plastic greenhouses enhance regional ecosystem services beyond the food supply? *Front. Ecol. Environ.* **2013**, *11*, 43–49.

30. Yuan, Y.; Liu, J.; Jin, Z. An integrated assessment of positive and negative effects of high-yielding cropland ecosyste*m services in* Luancheng County, Hebei Province of North China. *Chin. J. Ecol.* **2011**, *30*, 2809–2814. (In Chinese)

31. Millennium Ecosystem Assessment. *Ecosystems and Human Well-Being*; Island Press: Washington, DC, USA, 2005.

32. Wallace, K.J. Classification of ecosystem services: Problems and solutions. *Biol. Conserv.* **2007**, *139*, 235–246.

33. Fisher, B.; Kerry Turner, R. Ecosystem services: Classification for valuation. *Biol. Conserv.* **2008**, *141*, 1167–1169.

34. Boyd, J.; Banzhaf, S. What are ecosystem services? The need for standardized environmental accounting units. *Ecol. Econ.* **2007**, *63*, 616–626.

35. Fisher, B.; Turner, R.K.; Morling, P. Defining and classifying ecosystem services for decision making. *Ecol. Econ.* **2009**, *68*, 643–653.

36. Lyytimäki, J.; Petersen, L.K.; Normander, B.; Bezák, P. Nature as a nuisance? Ecosystem services and disservices to urban lifestyle. *Environ. Sci.* **2008**, *5*, 161–172.

37. Herendeen, R.A. Energy analysis and EMERGY analysis—A comparison. *Ecol. Model.* **2004**, *178*, 227–237.

38. Rugani, B.; Benetto, E.; Arbault, D.; Tiruta-Barna, L. Emergy-based mid-point valuation of ecosystem goods and services for life cycle impact assessment. *Rev. Métall.* **2013**, *110*, 249–264.

39. Hau, J.L.; Bakshi, B.R. Promise and problems of emergy analysis. *Ecol. Model.* **2004**, *178*, 215–225.

40. Voora, V.; Thrift, C. Using Emergy to Value Ecosystem Goods and Services. Available online: http://www.iisd.org/pdf/2010/using_emergy.pdf (accessed on 25 November 2014).

41. Sciubba, E.; Ulgiati, S. Emergy and exergy analyses: Complementary methods or irreducible ideological options? *Energy* **2005**, *30*, 1953–1988.

42. Cleveland, C.J.; Kaufmann, R.K.; Stern, D.I. Aggregation and the role of energy in the economy. *Ecol. Econ.* **2000**, *32*, 301–317.

43. Dong, X.; Yang, W.; Ulgiati, S.; Yan, M.; Zhang, X. The impact of human activities on natural capital and ecosystem services of natural pastures in North Xinjiang, China. *Ecol. Model.* **2012**, 225, 28–39.

44. Jørgensen, S.E.; Nielsen, S.N. Tool boxes for an integrated ecological and environmental management. *Ecol. Indic.* **2012**, *21*, 104–109.

45. Pulselli, F.M.; Coscieme, L.; Bastianoni, S. Ecosystem services as a counterpart of emergy flows to ecosystems. *Ecol. Model.* **2011**, *222*, 2924–2928.

46. Campbell, E.T.; Tilley, D.R. The eco-price: How environmental emergy equates to currency. *Ecosyst. Serv.* **2014**, *7*, 128–140.

47. Campbell, E.T.; Tilley, D.R. Valuing ecosystem services from Maryland forests using environmental accounting. *Ecosyst. Serv.* **2014**, *7*, 141–151.

48. Coscieme, L.; Pulselli, F.M.; Marchettini, N.; Sutton, P.C.; Anderson, S.; Sweeney, S. Emergy and ecosystem services: A national biogeographical assessment. *Ecosyst. Serv.* **2014**, *7*, 152–159.

49. Vassallo, P.; Paoli, C.; Rovere, A.; Montefalcone, M.; Morri, C.; Bianchi, C.N. The value of the seagrass Posidonia oceanica: A natural capital assessment. *Mar. Pollut. Bull.* **2013**, *75*, 157–167.

50. Watanabe, M.D.B.; Ortega, E. Dynamic emergy accounting of water and carbon ecosystem services: A model to simulate the impacts of land-use change. *Ecol. Model.* **2014**, *271*, doi:10.1016/j.ecolmodel.2013.03.006.

51. Zhang, L.X.; Ulgiati, S.; Yang, Z.F.; Chen, B. Emergy evaluation and economic analysis of three wetland fish farming systems in Nansi Lake area, China. *J. Environ. Manag.* **2011**, *92*, 683–694.

52. Díaz-Delgado, C.; Fonseca, C.R.; Esteller, M.V.; Guerra-Cobián, V.H.; Fall, C. The establishment of integrated water resources management based on emergy accounting. *Ecol. Eng.* **2014**, *63*, 72–87.

53. Ingwersen, W.W. Uncertainty characterization for emergy values. *Ecol. Model.* **2010**, *221*, 445–452.

54. Li, L.J.; Lu, H.F.; Campbell, D.E.; Ren, H. Methods for estimating the uncertainty in emergy table-form models. *Ecol. Model.* **2011**, *222*, 2615–2622.

55. Chen, G.Q.; Jiang, M.M.; Chen, B.; Yang, Z.F.; Lin, C. Emergy analysis of Chinese agriculture. *Agric. Ecosyst. Environ.* **2006**, *115*, 161–173.

56. Foley, J.A.; DeFries, R.; Asner, G.P.; Barford, C.; Bonan, G.; Carpenter, S.R.; Chapin, F.S.; Coe, M.T.; Daily, G.C.; Gibbs, H.K. Global consequences of land use. *Science* **2005**, *309*, 570–574.

57. Zhang, W.-G.; Hu, Y.-M.; Zhang, J.; Liu, M.; Yang, Z.-P. Assessment of land use change and potential eco-service value in the upper reaches of Minjiang River, China. *J. For. Res.* **2007**, *18*, 97–102.

58. Metzger, M.J.; Rounsevell, M.D.A.; Acosta-Michlik, L.; Leemans, R.; Schröter, D. The vulnerability of ecosystem services to land use change. *Agric. Ecosyst. Environ.* **2006**, *114*, 69–85.

59. Shaw, M.R.; Pendleton, L.; Cameron, D.R.; Morris, B.; Bachelet, D.; Klausmeyer, K.; MacKenzie, J.; Conklin, D.R.; Bratman, G.N.; Lenihan, J.; *et al.* The impact of climate change on California's ecosystem services. *Clim. Chang.* **2011**, *109*, 465–484.

60. Pan, Y.; Xu, Z.; Wu, J. Spatial differences of the supply of multiple ecosystem services and the environmental and land use factors affecting them. *Ecosyst. Serv.* **2013**, *5*, 4–10.

61. Yuan, Z.; Shen, Y. Estimation of Agricultural Water Consumption from Meteorological and Yield Data: A Case Study of Hebei, North China. *PLoS ONE* **2013**, *8*, doi:10.1371/journal.pone.0058685.

62. Lan, S.F.; Qin, P.; Lu, H.F. *Emergy Analysis of Ecological Economic Systems*; Chemical Industrial Press: Beijing, China, 2002. (In Chinese).

63. Wei, W. Groundwater age and recharge temperature in the Quantenary aquifers in North Chian Plain. Master's Thesis, Chinese Academy of Geological Sciences, Shijiazhuang, China, 2007. (In Chinese)

64. Zhao, G.S.; Jiang, H.R.; Wu, W.L. Sustainability of farmland ecosystem with high yield based on emergy analysis method. *Trans. Chin. Soc. Agric. Eng.* **2011**, *27*, 318–323. (In Chinese)

65. Liu, J.E.; Zhou, H.; Qin, P.; Zhou, J.; Wang, G. Comparisons of ecosystem services among three conversion systems in Yancheng National Nature Reserve. *Ecol. Eng.* **2009**, *35*, 609–629.

66. Du, F.Y. Emergy Analysis of Farmland system in Hebei Province. Mater's Thesis, Agricultural University of Hebei, Baoding, China, 2008. (In Chinese)

67. Hu, C.S.; Cheng, Y.S. *Chinese Ecosystem Observation and Research Data Sets, Agro-Ecosystem Volume, Hebei Luancheng Station*; China Agriculture Press: Beijing, China, 2011. (In Chinese).

68. Chen, D.D.; Gao, W.S.; Sui, P.; Wu, T.L. Dynamic Analysis on Energy Efficiency of Modern Planting System and Grain Production—A Case Study of Luancheng, Hebei. *Prog. Geogr.* **2008**, *27*, 99–104. (In Chinese)

69. Luo, S.M. *Agroecology*; China Agriculture Press: Beijing, China, 2001. (In Chinese)

70. Office of the People's Government of Hebei Province, Hebei Bureau of Statistics. *Hebei Rural Statistical Yearbook*; China Statistics Press: Beijing, China. (In Chinese)

71. Chinese Agricultural Yearbook Editorial Board. *China Agriculture Yearbook*; Chinese Agriculture Press: Beijing, China, 2006. (In Chinese)

72. Tang, H.; Zheng, Y.; Chen, F.; Yang, L.G.; Zhang, H.L. Ecosystem Services Valuation of Difference Croplands and Cropping Systems in Beijing Suburb. *Ecol. Econ.* **2008**, *7*, 86–89. (In Chinese)

73. State Forestry Administration. *Specification for Assessment of Forest Ecosystem Services in China*; China Forestry Industry Standard KT/T 1721–2008; China Forestry Industry Standard: Beijing, China, 2008. (In Chinese)

74. Ding, D.Z. *Hebei Soil Types*; Hebei Science and Technology Press: Shijiazhuang, China, 1992. (In Chinese)

75. Zeng, J.H.; Wang, Z.P.; Hu, C.S.; Zhang, Y.M. Soil organic matter decomposition and accumulation characteristics in North China. *Soil Fertil.* **1996**, *4*, 1–4. (In Chinese)

76. Zhang, X.Y.; Yuan, X.L. A Field Study on the Relationship of Soil Water Content and Water Uptake by Winter Wheat Root System. *Acta Agriculturae Boreali-Sinica* **1995**, *10*, 99–104. (In Chinese).

77. Zhang, X.Y. Variation of root, leaf water potentials and stomatal resistance among ten millet varieties. *Eco-Agric. Res.* **1997**, *5*, 37–39. (In Chinese)

78. Zhang, G.M. Study on emission fluxes of greenhouse gases from cropland soils and their regional estimation. Master's Thesis, Shanxi Agricultural University, Shanxi, China, 2003. (In Chinese)

79. Zhang, Y.M.; Hu, C.S.; Zhang, J.B.; Li, X.X.; Dong, W.X. Nitrogen cycling and balance in agricultural ecosystem in piedmont plain of Taihang Mountains. *Plant Nutr. Fertil. Sci.* **2006**, *12*, 5–11. (In Chinese)

80. Liu, G.D. Methods and Applications to Evaluate the Environmental Impacts of Regional Agriculture—A Case Study on High-Yielding Count, Huantai, North China. Ph.D. Thesis, China Agricultural University, Beijing, China, 2004. (In Chinese)

Biomass Power Generation Industry Efficiency Evaluation in China

Qingyou Yan and Jie Tao

Abstract: In this paper, we compare the properties of the traditional additive-based data envelopment analysis (hereafter, referred to as DEA) models and propose two generalized DEA models, *i.e.*, the big M additive-based DEA (hereafter, referred to as BMA) model and the big M additive-based super-efficiency DEA (hereafter, referred to as BMAS) model, to evaluate the performance of the biomass power plants in China in 2012. The virtues of the new models are two-fold: one is that they inherited the properties of the traditional additive-based DEA models and derived more new additive-based DEA forms; the other is that they can rank the efficient decision making units (hereafter, referred to as DMUs). Therefore, the new models have great potential to be applied in sustainable energy project evaluation. Then, we applied the two new DEA models to evaluate the performance of the biomass power plants in China and find that the efficiency of biomass power plants in the northern part of China is higher than that in the southern part of China. The only three efficient biomass power plants are all in the northern part of China. Furthermore, based on the results of the Wilcoxon-Mann-Whitney rank-sum test and the Kolmogorov-Smirnov test, there is a great technology gap between the biomass power plants in the northern part of China and those in the southern part of China.

Reprinted from *Sustainability*. Cite as: Yan, Q.; Tao, J. Biomass Power Generation Industry Efficiency Evaluation in China. *Sustainability* **2014**, *6*, 8720-8735.

1. Introduction

As a developing country with a huge population, China is always relatively short of energy storage. Now, with the rapid development of the economy, along with industrialization and urbanization, the energy consumption of China will increase to four billion Mtce, and the gap between energy production and consumption will reach half of the total energy volume. Meanwhile, China is still one of the few countries that rely on fossil energy, which has caused severe environmental problems. The energy shortage and defective infrastructure now is pushing China to seek out and develop substitution energies.

Being the one and only material and easily-stored renewable energy, biomass power generation is getting more and more attention globally. Till the end of 2012, the installation capacity of biomass power in the U.S. had exceeded 10,000 megawatts, and the capacity now has been planned to account for 50% of their total energy production. Furthermore, Germany aims to use biomass power to meet 16% of the whole country's electricity demand, 10% of the heating demand and 15% of electric vehicle (hereafter referred to as EV) power.

According to the publicized the International Energy Agency (IEA) data in 2012, China has abundant biomass resources with a productivity of about five billion tons per year, and this amount ranks only next to fossil resource. Hence, China can develop biomass power. However, compared with the development of hydro, nuclear, wind and solar power, biomass power in China had not fully started until the implementation of the Renewable Energy Act in 2006. Now, the installation capacity has increased by 30% each year, and according to the Long-Term Renewable Energy Development Plan, in 2020, the expected biomass power installation capacity will be 30 GW. With the industrial plan and policy incentives, biomass power in China has entered the track of high-speed development.

Under this situation, if enough data can be collected and used to evaluate the industry efficiency quantitatively, then the assessment can provide a basis for decision making and suggest industrial or policy measurements to improve efficiency, which, in return, will accelerate the industrialization of biomass power in China and keep the development leading in the optimal direction.

2. Literature Review and Emerged Concerns

Developed countries in Europe and the U.S. started their research on biomass power in the 1970s, and the early research commonly focused on the energy conversion technology. With the subsequent development of the biomass power industry, the studies and concerns shifted to the generating cost and industry efficiency. Hooper had come up with countermeasures from the view of investment for the development of the biomass power generation industry. In the research, technology needed to be advanced, and it was the best choice to industrialize and commercialize [1]. Biomass power took off late in China, and the studies about biomass generating technology were summed up and learned. Hence, the domestic research began with focusing on the economy and industry efficiency.

After the implementation of the Renewable Energy Act, the biomass power in China entered a phase with high-speed in industrial growth. It created profitability for the biomass power generation in China and encouraged researchers to perform comprehensive assessment methods to study the industry efficiency, not only to calculate the technical efficiency, but also to evaluate the scale and overall efficiency. Along this track, Christoph and Perrels used scenario analysis and an input-output model and studied the relationship between biomass power and CO_2 emissions [2]. Klevas and Denis used DEA to analyze the technical efficiency of some common renewable technologies, including biomass technology [3,4]. Based on their methods, Peng Zhou in China meliorated the DEA model and made it suitable for overall efficiency assessment [5]. DEA methods showed their advantage in evaluating the efficiency of the biomass power generation industry, and domestic scholars started to apply them to China's own data. Zhao chose SWOT to analyze the industry state of the biomass power generation of China and thought that the advantage lied in the increasing electricity demand and changing electricity price, yet, the industry was sensitive to local policy and the local industrial environment [6]. Kautto also compared the regional and national biomass power plan, finding that if the regional plan can cooperate with national development better, the whole biomass industry efficiency would take a step toward a higher level [7]. In all, the regional or local situation indeed impacts the industry efficiency of biomass power generation.

As for China, with multiple geographical features and an unbalanced economy, the regional difference is distinct, and the regional effect easily influences the biomass industry assessment; hence, the region needs to be considered as a key factor when performing an industry efficiency evaluation. Other than that, when Ramon used Multi-criteria data envelopment analysis model

(MCDEA) to assess the efficiencies of 13 projects located in different areas, some weights of the indexes were zero [8], which made the index lose the ability to contribute to the assessment and led to the inaccuracy of the efficiency ranking. Li *et al.* used a unified efficiency DEA model to evaluate the performance of 24 power companies in China [9].

In terms of methodology, DEA is widely used in evaluating the performance of biomass power generation plants. DEA, first proposed by Charnes *et al.* in 1978, is a methodology for evaluating and measuring the relative efficiencies of a set of decision making units (DMUs) that use multiple inputs to produce multiple outputs [10]. Subsequent to that pioneering study, the research work has been mainly focused on constructing different DEA models to compensate for the disadvantage of traditional DEA models, e.g., the additive model [11], slack-based model [12,13], cross-efficiency model [14] and assurance region model [15] have been proposed to solve a variety of problems [16].

Till now, there have still been two problems concerning the application of DEA models: one is that the traditional models are unable to deal with negative or nil data, and the other is that they cannot rank the efficient DMUs. Based on this, this paper presents two new models, the big M additive-based DEA (BMA) model and big M additive-based super-efficiency DEA (BMAS) model, to solve this problem. In addition, we apply the new DEA models to evaluate the performance of biomass generation plants in China and give the relevant policies based on the result of the model.

3. Industrial Characteristics of Biomass Power Generation in China

3.1. Conservative Rising Development Tendency

Since 2006, the biomass power generation industry has made huge progress. According to Zhao *et al.* [17], from 2008 to 2012, the installation capacity increased from 315 megawatts to 850 megawatts and the investment increased from 5.543 billion to 12.779 billion dollars; both of these two indicators' average growth rate per annum reached above 20%, which is shown in Figure 1.

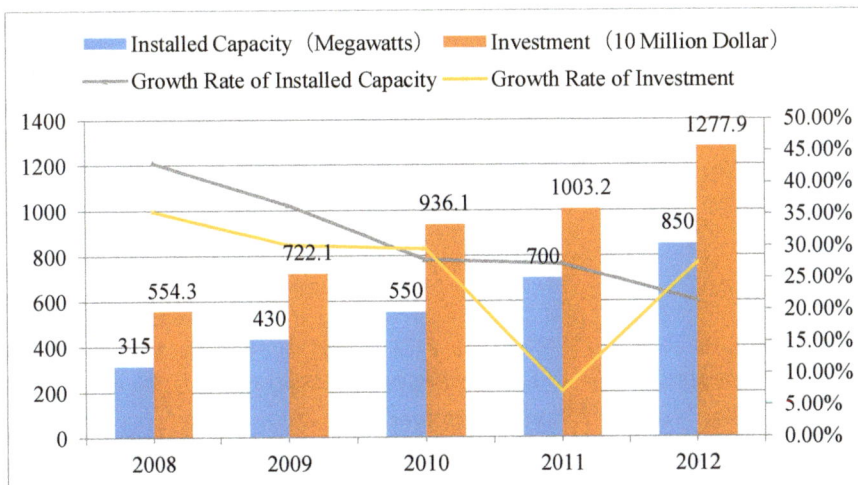

Figure 1. Industry scale and investment tendency chart of biomass power generation in China.

The average growth rate per annum of installation capacity and investment were both high during the recent past five years, but they were reducing each year, which showed that in China, the biomass power generation scale keeps growing, but the growth rate is sinking. On the one hand, this fits the growth law; on the other hand, this shows that China is developing biomass power with a positive, but conservative attitude.

According to "the 12th Five-Year Plan" and Long-Term Renewable Energy Development Plan, the Chinese government plans to develop the installation capacity of biomass power to 13 GW and invest over 90 billion RMB to achieve this goal.

3.2. Simple Impacts from the Industry Chain

In China, the industry chain of biomass power generation is simple. The up-stream companies are fuel suppliers and device manufactures, while the down-stream one is the power-grid company. Because the fuel cost accounts for almost 60% of the whole biomass power generation cost in China, the impact from the up-stream is the fuel cost. According to the IEA estimate (shown in Table 1), the potential biomass resource is abandoned; therefore, the fuel cost will not change dramatically in the near future.

Table 1. Potential biomass resource estimates in China (10^8 Mtce).

Year		2010	2020	2030	2050
Biomass resource type	Present biomass resource	2.8	2.8	2.8	2.8
	Newly-added organic waste	0.6	1.7	2.2	2.7
	Present woodland growth	0.05	0.3	0.7	1.37
	New ground marginal product	0.05	0.3	1	2
	Total potential	3.5	5.1	6.7	8.9

As for the down-stream company, because China now practices a "full amount buy in" policy for renewable energy and biomass power does not account for a large proportion, the electricity demand fluctuation has little impact on the biomass power generation industry, and the main impact is reflected by the feed-in price. According to the regulation "Agriculture and Forestry Biomass Price Policy" from the National Development and Reform Commission of China (NDRC), the feed-in price is set to 0.75 RMB per KWh (including the tax) for the new biomass project, which has not obtained its investor through biding.

3.3. Distinct Regional Difference

It is distinctive that the biomass power generation industry in China has a regional feature. This is partly because different regions have different types of biomass resources, and this is also due to the production characteristic of different biomass resources. For example, straw burning biomass power generation plants have been built in the south of China, because that area is rich in crop resources; while in the east of China, the city area produces lots of municipal waste and so garbage power plants are built.

Now, China's biomass power capacity is mainly located in eastern China and then ranks from mid-south, northeast, north, southwest to northwest China. Till the end of 2012, the biomass power installation capacity distributed in the above area is shown in Figure 2. This proportion will not change in the following years and so neither does the investment distribution.

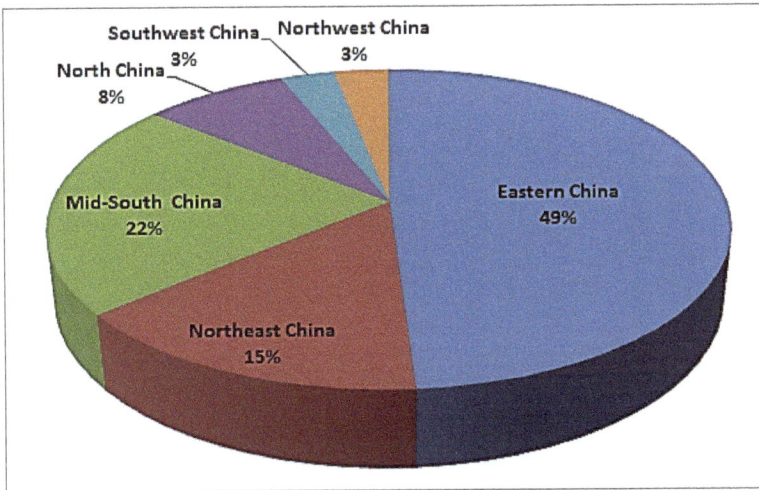

Figure 2. Biomass power distribution in China till 2012.

4. Methodology

Suppose there are n units being evaluated (in short, DMUs); each DMU has m inputs and s outputs. X denotes the $m \times n$ input indicator matrix, while Y denotes the $s \times n$ output indicator matrix. That is to say, x_{ij} denotes the i-th input indicator of the j-th DMU, while y_{rj} denotes the r-th output indicator of the j-th DMU. Finally, X_j denotes the input vector of the j-th DMU, while Y_j denotes the output vector of the j-th DMU.

4.1. Drawbacks of Traditional DEA Models

The CCR model [10] and BCC model [18] are the two main traditional DEA models. Compared to the CCR model, the BCC model takes into account the factors of returns to scale; thus, here, the BCC model is selected to show the drawbacks of classical DEA models when facing negative and nil data.

The BCC linear program (LP) model and its dual model can be illustrated as follows:

$$\max \mu^T Y_o + \mu_0$$
$$st. \begin{cases} \omega^T X_o = 1, \\ \mu^T Y_j - \omega^T X_j + \mu_0 \leq 0, j = 1,...,n, \\ \mu \geq 0, \omega \geq 0. \end{cases} \quad (1)$$

$$\min \theta$$

$$st.\begin{cases} \sum_{j=1}^{n} \lambda_j y_{rj} - s_r^+ = \theta y_{ro}, r = 1,...,s, \\ -\sum_{j=1}^{n} \lambda_j x_{ij} - s_i^- = -x_{io}, i = 1,...,m, \\ \sum_{j=1}^{n} \lambda_j = 1, \\ \lambda_j \geq 0, s_i^- \geq 0, s_r^+ \geq 0, j = 1,...,n. \end{cases} \tag{2}$$

here, ω and μ denote the weight vectors of the inputs and outputs, respectively, and "*st.*" denotes "subject to". Traditional DEA models supposed that each DMU's value is positive, which demonstrating that they are powerless when facing negative and nil data.

Lins *et al.* pointed out that, "In order to use the above amounts in the DEA it was necessary to alter the variables so as to avoid negative or nil values. To do so, all the cells in each column with negative or nil values were added to the lowest value in this column, with the addition of one unit" [19]. It is noted that, since the traditional DEA model does not have the translation invariance property, the method used by Lins *et al.* is not correct when facing negative and nil data.

4.2. Additive-Based DEA Models

In order to solve the negative data problem, many new DEA models were proposed, most of which were characterized as additive-based models, because of the translation invariance property of the additive model. The translation invariance, as is pointed out by Lovell and Pastor, is critical when the data contain zero or negative values and must be translated prior to analysis with available software packages [20]. The first additive-based DEA model, named the constant weighted additive model (CWA-DEA), was proposed by Pastor in 1994. It shared the translation invariance property with the original additive model, while neither of them was unit invariant [20]. In order to obtain a model that shared both the translation invariance and unit invariance properties, Lovell and Pastor in 1995 proposed the normalized weighted additive model (NWA-DEA), which was a great step forward in the history of additive-based DEA models [20]. They used the sample standard deviations of the output variables and the input variables, respectively, to replace the constant weight in the CWA-DEA model, and they pointed out that any first order dispersion measures can also be used to normalize the input excess and output slack variables. Apart from the translation invariance and unit invariance properties, three other important properties were proposed by Cooper *et al.* to testify to the quality of the additive-based DEA models [21]. Furthermore, based on the five properties, Cooper *et al.*, 1999, and Cooper *et al.*, 2011, extended the NWA-DEA model and proposed the famous RAM and BAM models [21,22]. The five properties were:

(P1) The optima is between 0 and 1;
(P2) The optima is 0 when DMU$_o$ is fully inefficient, while the optima is 1 when DMU$_o$ is fully efficient;

(P3) The optima is well defined and unit invariant;

(P4) The optima is strongly monotonic;

(P5) The optima is translation invariant.

According to Cooper *et al.*, 1999, the "strong monotonicity" property is described as follows: holding all other inputs and outputs constant, an increase in any of its inputs will increase the inefficiency score for an inefficient DMUo. The same is true for a decrease in any of its outputs [21].

Table 2 describes whether current DEA models satisfy these five properties. According to Lovell and Pastor, 1994, the CCR model and the normalized weighted CCR model were not translation invariant, while the BCC model and the normalized BCC model are translation invariant in a limited sense, being invariant with respect to the translation of inputs or outputs, but not both. The radial component of the efficiency measure obtained from the BCC model and the CCR model is unit invariant, but the slack component is not [20]. The additive model can only measure the inefficiencies of the DMUs, and the optima are not between 0 and 1. With respect to P4, it is obvious that only the RAM model satisfied the strongly monotonic property, while the others are all monotonic.

Table 2. Comparisons of traditional DEA models on the five properties.

Model	P1	P2	P3	P4	P5
CCR	Yes	Yes	Partially Units Invariance	Monotonic	NO
Normalized weighted CCR	Yes	Yes	Units Invariance	Monotonic	NO
BCC	Yes	Yes	Partially Units Invariance	Monotonic	Partially Translation Invariance
Normalized weighted BCC	Yes	Yes	Units Invariance	Monotonic	Partially Translation Invariance
Additive model	No	No	No	Monotonic	Yes
Normalized weighted Additive Model	No	No	Units Invariance	Monotonic	Yes
SBM	Yes	Yes	Yes	Monotonic	Yes
RAM	Yes	Yes	Yes	Strongly Monotonic	Yes
BAM	Yes	Yes	Yes	Monotonic	Yes

4.3. Generalized Additive-Based DEA Model-BMA Model

Based on the previous research, we proposed a generalized additive-based DEA model, which was called the big M additive-based DEA model. We here not only show that the previous additive-based DEA models were the particular form of the big M additive-based DEA model, but also show that other different forms of additive-based models can be derived from the big M additive-based DEA model.

Consider the following model:

$$\theta = \min \left[1 - \frac{\displaystyle\sum_{r=1}^{s} \frac{s_r^+}{\Psi(\Omega)} + \sum_{i=1}^{m} \frac{s_i^-}{\Phi(\Omega)}}{M} \right]$$

$$st. \begin{cases} \displaystyle\sum_{j=1}^{n} \lambda_j y_{rj} - s_r^+ = y_{ro}, r = 1,...,s, \\[2mm] -\displaystyle\sum_{j=1}^{n} \lambda_j x_{ij} - s_i^- = -x_{io}, i = 1,...,m, \\[2mm] \displaystyle\sum_{j=1}^{n} \lambda_j = 1, \\[2mm] \lambda_j \geq 0, s_i^- \geq 0, s_r^+ \geq 0, j = 1,...,n. \end{cases} \qquad (3)$$

here, Ω denotes the sample space, $\Psi(\Omega)$ and $\Phi(\Omega)$ are non-zero mappings from the sample space to $\Box$ with homogeneous properties, $i.e.$, $\theta\Phi(\Omega) = \Phi(\theta\Omega)$, $\theta\Psi(\Omega) = \Psi(\theta\Omega)$, $\theta \in \Box$, and big M is a large real number. The following theorem motivates the proposition of the homogeneity property of the mappings $\Psi(\Omega)$ and $\Phi(\Omega)$.

Theorem 1: (Lovell and Pastor [20]) In an additive DEA model, scaling an input (output) by multiplying it by a constant $\alpha > 0$ is equivalent to leaving the input (output) unscaled and multiplying the corresponding input excess (output slack) variable in the objective function by the same constant.

Therefore, Model (3) is unit invariant if and only if $\Psi(\Omega)$ and $\Phi(\Omega)$ satisfy the homogeneity property. In this sense, the additive-based DEA models mentioned above are only the particular form of the big M additive-based DEA model. Obviously, Model (3) can be transformed into the additive model, if we set $\Psi(\Omega) = 1$, $\Phi(\Omega) = 1$ and $M = 1$. It can be transformed into the normalized weighted additive model when we set $\Psi(\Omega)_r = \sigma_r$, $\Phi(\Omega)_i = \sigma_i$ and $M = 1$. Here, σ_r, σ_i denote the sample standard deviation of the r-th output variable and the i-th input variable. It can be transformed into the SBM model when we set $\Psi(\Omega)_r = s \cdot y_{ro}$, $\Phi(\Omega)_i = m \cdot x_{io}$ and $M = 1$. It can be transformed into the RAM model when we set $\Psi(\Omega)_r = R_r^+$, $\Phi(\Omega)_i = R_i^-$ and $M = (m + s)$, where $R_i^- = \bar{x}_i - \underline{x}_i$ with $\bar{x}_i = \max\{x_{ij}, j = 1,...,n\}$, $\underline{x}_i = \min\{x_{ij}, j = 1,...,n\}$ and $R_r^+ = \bar{y}_r - \underline{y}_r$ with $\bar{y}_r = \max\{y_{rj}, j = 1,...,n\}$, $\underline{y}_r = \min\{y_{rj}, j = 1,...,n\}$. It can be transformed into the BAM model when we set $\Psi(\Omega)_r = U_{ro}^+$, $\Phi(\Omega) = L_{io}^-$ and $M = (m + s)$, where $L_{io}^- = x_{io} - \underline{x}_i$ and $U_{ro}^+ = \bar{y}_r - y_{ro}$. Moreover, it is obvious that the mappings $\Psi(\Omega)$ and $\Phi(\Omega)$ control the unit invariance property, $i.e.$, property (P2) and M control the properties of (P1), (P2) and (P4).

Furthermore, we can get various additive-based DEA models when we set different types of $\Psi(\Omega)$, $\Phi(\Omega)$ and M. We listed several types of $\Psi(\Omega)$, $\Phi(\Omega)$ and M as follows:

Case 1: p, q order geometric moment. $\Psi(\Omega) = (\sum_{j=1}^{n} y_{rj}^{p})^{\frac{1}{p}}, p = 1,...,\infty$, and

$\Phi(\Omega) = (\sum_{j=1}^{n} x_{ij}^{q})^{\frac{1}{q}}, q = 1,...,\infty$. It is noted that $\Psi(\Omega) = \max\{y_{rj}, j = 1, ..., n\}$ and $\Phi(\Omega) = \max\{x_{ij}, j = 1,$

..., n\} when p,q = ∞.

Case 2: p, q order central moment. $\Psi(\Omega) = (\sum_{j=1}^{n} (y_{rj} - \bar{y}_r)^{p})^{\frac{1}{p}}, p = 1,...,\infty$, and

$\Phi(\Omega) = (\sum_{j=1}^{n} (x_{ij} - \bar{x}_i)^{q})^{\frac{1}{q}}, q = 1,...,\infty$, where $\bar{x}_i, \bar{y}_r$ denote the mean value of the i-th input and the r-th

output, respectively.

Then, we researched whether the generalized additive-based DEA model satisfies the five properties mentioned above.

Theorem 2: (P1) The optimum of Model (3) is between 0 and 1 only when M is big enough.

Proof: Since whatever $\Psi(\Omega)$ and $\Phi(\Omega)$ are, we could always find an M big enough, such that

$$\sum_{r=1}^{s} \frac{s_r^+}{\Psi(\Omega)} + \sum_{i=1}^{m} \frac{s_i^-}{\Phi(\Omega)} = \sum_{r=1}^{s} \frac{\sum_{j=1}^{n} \lambda_j y_{rj} - y_{ro}}{\Psi(\Omega)} + \sum_{i=1}^{m} \frac{x_{io} - \sum_{j=1}^{n} \lambda_j x_{ij}}{\Phi(\Omega)} \leq M \quad ; \quad \text{thus}$$

$$1 - \frac{\sum_{r=1}^{s} \frac{s_r^+}{\Psi(\Omega)} + \sum_{i=1}^{m} \frac{s_i^-}{\Phi(\Omega)}}{M} \geq 0 \quad . \quad \text{Furthermore,} \quad \text{since} \quad \sum_{r=1}^{s} \frac{s_r^+}{\Psi(\Omega)} + \sum_{i=1}^{m} \frac{s_i^-}{\Phi(\Omega)} \geq 0 \quad ,$$

$$1 - \frac{\sum_{r=1}^{s} \frac{s_r^+}{\Psi(\Omega)} + \sum_{i=1}^{m} \frac{s_i^-}{\Phi(\Omega)}}{M} \leq 1 .$$

Theorem 3: (P2) The optimum of Model (3) is 1 when DMU$_o$ is fully efficient, while the optimum of Model (3) is not 0 when DMU$_o$ is fully inefficient.

Proof: It is obvious that s_i^-, s_r^+ are all equal to zero when DMU$_o$ is fully efficient; thus, the optimum is one. However, when DMU$_o$ is fully inefficient, the optimum varies when different M are chosen.

Theorem 4: (P3) The optimum of Model (3) is well defined and unit invariant.

Proof: Based on Theorem 1, it is obvious that Model (3) is unit invariant if and only if $\Psi(\Omega)$ and $\Phi(\Omega)$ satisfy the homogeneity property. Moreover, since $\Psi(\Omega)$ and $\Phi(\Omega)$ are non-zero mappings, Model (3) is well defined.

With respect to (P4) and (P5), it is obvious that the optima of Model (3) is not necessarily monotonic (the proof is similar to that of (P4') in Cooper *et al.*, 2011, and the optimum is not monotonic when we simply choose $\Psi(\Omega) = x_i$.), and it satisfies the translation invariance property.

4.4. BMAS Model

Since classical DEA models can only recognize the efficiency of each unit, when there are many efficient DMUs, they cannot be ranked. Classic super-efficiency DEA models can effectively solve the problem of ranking effective decision making units, but they cannot deal with the negative and nil data problem. Based on this, this paper presents the BMAS model to solve these problems.

Consider the model below:

$$\theta = \min\left[1 - \frac{\displaystyle\sum_{r=1}^{s}\frac{s_r^+}{\Psi(\Omega)} + \sum_{i=1}^{m}\frac{s_i^-}{\Phi(\Omega)}}{M}\right]$$

$$st.\begin{cases}\displaystyle\sum_{\substack{j=1 \\ j\neq j_0}}^{n}\lambda_j Y_{rj} - s_r^+ = Y_{rj_0}, r = 1,...,s, \\[2mm] -\displaystyle\sum_{\substack{j=1 \\ j\neq j_0}}^{n}\lambda_j X_{ij} - s_i^- = -X_{ij_0}, i = 1,...,m, \\[2mm] \displaystyle\sum_{\substack{j=1 \\ j\neq j_0}}^{n}\lambda_j = 1, \\[2mm] \lambda_j \geq 0, s_i^- \geq 0, s_r^+ \geq 0, j = 1,...,n.\end{cases} \tag{4}$$

The difference between the BMA model and BMAS model is only in the construction of the production possibility set. The BMAS model excludes the DMU evaluated from the production possibility set and only considers the new production possibility set constructed by the remaining DMUs. Then, we could evaluate this DMU based on the new production possibility set.

Thus, when the DMU being evaluated is efficient in the original DEA model, it would be outside of the new production possibility set which is constructed by the remaining DMUs. As a result, s_i^-, s_r^+ would be less than zero, so that the value of the objective function would be greater than one. In addition, since the distance between DMUs and the frontier varies from different DMUs, the efficiency would be measured by these distances. If the distance is short, the efficiency is relatively low, while if the distance is long, the efficiency is relatively high. Figure 1 shows that the production possibility set of the original DEA models (BCC) is the district constructed by EBACF, while the new production possibility set of the super-efficiency DEA models is the district constructed by EBCF. With respect to A, Figure 3 shows that s_i^-, s_r^+ are less than zero.

5. Empirical Analysis

5.1. Efficiency Analysis

In China, the efficiency of biomass power plants can be seen as the efficiency of the entire biomass generation industry. We employ the methodology in Section 4 to evaluate the efficiency of biomass power plants in China. The data consist of 11 biomass power plants in 2012 (it is noted that in China, most biomass power plants' data are classified information and are unavailable to the public; therefore, we here only obtained the data of 11 biomass power plants by making surveys). The input variables selected are: greenhouse gases emission, investment costs, operation and maintenance costs. The output variables selected are: potential job creation and potential distributed power generation. Here, greenhouse gas emissions were calculated based on the documents of the Methods and Guidelines of Calculating the Greenhouse Gases Emissions of Chinese Power Plants. Data of investment costs, operation and maintenance costs were obtained from the financial statements of each biomass power plant. Potential job creation means the maximum amount of people who are willing to work at this biomass power plant, and these data were obtained by surveys to the 11 biomass power plants. Potential distributed power generation means the maximum power generated by this biomass power plant per year (or, in other words, power generation capacity (it is noted that the power generation capacity is not equal to the electric power production, and it is better to use the power generation capacity to evaluate the biomass power plants). Table 3 provides the description statistics of these 11 biomass power plants. We can see that the minimum value of operation and maintenance costs and the mean value of green gases emission are negative.

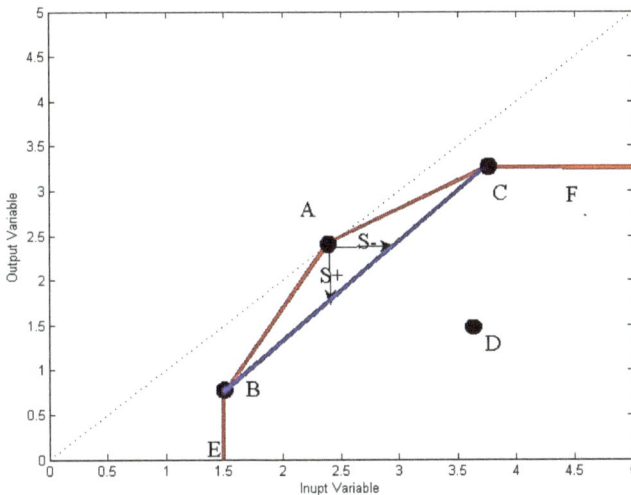

Figure 3. The improvement of inefficient DMU on the production possibility set.

Table 3. Description statistics of the inputs and outputs of 11 biomass power plants in China.

Inputs/Outputs	Unit	Average	Std	Max	Min
Greenhouse Gas Emissions (I)	tCO$_2$/GWh	−203.961	542.0662	600	−1223
O&M + CC Costs (I)	RMB/MWh	14.69	25.4917	62.53	−26.52
Investment Costs (I)	RMB/MWh	38.15727	18.70269	76	14.96
Potential Job Creation (O)	Job/TWh	7811.348	11,466.02	35,347.69	1.88
Potential Distributed Power Generation (O)	GWH/year	54,870.36	35,121.65	133,296	6833

Here, we set $\Psi(\Omega)_r = \sigma_r$, $\Phi(\Omega)_i = \sigma_i$ and M equal to 1, and the BMA model was transformed as the normalized weighted additive-based model, which is proposed by Lovell and Pastor [20]. In order to compare the results of the BMA model with those of the traditional DEA models, we use the BCC model, the BMA model and the BMAS model to evaluate the performance of the biomass power plants in China.

Table 4 shows the result of the efficiencies obtained by the BCC model, the BMA model (normalized weighted additive-based model) and the BMAS model. The second column of Table 4 gives the efficiency of each biomass power plant, which is obtained from the BCC model, while the third column of Table 4 gives the efficiency of each biomass power plant, which is obtained by the BMA model. In comparison, the efficient biomass power plants whose efficiencies are equal to one are altered when using the BMA model. In specific, there are 10 efficient biomass power plants when using the BCC model to evaluate the performance of the 11 biomass power plants, and there are only three efficient power plants when using the BMA model, *i.e.*, Jilin Changling, Neimeng Zhaoxin and Jinlin Gongzhuling. The main reason here is that the traditional DEA model does not have the translation invariance property; consequently, it will draw the wrong conclusion if all of the same indicators of the decision-making units increase a constant at the same time. It is noted that the function of the BMA model is to find out the efficient DMUs when facing the negative and nil data problem. However, the efficiency value of each DMU does not make any sense; it is altered when the value of the constant M alters. Meanwhile, since three DMUs' efficiency reaches one, it is necessary to use the BMAS model to rank these efficient DMUs. The fourth column of Table 4 illustrates the super efficiency of the three efficient DMUs, which are Jilin Changling, Neimeng Zhaoxin and Jinlin Gongzhuling. From the fourth column of Table 4, it could be seen that the super efficiency of Jinlin Gongzhuling is the highest, which is 1.033233, while Neimeng Zhaoxin, with a super efficiency of 1.012733, comes in at second place, and Jilin Changling, with a super efficiency of 1.010540, in third place. Thus, these efficient DMUs, whose efficiencies are equal to one, could be ranked through their super efficiency. The fifth column of Table 4 illustrates the ranking order of each biomass power plant based on the BCC model, while the sixth column of Table 4 illustrates the ranking order of each biomass power plant given by the BMA model and the BMAS DEA model. In comparison, it could be found that the ranking order of most biomass power plants changes.

Table 4. Efficiency and ranking of each biomass power plant.

Biomass Power Plants	BCC Efficiency	BMA Efficiency	BMAS Efficiency	BCC Rank	BMA Rank	Location
Shangdong Pingyuan	1.00	0.882389		7	9	South
Hebei Wuqiao	1.00	0.984928		9	8	North
Hebei Yuanshi	1.00	0.981673		10	10	North
Anhui Shouxian	1.00	0.859524		11	11	South
Jilin Changling	1.00	1.000000	1.010540	6	3	North
Neimeng Zhaoxin	1.00	1.000000	1.012733	3	2	North
Hengshui Taida	1.00	0.988927		2	5	North
Jilin Gongzhuling	1.00	1.000000	1.033233	1	1	North
Dongping Guangyuan	1.00	0.887661		5	6	South
Shandong Pingquan	1.00	0.886453		8	7	South
Jiangxi Ganxian	0.331	0.888768		4	4	South

5.2. Group Analysis

Based on the distinct regional difference in Section 3.3, we found that the biomass power generation industry in China has a regional distribution feature, *i.e.*, China's biomass power capacity is mainly distributed in eastern China and then ranks from mid-south, northeast, north, southwest to northwest China. Therefore, it is necessary to analyze the regional differences of efficiency. Generally speaking, the investment in the biomass power generation industry of the southern part of China is much higher than that of the northern part. Therefore, is it true that the efficiencies of the southern part are significantly higher than those of the northern part?

Assumption: The efficiencies of the biomass power plants in the southern part of China are higher than those in the northern part of China.

In terms of location, from Table 5 we can see that there are five biomass power plants in the south part of China and six biomass power plants in the north part of China. In terms of the mean value, the efficiency of the northern group is 0.9923, a little higher than that of the southern group, indicating that, on average, the performance of the biomass power plants in the northern part of China is better than those in the southern part of China. In terms of the standard deviation, the value of the northern part is lower than that of the southern part, indicating that there is a great efficiency gap among the performance of biomass power plants in the southern part of China. It is noted that the maximum value of the efficiency in the southern part is 0.9888, indicating that the three efficient biomass power plants are all in the northern part of China.

Table 5. Description statistics of the efficiency between the northern group and southern group.

Group	Mean	Std	Max	Min
North	0.9923	0.0079	1.0000	0.9814
South	0.8009	0.0421	0.8888	0.7595

Furthermore, we conduct the Wilcoxon-Mann-Whitney rank-sum test and the Kolmogorov-Smirnov (here after referred to as K-S) test to determine the significant differences in the efficiencies between the north and south group. Figure 4 shows the distribution kernel density of the two groups, and Table 6 shows the results of the test. In terms of the Wilcoxon-Mann-Whitney rank-sum test, all null hypotheses are rejected at the 1% level, indicating that the rank differences in the efficiencies between the two groups are significant and provide support for the technological heterogeneity of the two groups. In terms of the Kolmogorov–Smirnov test, all null hypotheses are rejected at the 1% level, indicating that kernel density distribution differences exist in the efficiencies between the two groups.

From the analysis above, it is obvious that although the biomass power capacity in the southern part of China is much larger than that in the northern part, the efficiency of the northern part is significantly higher than that of the southern part. The main reason is that in the southern part, the government invested more resources (such as capital, manpower, and so on) in order to get more returns. However, the biomass power industry in China is still in the early stage, and the technology and management level are relatively lower than that of the developed countries. Therefore, there exists a large number of redundant capital investments. Therefore, the efficiencies of the southern part of China are relatively lower than those of the northern part of China. As for policy makers, they should be sensitive to the differences between the two regions. Moreover, biomass power plants should pay more attention to improving their technology level and management level when making massive investments. Finally, in order to improve the efficiencies of the southern part of China, policy makers should provide more incentives to the biomass power plants in the southern part of the China, so as to catch up with those in the northern part of China and promote the entire biomass generation industry.

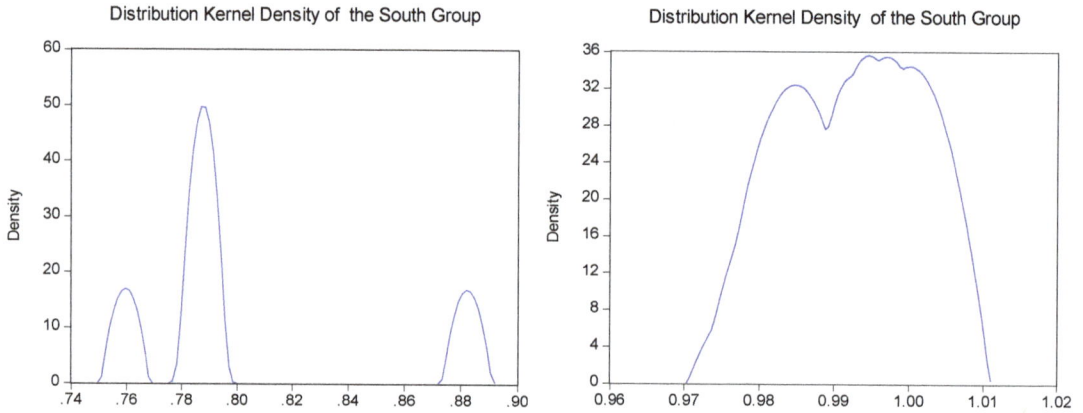

Figure 4. Distribution kernel density of the two group's efficiencies.

Table 6. The Mann–Whitney rank-sum test and the Kolmogorov–Smirnov test.

Variable	Mann–Whitney U	Prob	K-S	Prob
efficiency	15	0.006	1.651	0.009

6. Conclusions

In this paper, we have compared the properties of the existing additive-based DEA models and proposed two generalized additive-based models, *i.e.*, the BMA model and the BMAS model, to evaluate the performance of the biomass power plants in China in 2012. Compared to the traditional DEA models, there are two virtues of the new models that we proposed: one is that they inherited the properties of the traditional additive-based DEA models and derived more new additive-based DEA forms; the other is that they can rank the efficient DMUs. Then, we applied the two new DEA models to the evaluation of the biomass power plants in China. The conclusions are three-folds: first, the BMA and the BMAS model have great potential to be applied in sustainable energy project evaluation; second, the efficiency of biomass power plants in the northern part of China is higher than that in the southern part of China (the only three efficient biomass power plants are all in the northern part of China); finally, based on the results of the Wilcoxon–Mann–Whitney rank-sum test and the Kolmogorov–Smirnov test, there is a great technology gap between the biomass power plants in the northern part of China and those in the southern part of China. Based on the results of the model, policy makers should provide more incentives to the biomass power plants in the southern part of the China, so as to catch up with those in the northern part of China and promote the entire biomass generation industry.

Acknowledgments

This work is financially supported by the Fundamental Research Funds for the Central Universities (13XS26).

Author Contributions

Jie Tao completed the paper. Qingyou Yan gave much good research advice.

Conflicts of Interest

The authors declare no conflict of interest.

References

1. Hooper, R.; Li, J. Summary of the factors critical to the commercial application of bioenergy technologies. *Biomass Bioenerg.* **1996**, *11*, 469–474.
2. Weber, C.; Perrels, A. Modelling lifestyle effects on energy demand and related emissions. *Energ. Policy* **2000**, *28*, 549–566.
3. Klevas, V.; Streimikiene, D.; Grikstaite, R. Sustainable energy in Baltic States. *Energ. Policy* **2007**, *35*, 76–90.
4. St. Denis, G.; Parker, P. Community energy planning in Canada: The role of renewable energy. *Renew. Sustain. Energ. Rev.* **2009**, *13*, 2088–2095.
5. Zhou, P.; Ang, B.W.; Wang, H. Energy and CO_2 emission performance in electricity generation: A non-radial directional distance function approach. *Eur. J. Oper. Res.* **2012**, *221*, 625–635.
6. Zhao, Z.Y.; Yan, H. Assessment of the biomass power generation industry in China. *Renew. Energy* **2012**, *37*, 53–60.
7. Kautto, N.; Peck, P. Regional biomass planning—Helping to realise national renewable energy goals? *Renew. Energy* **2012**, *46*, 23–30.
8. San Cristóbal, J.R. A multi criteria data envelopment analysis model to evaluate the efficiency of the Renewable Energy technologies. *Renew. Energy* **2011**, *36*, 2742–2746.
9. Li, J.; Li, J.; Zheng, F. Unified Efficiency Measurement of Electric Power Supply Companies in China. *Sustainability* **2014**, *6*, 779–793.
10. Charnes, A.; Cooper, W.W.; Rhodes, E. Measuring the efficiency of decision making units. *Eur. J. Oper. Res.* **1978**, *2*, 429–444.
11. Ali, A.I.; Seiford, L.M. The Mathematical Programming Approach to Efficiency Analysis. In *The Measurement of Productive Efficiency: Techniques and Applications*; Fried, H.O., Schmidt, S.S., Eds.; Oxford University Press: London, UK, 1993; pp. 120–159.

12. Du, J.; Liang, L.; Zhu, J. A slacks-based measure of super-efficiency in data envelopment analysis: A comment. *Eur. J. Oper. Res.* **2010**, *204*, 694–697.
13. Tone, K. A slacks-based measure of efficiency in data envelopment analysis. *Eur. J. Oper. Res.* **2001**, *130*, 498–509.
14. Oral, M.; Kettani, O.; Lang, P. A methodology for collective evaluation and selection of industrial R&D projects. *Manag. Sci.* **1991**, *37*, 871–885.
15. Thompson, R.G.; Langemeier, L.N.; Lee, C.T.; Lee, E. The role of multiplier bounds in efficiency analysis with application to Kansas farming. *J. Econom.* **1990**, *46*, 93–108.
16. Liu, J.S.; Lu, L.Y.Y.; Lu, W.M.; Lin, B.Y.J. A survey of DEA applications. *Omega* **2013**, *41*, 893–902.
17. Xingang, Z.; Zhongfu, T.; Pingkuo, L. Development goal of 30 GW for China's biomass power generation: Will it be achieved? *Renew. Sustain. Energ. Rev.* **2013**, *25*, 310–317.
18. Banker, R.D.; Charnes, A.; Cooper, W.W. Some Models for Estimating Technical and Scale Inefficiencies in Data Envelopment Analysis. *Manag. Sci.* **1984**, *30*, 1078–1092.
19. Lins, M.E.; Oliveira, L.B.; da Silva, A.C.M.; Rosa, L.P. Performance assessment of Alternative Energy Resources in Brazilian power sector using Data Envelopment Analysis. *Renew. Sustain. Energy Rev.* **2012**, *16*, 898–903.
20. Lovell, C.A.K.; Pastor, J.T. Units invariant and translation invariant DEA models. *Oper. Res. Lett.* **1995**, *18*, 147–151.
21. Cooper, W.; Park, K.; Pastor, J. RAM: A Range Adjusted Measure of Inefficiency for Use with Additive Models, and Relations to Other Models and Measures in DEA. *J. Product. Anal.* **1999**, *11*, 5–42.
22. Cooper, W.W.; Pastor, J.T.; Borras, F.; Aparicio, J. BAM: A bounded adjusted measure of efficiency for use with bounded additive models. *J. Product. Anal.* **2011**, *35*, 85–94.

Environmental Justice and Sustainability Impact Assessment: In Search of Solutions to Ethnic Conflicts Caused by Coal Mining in Inner Mongolia, China

Lee Liu, Jie Liu and Zhenguo Zhang

Abstract: The Chinese government adopted more specific and stringent environmental impact assessment (EIA) guidelines in 2011, soon after the widespread ethnic protests against coal mining in Inner Mongolia. However, our research suggests that the root of the ethnic tension is a sustainability problem, in addition to environmental issues. In particular, the Mongolians do not feel they have benefited from the mining of their resources. Existing environmental assessment tools are inadequate to address sustainability, which is concerned with environmental protection, social justice and economic equity. Thus, it is necessary to develop a sustainability impact assessment (SIA) to fill in the gap. SIA would be in theory and practice a better tool than EIA for assessing sustainability impact. However, China's political system presents a major challenge to promoting social and economic equity. Another practical challenge for SIA is corruption which has been also responsible for the failing of EIA in assessing environmental impacts of coal mining in Inner Mongolia. Under the current political system, China should adopt the SIA while continuing its fight against corruption.

Reprinted from *Sustainability*. Cite as: Liu, L.; Liu, J.; Zhang, Z. Environmental Justice and Sustainability Impact Assessment: In Search of Solutions to Ethnic Conflicts Caused by Coal Mining in Inner Mongolia, China. *Sustainability* **2014**, *6*, 8756-8774.

1. Introduction

The Inner Mongolia Autonomous Region forms much of China's strategic northern frontier bordering Mongolia and Russia. In a region where protest is rare, a series of Mongolian demonstrations across the region, including these in the capital Hohhot, took the world by surprise in the spring of 2011. Students demonstrated and clashed with police, demanding justice. The events were triggered by an incident near Xilinhot (Figure 1). A Chinese truck driver killed a Mongolian herdsman who was blocking a convoy of coal trucks from driving through his pastureland. Chinese and international media widely reported the protests that underscored simmering discontent over environmental damage from mining in this resource-rich region [1–3]. To quell the demonstrations, the government declared martial law and cracked down on the activists while pledging to look into the impact of the mining industry on the environment and local culture.

We were curious how mining-related environmental problems led to ethnic conflicts in a region that had been relatively free of ethnic tensions in recent history. Our initial investigation indicated that mining caused serious environmental and economic injustice to the Mongolian herdsmen. We found earlier reports on mining pollution in Inner Mongolia. For example, the *Beijing Youth Daily* reported that a few rare-earth refineries polluted the grassland and killed 60,000 livestock that belonged to 190 herdsmen from 1996–2003 [4]. Another report found that arsenic poisoning was threatening the lives of the nearly 300,000 people in the Ordos Region; 2000 were already sick and many died of cancer, producing cancer villages [5]. China has hundreds of cancer villages, places where cancer rates are unexpectedly high and industrial pollution is suspected as the main cause [6]. However, most of them are in the more developed regions on the eastern coast. In Inner Mongolia, the main cause is suspected to be water pollution caused by mining.

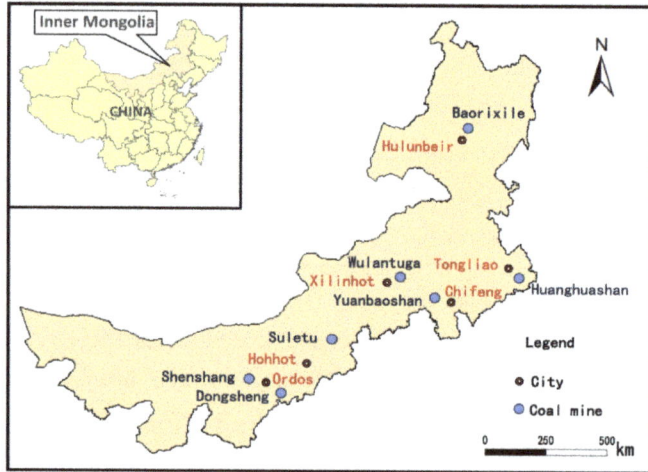

Figure 1. Location of the studied cities and coal mines in Inner Mongolia.

Mongolians have many long-established grievances, such as those reported by Jacobs: the ecological destruction wrought by an unprecedented mining boom, a perception that economic growth disproportionately benefits the Chinese and the rapid disappearance of Inner Mongolia's pastoral tradition [7].

Qian *et al.* find quantitative evidence to support the conclusion that the expansion of coal mining and associated industry and population increase was the major cause of grassland degradation in the Holingol region of Tongliao City, Inner Mongolia [8]. While mines are expanding, underground water is being over-extracted and coal-fired power plants as well as chemical plants are being established [9]. Coal mining and associated electricity generation have seriously degraded the water resource and the livelihood of local people in Inner Mongolia [8].

Greenpeace reports that in China, a coal chemical project in the dry Inner Mongolia region, part of a new mega coal power base, had extracted so much water in 8 years of operation that it caused the local water table to drop by up to 100 m, and the local lake to shrink by 62%. Due to lowering of water table, large areas of grassland have subsided (Figure 2). The drastic ecological impacts have forced thousands of local residents to become 'ecological migrants' [10].

At the costs of the environment and local residents' livelihood, Inner Mongolia has since 2002 experienced an economic boom based on mining. The wealth from the economic boom has not been fairly distributed. Many Chinese investors have benefited from the mining operations and become billionaires. Ordos became one of the wealthiest cities in China. However, ordinary Mongolian herdsmen are not benefiting from that boom, which is based on exploitation of what they view as their resources. Coal development on the grasslands does not increase the herdsmen's income or materially improve their life but instead has dampened their future by degrading the environment [11], causing injustice and sustainability disparities [12,13].

Figure 2. Baorixile, Hulunbeir, Inner Mongolia: grassland subsidence due to lowering of water table caused by coal mining [10]. © Lu Guang/Greenpeace.

The paper draws from global knowledge of environmental justice and assessment approaches and applies it to Inner Mongolia. It argues for the need of developing a sustainability impact assessment (SIA) and demonstrates that such a need is particularly urgent for subtle ethnic regions such as Inner Mongolia. We explore answers to five related questions: (1) What are the theoretical bases for developing an SIA that emphasizes justice? (2) How have assessment approaches been practiced in China? (3) Why has environmental impact assessment (EIA) not worked for Inner Mongolia in the current EIA system? (4) How and why do we need to explore an SIA that supports environmental justice in order to help with sustainability? (5) What should China do in search of solutions to ethnic conflicts in Inner Mongolia?

The analyses were based on data collected during fieldwork through qualitative research methods including site inspections and semi-structured interviews and discussions with local officials and scholars concerning environmental and economic issues. The study covers seven major coal-mining areas: Dongsheng, Shenshang, Suletu, Yuanbaoshan, Wulantuga, Baorixile, and Huanghuashan, in six associated city regions: Ordos, Hohhot, Xilinhot, Chifeng, Tongliao, and Hulunbeir (Figure 1). Primary and secondary data were collected during fieldwork in the summers from 2011–2013. The initial report was presented and discussed at the International Conference on Sustainability Assessment at Dalian Nationalities University. Follow-up fieldwork and research was conducted after the conference to further verify and interpret the research findings. We realize that our study areas were limited to only a few places. Due to lack of time, financial support, and availability of data and information, we were not able to obtain quantitative data or conduct more in-depth investigations. Environmental justice, sustainability impact assessment and ethnic conflicts in China are topics that are contested and require more systematic research. As a result, caution is needed when drawing conclusions from our findings.

In search of solutions to environmental degradation, injustice, and ethnic conflicts in the region, we first examine how project assessment tools could help. For example, environmental impact assessment (EIA) has been regarded as an important measure to control environmental impact in many countries, and some governments, such as those of the United States and Scotland, have attempted to use environmental assessment tools to deliver environmental justice [14]. However, environmental assessment tools in theory and practice appear to be inadequate when sustainability, not just the environment, is the subject for assessment.

2. The Theoretical Basis for Sustainability Impact Assessment that Supports Environmental Justice

This section starts with an overview of the meaning of sustainability and its indicators, and criteria that have been used or proposed for EIA. Following discussions over the use of existing assessment approaches to assess sustainability, the focus is on exploring the possibility of incorporating justice and equity into existing assessments and SIA.

2.1. Sustainability: The Three Pillars

Sustainability means meeting the needs of the present without sacrificing the ability of future generations to meet their own needs [15]. It has been illustrated as having three overlapping dimensions: the simultaneous pursuit of economic prosperity, environmental quality, and social equity, also known as the "three pillars" of sustainability [16–18]. In addition, cultural sustainability is widely regarded as an important element for people to achieve a more satisfactory intellectual, emotional, moral and spiritual existence. Recent holistic and inclusive thinking of sustainability emphasizes overlapping dimensions and the interaction among them [18]. In addition to environmental and material needs that may be fulfilled through economic development, humans also need social development to improve social justice, equality, and security. While acknowledging the interactions among different dimensions, Gibson et al. caution against a simplistic application of the three pillar model, pointing out that it serves to emphasize tensions among competing interests [19]. In contrast, their criteria cross the traditional and limiting divides to provide a more holistic conceptualization [19]. They also criticizes approaches to sustainability that over-emphasize local considerations or that focus too strongly on efficiency measures, recalling that the sustainability discourse is essentially global and the West must challenge some fundamental cornerstones of its way of life, and particularly the obsession with economic growth [19].

2.2. Sustainability and Justice

Dobson provides a detailed discussion of the relationship between environmental justice and sustainability. He argues that "the discourses of sustainability and justice may be related" but "the question of whether sustainability and justice are compatible objectives can only be resolved empirically, and the range and depth of empirical research required in resolving this question has not been done" [20]. We argue that SIA for subtle ethnic regions such as Inner Mongolia should stress justice, including environmental, social, and economic justice, and equity, which has been recognized as a key element of sustainability. Sustainability is about meeting needs. Justice has increasingly been recognized as one of such needs. There is no sustainability without justice. Furthermore, the United Nations resolution 66/197 on sustainable development pays special attention to the welfare of ethnic minorities: recognizing and supporting their identity, culture and interests; avoiding endangering their cultural heritage, practices and traditional knowledge; and preserving and respecting non-market approaches that contribute to the eradication of poverty [21]. Iris Marion Young has also questioned the common practice of reducing social justice to distributive justice and argued for group-differentiated policies and a principle of group representation [22]. Using justice as an overarching element can help develop a more holistic SIA for ethnic regions such as Inner Mongolia.

The concept of environmental justice was first developed in the early 1980s during the social movement in the United States on the fair distribution of environmental benefits and burdens. The United States Environmental Protection Agency defines environmental justice as "the fair treatment and meaningful involvement of all people regardless of race, color, sex, national origin, or income with respect to the development, implementation and enforcement of environmental laws, regulations, and policies" [23]. Three different notions of justice have been applied, including distribution, recognition, and procedure (or participation) [24]. Procedural justice means that those who are most affected by decisions should have particular rights to be involved and have their voices heard on a fully informed basis [25]. Participation has also been demanded as an instrument of EIA. Since ex ante analysis of potential impacts of planned projects on the environment is difficult, participation is intended to reduce uncertainty by intra-subjective judgment; furthermore, participation increases the transparency of the decision-making process [26]. From a social science point of view, participation is a central element of sustainability [26]. Participation, however, is difficult to translate meaningfully into quantitative terms as a social indicator [26]. Direct and open debates among the people who will be affected by the development lay the foundation for conflict resolution in Inner Mongolia, if SIA can be incorporated at the planning level in order to influence decision making and support policies that affect regional sustainability [27].

2.3. Sustainability: Indicators and Criteria for Assessments

In practice, economic and social indicators and criteria have been used, in addition to environmental ones, for assessing sustainability [28]. For example, Becker presents an overview on sustainability indicators for assessing economic, environmental, and social sustainability which includes "equity coefficients (Gini coefficient, Atkinson's weighted index of income distribution), disposable family income, and social costs, participation, and tenure rights [26]. Herder *et al.* used "production costs" and "local value added" as economic indicators and "employment" as a social indicator [29]. However, the incorporation of these sustainability truths into assessment and decision-making processes remains somewhat daunting in practice. Lamorgese and Geneletti developed a framework for evaluating planning against sustainability criteria and found that criteria explicitly linked to intra- and inter-generational equity is rarely addressed [30]. Jain and Jain emphasize the need for an alternative index which considers sustainability of human development and formulates an index based on strong sustainability [31]. Shah and Gibson have developed a set of 12 core procedural and substantive-level sustainability criteria to be used as a guide for clarifying development purposes, identifying potentially desirable options, comparing alternatives and monitoring implementation for infrastructure at the water-agriculture-energy nexus in India [32]. They believe that sustainability-based tools encourage comprehensive attention to issues at the core of sustainability thinking and application: relative to conventional assessment approaches, assessments applying explicit sustainability criteria encourage lasting benefits within complex socio-ecological systems through assessing interdependencies and opportunities, sensitivities and vulnerabilities of regional ecologies, incorporating systems, resiliency and complexity frameworks. SIA for Inner Mongolia should learn from the international experience to develop specific indicators and criteria that help with ethnic equality and harmony.

2.4. The Debate over the Use of Existing Assessment Approaches to Assess Sustainability

Researchers have been debating over the use of existing assessment approaches such as environmental impact assessment (EIA) and strategic environmental assessment (SEA) to assess sustainability. For examples, Zhu *et al.* advocate an impact-centered SEA with institutional components as an alternative to the impact-based approach which seems unable to address institutional weaknesses in most conventional SEA cases in China [33]. Lam, Chen, and Wu affirm the potential role of SEA in fostering a sustainable and harmonious society and the need to mainstream sustainability considerations in the formulation of national plans and strategies [34]. Hacking and Guthrie identify the features that are typically promoted for improving the sustainable development directedness of assessments and a framework which reconciles the broad range of emerging approaches and tackles the inconsistent use of terminology [35]. Morrison-Saunders and Retief assert that internationally there is a growing demand for EIA to move away from its traditional focus towards delivering more sustainable outcomes [36]. They argue that it is possible to use EIA to deliver some sustainability objectives in South Africa, if EIA practices strictly follow a strong and explicit sustainability mandate [36]. To advance SEA for sustainability, White and Noble examined the incorporation of sustainability in SEA and identified several common themes by which SEA can support sustainability, as well as "many underlying barriers that challenge SEA for sustainability, including the variable interpretations of the scope of sustainability in SEA; the limited use of assessment criteria directly linked to sustainability objectives; and challenges for decision-makers in operationalizing sustainability in SEA and adapting PPP (policy, plan, and program) development decision-making processes to include sustainability issues" [37].

2.5. The Possibility of Incorporating Environmental Justice into Environmental Assessments

Jackson and Illsley proposed that SEA could be used to help deliver environmental justice [38]. Krieg and Faber suggest that environmental injustices exist on a remarkably consistent continuum for nearly all communities and a cumulative environmental justice impact assessment should take into account the total environmental burden and related health impacts upon residents [39]. Connelly and Richardson argue that "we cannot debate SEA procedures in isolation from questions of value, and that these debates should foreground qualities of outcomes rather than become preoccupied with qualities of process" [40]. They "explore how theories of environmental justice could provide a useful basis for establishing how to deal with questions of value in SEA, and help in understanding when SEA is successful and when it is not" [40]. They assert that "Good SEA must be able to take into account the distributional consequences of policies, plans, or programs, with decisions driven by the recognition that certain groups tend to systematically lose out in the distribution of environmental goods and bads" [40]. Walker finds that although practices are evolving there is a little routine assessment of distributional inequalities, which should become part of established practice to ensure that inequalities are revealed and matters of justice are given a higher profile [41]. On the other hand, Mclauchlan and Joao oppose the use of strategic environmental assessment (SEA) to deliver environmental justice, partly because "a direct focus on the environment requires that factors associated with environmental justice are not central to SEA" [14].

The literature indicates that it is possible to use environmental assessment to incorporate environmental justice criteria such as public participation. In fact, public participation is considered as an integral part of the EIA procedure [42]. A major challenge is that environmental justice is a social factor, which is not central to environmental assessments. In China, EIA is often inadequately implemented and social factors tend to be neglected. For examples, Ren finds that "EIA in China has evolved into a fairly comprehensive and technically adequate system, but the problem lies in its poor enforcement and implementation, due to the political system and incentive mechanisms, institutional arrangements, and regulatory and methodological shortcomings" [43]. Yang criticizes that "public participation in the Chinese EIA system has not been effectively carried out" [44]. These problems have significant implications to EIA in Inner Mongolia.

2.6. The Possibility of Developing Sustainability Impact Assessments with Stress on Justice

The literature on EIA, SEA, and environmental justice may provide a theoretical context for developing SIA. The theory and practice of SIA have been discussed with case studies from different parts of the world. For example, Gibson *et al.* conceptualize sustainability assessment as a marriage between sustainable development and environmental assessment [19]. Huber made the distinction of social justice based on need, on performance, and on property as different dimensions of equity, which are not taken into account in static, target-oriented sustainability policies [26]. Bond *et al.* point out that sustainability assessment is an increasingly important tool for informing planning and development decisions across the globe [45]. Required by law in some countries, strongly recommended in others, a comprehensive analysis of why sustainability assessment is needed and clarification of the value-laden and political nature of assessments is long overdue [45]. The remaining of the paper will attempt to demonstrate the need to develop an SIA that stresses justice in order to reduce ethnic tensions in Inner Mongolia.

3. Assessment Practices in China

China faces a daunting task for improving its environmental performance, particularly in the ethnic regions where the environment is fragile, ecological systems are sensitive, the economy is underdeveloped, and ethnic relations are subtle. Different approaches have been proposed to deal with the task. Many believe that economic growth is the key for environmental improvement and social political stability [46]. This belief supports China's Go West policy, which covers all provincial level ethnic regions. While that policy has resulted in economic growth in some areas, there are indications that the environmental costs have been enormous and ethnic relations are getting worse. Exploitation of natural resources in the ethnic regions is followed by rapid environmental degradation. "Go West" has in some way become "Pollute West" under the "grow first, clean up later" approach to development. Inner Mongolia is a good example. It was once an endless field of grassland, punctuated by mountains and the occasional yurt. Now Inner Mongolia is the country's top coal producer, accounting for about a quarter of all domestic supply—doubling what it was in 2005 [1].

On the other hand, sustainable development has also been the view of some top Chinese officials such as the former premier Wen Jiabao. Sustainability management has shown that environmental problems and social problems are closely related [47], especially in the case of China [48]. Among the many possible methods for improving environmental performance, EIA has been used in China, including its ethnic regions. For examples, the Asian Development Bank (ADB) was particularly comprehensive in its assessment of Inner Mongolia Environment Improvement Project (Phase II) [49], following ADB's Environmental Assessment Guidelines [50]. The ADB report recommended that Inner Mongolia install "clean coal" technologies now to reduce global warming and reverse the climate change caused by current coal mining [51]. Many coal mining companies in Inner Mongolia have drafted EIA and posted notifications for the public to provide feedback.

However, EIAs seem to have not had any significant impact, as coal mining and associated industries continue to expand. Mining pollution causes local herders to lose their sheep and cattle and thousands of pits left behind by the mining companies cause fatalities to the herds [9]. Consequently, coal mining has contributed to increased ethnic tension and conflicts in Inner Mongolia. Investigations found that the common people have got poorer in natural-resource rich places such as Inner Mongolia, while government officials and mine bosses got extremely wealthy, increasing social unrest [52]. The Inner Mongolia Government issued a document asking local governments and agencies to follow governmental directives to adequately protect the environment and people's livelihood [53]. The document clearly stated that promised compensations to herders who lost land due to mining should be honored. Wealth from the mining should be partially used to help improve local infrastructure and living conditions. The document, however, fails to recommend any concrete procedures to insure the local residents receive their share of the mining wealth. The document encourages coal companies to invest in non-coal industries locally. This kind of investment helps to diversify the economy and increase government revenues and GDP. However, further industrialization has been accompanied by worsening environmental degradation and damage to the agricultural environment needed to support the livelihood of the Mongolian herders.

Nevertheless, progress was made. The Inner Mongolia Government claimed to have halted 476 illegal mining projects, ordered 887 mines to suspend operations, permanently shut down 73 mines, intervened in 100 disputes between local herders and mining companies, and established a mechanism involving the government, miners and local residents to resolve disputes through dialogue [54]. However, new protests continue to be reported by the international media [55]. Tang suggests that a rise in public protests in China signals a failure of environmental governance, where officials use legal threats to extract benefit from polluters, but the power of developers in China remains untouched, despite widespread protests against polluting projects [56].

4. Why Environmental Impact Assessments Have Failed for Inner Mongolia

The last section elaborates the failure of environmental assessments to do their job for Inner Mongolian mining projects. This section will specifically answer three questions: (1) why did the mining projects fail to conduct environmental assessments when they would be expected? (2) Would environmental assessments have had an impact on the projects if they were conducted? (3) Would environmental impact assessments have addressed the questions of sustainability and environmental justice adequately, even if they were conducted?

4.1. Why Did Projects Fail to Conduct Environmental Impact Assessments?

The failure of EIA may be one of the many factors for explaining environmental degradation caused by coal mining in Inner Mongolia. Here are a few scenarios based on our investigation. First, an EIA is not conducted at all. This applies to the many small scale mining operations. Many are "illegal" as they do not have any permit. These operations tend to pay no attention to the environment. They are allowed to be in operation mainly through bribing the government officials who will then turn a blind eye on the environmental destructions. Under the pressure from repeated local protests, there has been a tightening of regulations and cracking down on these operations. However, they continue to be a major threat as corruption will continue to be severe. A second scenario is that an EIA is conducted, but is falsified as the required criteria were not followed. This is concerned with the legal operations. Again, official corruption is involved, which is the main reason EIAs are not conducted or are falsified.

4.2. Would Environmental Impact Assessments Have Had an Impact?

We find that large state coal mines often had an EIA conducted. We examined key government directives that provide technical guidelines for EIA for coal mines. China passed its EIA laws in 2002. In 2006, Technical Guidelines for EIA Coal Mine Master Plans was drafted. The guidelines did not include any mandatory requirements in terms of EIA. That left much room for interpretation of activities as to what was appropriate. It stated that the plan should include descriptions concerning water, air pollution, land restoration, and public participation [57]. It is unclear how many EIAs were done. However, an online search found four Master Plan EIAs, which were posted for public notification as required by the EIA laws, an indication of implementation of the EIA laws and the 2006 Guidelines. These four EIAs, three from Ordos [58–60] and one from Hulunbeir [61] are identical in terms of structure and contents, suggesting that they followed the same standard format and guidelines used in coal mining in Inner Mongolia and possibly nationwide.

The public notifications are very superficial, mainly an overview of the planned project which follows the guidelines but lack any specifics. Accompanying the notifications, a survey form asks questions such as: What do you think of the current environmental conditions in your area? What impact will the project have on the environment? Do you support the implementation of the project or not? The notifications were published in local newspapers or government websites. The public were given 10 business days to respond, which was too short by international standards. A search did not find any cases where public feedback was publicized or had any effect on the plans. That might suggest that public participation did not play any role in the plan and the EIAs were done superficially. Few EIA notifications were found online for the period between 2008 and 2011, only one for 2010 [62] and one for 2011 [63].

The Technical Audit Points for Coal Mine Master Plan EIA Report was published in October 2011 by the Ministry of Environmental Protection [64], after the widespread protests in Inner Mongolia in May. This is a comprehensive directive that provides detailed requirements for coal mine planning and EIA. The Circumstances for Rejecting and Requiring Revisions of the Plan include six items and three of them are:

A. The project may cause major impact to the ecology or underground water (quantity or quality) but the plan does not provide mature and practical ecological recovery and protection measures;
B. The local resource and environment is unable to provide the capacity for the possible direct and indirect impact of urbanization and industrialization due to coal mining;
C. The majority of the public participants do not support the implementation of the project plan.

One of the Circumstances for Requiring Revisions of the Plan for reevaluation includes irregularities of public participation, no explanations for accepting or rejecting public suggestions, or obviously unreasonable rejection of public suggestions. The directive also states that the master plan should ensure that the mining operation will protect the ecological integrity and biodiversity and prevent desertification. Air pollution needs to be controlled during mining, transportation, and storage, consumption, and waste management. However, many of the requirements are still vague due to lack of specifics.

The latest EIA documents we found online include two EIA notifications [65,66] and one EIA report [67]. They reflect the more stringent guidelines and contain more specifics than the 2006 and 2010 ones. The posting of an EIA Report provides information to the public. Interestingly, however, the lead author of both the 2006 and 2011 guidelines was Beijing Huayu Engineering Co., Ltd. of the Sino-Coal International Engineering Group, in cooperation with the State Environmental Protection Bureau of China (now Ministry of Environmental Protection). Huayu or some other firms within Sino-Coal have been the sole authors for the EIAs. So the guidelines and EIAs are likely to be on the side of the coal industry, rather than the affected communities. The number of EIAs available online is very small, compared to the number of mines in the region, possibly over 100. According to the Chinese search engine Baidu, there were 82 state-owned mines in Inner Mongolia in 2009, including five state-owned enterprises, 42 state-owned major mines, and 36 state-owned local mines [68]. The number should have increased, judging from the increased coal output in the region. More importantly, the new EIA requirements were probably not followed in Inner Mongolia coal mine planning and operations, judging by the high level of environmental degradation due to coal mining, as reported in Chinese official and international media. Consequently, EIAs have had only a limited impact in protecting the environment.

4.3. Would Environmental Impact Assessments Have Addressed Sustainability and Environmental Justice Adequately?

Furthermore, even if the more stringent EIA requirements were closely followed, many social problems caused by the exploitation of natural resources in the ethnic regions were not going to go away, as many Mongolians are likely to view the coal and the land as theirs, that they inherited from their ancestors, and that the Chinese are outsiders coming in to take their resources away and destroy their land and lifestyle. The central government may claim that the resources are national property. Many Mongolians may believe that the nation should be the people instead of some state officials. Considering the history of settlement and the Mongolian way of thinking, the mining plans may have to incorporate the concept of environmental justice and respect the view of the local people and culture, in addition to protecting the environment. Improvement in EIAs is needed for the environment, but EIAs are inadequate for dealing with these social problems caused by resource exploitation in the ethnic regions.

5. The Need to Stress Justice and Equity in Sustainability Impact Assessment

The above discussed problems in EIA practices in China need to be dealt with. For example, public participation needs to be strengthened to allow full involvement from the beginning of the project planning. It would be worthwhile to explore ways to have EIAs conducted by an independent third party rather than by an affiliate of the coal companies, even though that may increase the operational cost of the assessment. EIA has not been taken seriously because it concerns only the environment, which is considered as a public good in China. The government, which is supposed to take care of the environment and the resources, puts economic growth first. Public participation has not been regarded as a key element in resource development as natural resources belong to the government (Officially they belong to the state, but in reality the government, rather than the people, is the state in China). Wealth from mining is mainly taken by the central government, with the rest of the wealth taken by different levels of the governments. The local government, which usually receives a third of the wealth, is left to take care of the social and economic welfare [52]. Corruption and lack of funding have meant little is done for the common people. Such injustice and inequity has led local people to organize to open illegal mines to steal and rob the resources, which they think should belong to them [52]. Li Bo, head of Friends of Nature, an environmental non-government organization (NGO) in China, believes that:

> The environmental assessment of development projects should be much more open. The possible existence of risk for any project—technological and economic, or social and political—should be fully discussed before the project is implemented. Right now, according to the law, there is a process for EIA. But the people who are in charge of executing these are only responsible to their seniors, not to the people under them. So these processes aren't very open, and their discussions aren't transparent. Because of this

many projects are approved, and then their problems are only discovered afterwards. An example is the recent PX incident—there's a lot of fear and rage. These things can tear a society apart [69].

Morrison-Saunders and Pope argue that there is inadequate consideration of trade-offs throughout the sustainability assessment process and insufficient considerations of how process decisions and compromises influence substantive outcomes [67]. If properly done, sustainability assessment should indicate who gets what, who loses what, how, when and why [70]. Current EIAs in China are concerned with trade-offs between the economy and the environment. They are not concerned with trade-offs among different social groups. We argue that SIA should be adopted for subtle ethnic regions in order to adequately evaluate economic, environmental and social impacts to help reduce ethnic tensions. These impacts are interrelated and cannot be mitigated successfully unless they are dealt with together. If fully enforced, SIA will ensure that the public is more involved and their interest is better taken care of when justice and equity are a matter of concern in the assessment.

As sustainability assessment is new in China, we draw below some international experiences to help with the discussion. Gibson *et al.* present the case of the assessment of the proposed major nickel mining project near Voisey's Bay on the north coast of Labrador (Canada), which is often considered the first attempt to conduct sustainability assessment within a project approval context [19]. They challenge the common conceptualization of sustainability of three intersecting pillars representing environmental, social and economic concerns, on which most practice of sustainability assessment is based [19]. Gibson reports that an innovative environmental assessment and a set of surrounding and consequential negotiations were conducted between 1997 and 2002 on the proposed project:

> The proponent and other participants wrestled directly and often openly with the project's potential contribution to local and regional sustainability. The resulting agreements to proceed were heavily influenced by the precedent-setting assessment, which imposed a "contribution to sustainability" test on the proposed undertaking. Given the profound differences in background, culture, priorities and formal power involved, as well as the record of tensions in the history of this case and before, the agreements also represent a considerable achievement in conflict resolution [71].

Faced with growing environmental and social crises, China's new leader, Xi Jinping, has criticized the "grow first, clean up later" approach and given more emphasis on ecological development than his predecessors. Among other things, he recently called for stopping the GDP-based promotion of government leaders [72]. Consequently, several provinces have lowered or abandoned using GDP as the only measure of success for city or county leaders, affecting over 70 of China's poorest cities or counties [73]. Evaluation will instead be based on poverty reduction and environmental protection [74]. It remains to be seen if the new policy will be applied to larger, wealthier cities. Nevertheless, cleanup efforts have been increasing. Many interviewed officials cared about the environment and were sympathetic for the Mongolian herders. There are indications that EIA will be more stringently and widely implemented. Kahya reports that:

> Concerns over water use from coal mining and gasification projects have led the Chinese government to change the rules for new schemes. Mirroring recent "national plans" to

tackle air pollution the Ministry of Water resources has announced a plan to limit coal expansion based on regional water capacity. The rules mean the approval process for large-scale projects must now include an appraisal of the available water [75].

Mclauchlan and Joao oppose the use of strategic environmental assessment (SEA) to deliver environmental justice, partly because "a direct focus on the environment requires that factors associated with environmental justice are not central to SEA [14]." That was exactly the case with some projects that conducted EIAs. China should borrow global knowledge in environmental justice and SIA to help with local practices and leapfrog the EIA stage to start SIA instead. Less-developed ethnic regions should leapfrog the "grow first, clean up later" stage and start practicing sustainability, so that further deepening of injustice and sustainability disparities might be avoided [12,13]. EIA has been useful in some countries. An important reason is that these countries tend to follow the rule of law and have an independent media and democratic government. Environmental injustice is partly inherited in the undemocratic system. There are limited options China has as major political reforms are unlikely to happen soon. Within the current political system, SIA certainly seems to be more useful than EIA in dealing with justice and equity problems.

6. Conclusions

In this paper, we have explored the theoretical basis and possibility of developing SIA with an emphasis on justice and equity to meet an urgent need in subtle ethnic regions such as Inner Mongolia, China. In coal mining practices in Inner Mongolia, an EIA was often not conducted for the large number of small scale so called "illegal" mines, or might have been falsified for many other mines through corrupted officials. Our focus, however, has mainly been on those that have conducted official EIAs following government guidelines but still fail to protect the environment partly because the guidelines are inadequate. The government has tightened control over EIA along with more specific and stringent guidelines.

Our research indicates that even if the new EIA guidelines are closely followed social justice and economic equity problems will continue to exist, as EIAs do not deal with any ethnic social problems. The assessment needs to include guidelines for justice and equity, in addition to protect the physical environment. EIAs appear to be inadequate for that. Even though certain elements of environmental justice such as participation can be incorporated into EIA/SEA, these elements are not central to environmental assessments. Environmental assessments are concerned with the environment while sustainability has addition concerns such as social justice and economic equity. Consequently, EIA/SEA is in theory and practice inadequate as a tool in meeting sustainability challenges.

SIA would be in theory and practice a better tool than EIA/SIA for assessing sustainability impact. The assessment needs to involve the effected ethnic groups at the very beginning and careful negotiations are needed so that agreements can be reached. This would be an appropriate approach to conflict resolution to take care of the profound differences and complex relations in the ethnic regions. Public participation in SIA is an effective measure to ensure social and economic justice and equity. SIA should recognize and respect traditional ethnic way of life, which has often been found to be environmentally sustainable. It is important to let the local people make their own decisions concerning the use of their resources. They tend to be the people who care about the environment the most and have the knowledge for sustainability. Social and economic equity, protecting the environment, and respect for nature, culture, and autonomy of local ethnic groups should all be key elements for SIA in the ethnic regions.

However, one practical challenge for SIA is corruption which has been also responsible for the failing of EIA in Inner Mongolia. China's political system presents another challenge to promoting social and economic equity. Political reforms are necessary to enhance ethnic justice and equity. Under the current political system, China should adopt the SIA for ethnic regions while continuing its fight against corruption.

Many of the concepts discussed in this paper are contested, such as sustainability, justice, and even participation. For example, Cooke and Kothari criticize "participatory development's potential for tyranny" as "it can lead to the unjust and illegitimate exercise of power" [76]. On the other hand, Hickey and Mohan argue for transforming problematic traditional practices to citizen participation. The contested nature of the terms shows the complexity of the issues and cautions us to avoid drawing simplistic conclusions [77]. We hope that our report will provide the initial information and stimulate future research into developing an SIA with justice and equity emphasized for easing ethnic conflicts in ethnic regions.

Acknowledgments

This research was based on fieldwork partly funded by grants from the National Geographic Society (Grant # 8980-1) and University of Central Missouri. The research benefited from expert discussions at the International Conference on Sustainability Assessment, 20–21 July 2013, Dalian Nationalities University, Dalian, China. The authors also wish to thank the anonymous reviewers for their constructive comments and suggestions on the earlier versions of the manuscript.

Author Contributions

Lee Liu designed the research, conducted literature research and fieldwork, and wrote the article; Jie Liu conducted literature research and fieldwork; Zhenguo Zhang conducted fieldwork and cartographic work.

Conflicts of Interest

The authors declare no conflict of interest.

References

1. Pierson, D. Coal Mining in China's Inner Mongolia Fuels Tensions. Available online: http://articles.latimes.com/2011/jun/02/business/la-fi-china-coal-20110602 (accessed on 20 June 2014).
2. Jacobs, A.; Shi, D. Anger over Protesters' Deaths Leads to Intensified Demonstrations by Mongolians. *New York Times* 31 May 2011.

3. Unrest in China: No Pastoral Idyll. Available online: http://www.economist.com/node/18775303 (accessed on 22 October 2014).

4. Wen, S. Illegal Emissions from Rare Earth Refineries Polluted Inner Mongolian Grassland and Killed 60,000 Livestock. Available online: http://news.sina.com.cn/c/2003-09-20/0524784605s.shtml (accessed on 28 August 2014). (In Chinese)

5. China Light & Power Company Syndicate (CLP) Netcom. Inner Mongolian Grassland Polluted, Cancer Villages Burst Out. Available online: http://www.china125.com/news/it/1837.htm (accessed on 28 August 2014). (In Chinese)

6. Liu, L. Made in China: Cancer Villages. *Environ. Sci. Policy Sustain. Dev.* **2010**, *52*, 8–21.

7. Jacobs, A. Ethnic Protests in China Have Lengthy Roots. Available online: http://www.nytimes.com/2011/06/11/world/asia/11mongolia.html?pagewanted=all&_r=0 (accessed on 23 October 2014).

8. Qian, T.; Bagan, H.; Kinoshita, T.; Yamagata, Y. Spatial-temporal analyses of surface coal mining dominated land degradation in Holingol, Inner Mongolia. *IEEE J. Sel. Topics Appl. Earth Obs. Rem. Sens.* **2014**, *7*, 1675–1687.

9. Yang, S. Inner Mongolia Sinking under the Weight of its Mining Industry. Available online: http://www.globaltimes.cn/content/734241.shtml (accessed on 20 June 2014).

10. Greenpeace. Coal Impacts on Water, 21 March 2014. Available online: http://www.greenpeace.org/international/en/campaigns/climate-change/coal/Water-impacts/ (accessed on 28 August 2014).

11. Dai, G.S.; Ulgiati, S.; Zhang, Y.S.; Yu, B.H.; Kang, M.Y.; Jin, Y.; Dong, X.B.; Zhang, X.S. The false promises of coal exploitation: How mining affects herdsmen well-being in the grassland ecosystems of Inner Mongolia. *Energy Policy* **2014**, *67*, 146–153.

12. Liu, L. Environmental poverty, a decomposed environmental Kuznets curve, and alternatives: Sustainability lessons from China. *Ecol. Econ.* **2012**, *73*, 86–92.

13. Liu, L. Geographic Approaches to Resolving Environmental Problems in Search of the Path to Sustainability: The Case of Polluting Plant Relocation in China. *Appl. Geogr.* **2013**, *45*, 138–146.

14. Mclauchlan, A.; Joao, E. The utopian goal of attempting to deliver environmental justice using SEA. *J. Environ. Assess. Policy Manag.* **2011**, *13*, 129–158.

15. World Commission on Environment and Development. *Our Common Future*; Oxford University Press: New York, NY, USA, 1987.

16. United Nations. 2005 World Summit Outcome, Resolution A/60/1. Available online: http://www.un.org/womenwatch/ods/A-RES-60-1-E.pdf (accessed on 22 October 2014).

17. Adams, W.M. The future of sustainability: Re-thinking environment and development in the twenty-first century. In Proceedings of the IUCN Renowned Thinkers Meeting, Gland, Switzerland, 29–31 January 2006.

18. Liu, L. Sustainability: Living within One's Own Ecological Means. *Sustainability* **2009**, *1*, 1412–1430.

19. Gibson, R.B.; Hassan, S.; Holtz, S.; Tansey, J.; Whitelaw, G. *Sustainability Assessment: Criteria and Processes*; Earthscan: London, UK, 2005.

20. Dobson, A. *Justice and the Environment: Conceptions of Environmental Sustainability and Theories of Distributive Justice*; Oxford University Press: Oxford, UK, 1998.

21. United Nations. Resolution 66/197 on Sustainable Development. Available online: http://www.un.org/ga/search/view_doc.asp?symbol=%20A/RES/66/197 (accessed on 23 June 2014).

22. Young, I.M. *Justice and the Politics of Difference*; Princeton University Press: Princeton, NJ, USA, 1990.

23. United States Environmental Protection Agency (USEPA). Environmental Justice. Available online: http://www.epa.gov/environmentaljustice (accessed on 20 June 2014).

24. Walker, G. Beyond distribution and proximity: Exploring the multiple spatialities of environmental justice. *Antipode* **2009**, *41*, 614–636.

25. Hampton, G. Environmental equity and public participation. *Policy Sci.* **1999**, *32*, 163–174.

26. Becker, B. Sustainability Assessment: A Review of Values, Concepts, and Methodological Approaches. Available online: http://www.worldbank.org/html/cgiar/publications/issues/issues10.pdf (accessed on 22 October 2014).

27. Jeon, C.M.; Amekudzi, A.A.; Guensler, R.L. Sustainability assessment at the transportation planning level: Performance measures and indexes. *Transp. Policy* **2013**, *25*, 10–21.

28. Carnus, J.; Hengeveld, G.; Mason, B. Sustainability Impact Assessment of Forest Management Alternatives in Europe: An Introductory Background and Framework. *Ecol. Soc.* **2012**, *17*, 35–39.

29. Herder, M.; Kolström, M.; Lindner, M.; Suominen, T.; Tuomasjukka, D.; Pekkanen, M. Sustainability Impact Assessment on the Production and Use of Different Wood and Fossil Fuels Employed for Energy Production in North Karelia, Finland. *Energies* **2012**, *5*, 4870–4891.

30. Lamorgese, L.; Geneletti, D. Sustainability principles in strategic environmental assessment: A framework for analysis and examples from Italian urban planning. *Environ. Impact Assess. Rev.* **2013**, *42*, 116–126.

31. Jain, P.; Jain, P. Sustainability assessment index: A strong sustainability approach to measure sustainable human development. *Int. J. Sustain. Dev. World Ecol.* **2013**, *20*, 116–122.

32. Shah, S.H.; Gibson, R.B. Large dam development in India: Sustainability criteria for the assessment of critical river basin infrastructure. *Int. J. River Basin Manag.* **2013**, *11*, 33–53.

33. Zhu, Z.; Wang, H.; Xu, H.; Bai, H. An alternative approach to institutional analysis in strategic environmental assessment in China. *J. Environ. Assess. Policy Manag.* **2010**, *12*, 155–183.

34. Lam, K.; Chen, Y.D.; Wu, J. Strategic environmental assessment in China: Opportunities, issues and challenges. *J. Environ. Assess. Policy Manag.* **2009**, *11*, 369–385.

35. Hacking, T.; Guthrie, P. A framework for clarifying the meaning of Triple Bottom-Line, Integrated and sustainability assessment. *Environ. Impact Assess. Rev.* **2008**, *28*, 73–89.

36. Morrison-Saunders, A.; Retief, F. Walking the sustainability assessment talk—Progressing the practice of environmental impact assessment (EIA). *Environ. Impact Assess. Rev.* **2012**, *36*, 34–41.

37. White, L.; Noble, B.F. Strategic environmental assessment for sustainability: A review of a decade of academic research. *Environ. Impact Assess. Rev.* **2013**, *42*, 60–65.

38. Jackson, T.; Illsley, B. An analysis of the theoretical rationale for using strategic environmental assessment to deliver environmental justice in the light of the Scottish Environmental Assessment Act. *Environ. Impact Assess. Rev.* **2007**, *27*, 607–623.

39. Krieg, E.J.; Faber, D.R. Not so Black and White: Environmental justice and cumulative impact assessments. *Environ. Impact Assess. Rev.* **2004**, *24*, 667–694.

40. Connelly, S.; Richardson, T. Value-driven SEA: Time for an environmental justice perspective? *Environ. Impact Assess. Rev.* **2005**, *25*, 391–409.

41. Walker, G. Environmental justice, impact assessment and the politics of knowledge: The implications of assessing the social distribution of environmental outcomes. *Environ. Impact Assess. Rev.* **2010**, *30*, 312–318.

42. Glucker, A.N.; Driessen, P.P.J.; Kolhoff, A.; Runhaar, H.A.C. Public participation in environmental impact assessment: Why, who and how? *Environ. Impact Assess. Rev.* **2013**, *43*, 104–111.

43. Ren, X. Implementation of environmental impact assessment in China. *J. Environ. Assess. Policy Manag.* **2013**, doi: 10.1142/S1464333213500099.

44. Yang, S. Public participation in the Chinese environmental impact assessment (EIA) system. *J. Environ. Assess. Policy Manag.* **2008**, *10*, 91–113.

45. Bond, A.J.; Morrison-Saunders, A.; Howitt, R. *Sustainability Assessment: Pluralism, Practice and Progress*; Routledge: New York, NY, USA, 2013.

46. Rock, M.; Angel, D. Grow first, clean up later? Industrial transformation in East Asia. *Environ. Sci. Policy Sustain. Dev.* **2007**, *49*, 8–19.

47. Wallimann, I. *Environmental Policy is Social Policy—Social Policy is Environmental Policy: Toward Sustainability Policy*; Springer: New York, NY, USA, 2013.

48. Liu, L. Chinese Model Cities and Cancer Villages: Where Environmental Policy is Social Policy. In *Environmental Policy is Social Policy—Social Policy is Environmental Policy: Toward Sustainability Policy*; Springer: New York, NY, USA, 2013; pp. 121–134.

49. Asian Development Bank (ADB). Asian Development Bank Inner Mongolia Autonomous Region Environment Improvement Project (Phase II), Environmental Assessment and Measures. Available online: http://www.adb.org/projects/documents/inner-mongolia-autonomous-region-environment-improvement-project-phase-ii-environ (accessed on 20 June 2014).

50. Asian Development Bank (ADB). ADB Environmental Assessment Guidelines. Available online: http://www.adb.org/documents/adb-environmental-assessment-guidelines (accessed on 20 June 2014).

51. Clark, W.; Isherwood, W. Inner Mongolia must "leapfrog" the energy mistakes of the western developed nations. *Util. Policy* **2010**, *18*, 29–45.

52. Liu, J.; Qin, Y.; Liu, J.; Ding, J. Rich Revenue and Poor Residents: Top 100 County is Poor. Available online: http://politics.people.com.cn/GB/30178/4372944.html (accessed on 20 June 2014). (In Chinese)

53. Inner Mongolia Government. Inner Mongolia Government on Further Regulate Mining Development in order to Protect the Environment and People's Livelihood. Available online: http://www.nmgmt.gov.cn/xxgk_show.aspx?id=23166&classid=15 (accessed on 20 June 2014).

54. Kosich, D. Chinese Officials Hail Inner Mongolian Mining Crackdown a Success. Available online: http://www.mineweb.com/mineweb/content/en/mineweb-political-economy?oid=145681&sn=Detail (accessed on 20 June 2014).

55. Earth First. Chinese authorities squash protests against land grabs in Inner Mongolia. *Newswire* 15 April 2013.

56. Tang, H. China's Street Protests Won't Change Failing System. Available online: https://www.chinadialogue.net/article/show/single/en/5660-China-s-street-protests-won-t-change-failing-system (accessed on 22 October 2014).

57. State Environmental Protection Bureau. Office Letter (2006) #860: Technical Guidelines for EIA Coal Mine Master Plans (Draft), Attachment 2, 2006. Available online: http://www.zhb.gov.cn/info/gw/bgth/200612/t20061220_97529.htm (accessed on 20 June 2014).

58. EIA Public Participation Notification for Inner Mongolia Ordos Xinjie Mine Master Plan. Available online: http://www.zge.gov.cn/zwgk_1/gggs/201106/t20110622_352209.html (accessed on 22 October 2014). (In Chinese)

59. EIA Public Participation Notification for Inner Mongolia Ordos Wanli Mine Master Plan. Available online: http://www.zge.gov.cn/zwgk_1/gggs/201106/t20110622_352214.html (accessed on 22 October 2014). (In Chinese)

60. EIA Public Participation Notification for Inner Mongolia Ordos Hujiert Mine Master Plan. Available online: http://www.zge.gov.cn/zwgk_1/gggs/201106/t20110622_352217.html (accessed on 22 October 2014). (In Chinese)

61. Xian, H. EIA Public Participation Notification for Inner Mongolia Huaneng Group Yimin Mine (Hexi District) Master Plan. Available online: http://www.hlbrdaily.com.cn/news/2/html/4209.html (accessed on 20 June 2014).

62. Guo, J. EIA Public Participation Notification for Inner Mongolia Baiyanhua Mine Master Plan. Available online: http://www.dmlhq.gov.cn/html/41/20109/1554.html (accessed on 22 October 2014). (In Chinese)

63. Beijing Coal Group. EIA for Inner Mongolia Ordos Gaojialiang Mine Resource Consolidation. Available online: http://www.beijingcoal.com/edit/UploadFile/201131692232642.pdf (accessed on 23 June 2014).

64. Ministry of Environmental Protection of PRC (MEP). Technical Audit Points for Coal Mine Master Plan Environmental Impact Assessment (EIA) Report. Available online: http://www.china-eia.com/docs/2011-10/20111026090001616628.pdf (accessed on 23 June 2014).

65. Ordos Government. EIA Public Participation Notification for Inner Mongolia Huineng Coal Electricity Group Changtanlu Open-Pit Mine and Coal Preparation Plant. Available online: http://wap.ordos.gov.cn/xxgk/tggs/201209/t20120907_677682.html (accessed on 22 October 2014). (In Chinese)

66. Hulunbeir Government. EIA Public Participation Notification for Inner Mongolia Xinbaerhu Wuyi Ranch Mine Master Plan. Available online: http://blog.renren.com/share/233586009/15236078100 (accessed on 22 October 2014). (In Chinese)

67. Ordos Coal Bureau. Tongjiangchuan Mine EIA Report. Available online: http://www.ordosmt.gov.cn/gzcy/201212/P020121203388094345155.pdf (accessed on 22 October 2014). (In Chinese)

68. Inner Mongolia Coal Mine Distribution, 2013. Available online: http://zhidao.baidu.com/question/126121010.html (accessed on 22 October 2014). (In Chinese)

69. Kaiman, J. China: If You Were Xi Jinping. Available online: http://www.theguardian.com/world/2012/nov/08/china-xi-jinping-expert-view (accessed on 23 June 2014).

70. Morrison-Saunders, A.; Pope, J. Conceptualising and managing trade-offs in sustainability assessment. *Environ. Impact Assess. Rev.* **2013**, *38*, 54–63.

71. Gibson, R.B. Sustainability assessment and conflict resolution: Reaching agreement to proceed with the Voisey's Bay nickel mine. *J. Clean. Prod.* **2006**, *14*, 334–348.

72. Wang, J. President Xi Jinping on Environmental Protection and Development: Clean Water and Green Mountains Are Productivity. Available online: http://www.ce.cn/xwzx/gnsz/szyw/201408/15/t20140815_3360500.shtml (accessed on 28 August 2014). (In Chinese)

73. Li, J. China Stops Using GDP as Single Measure of Success for over 70 Counties/Cities. Available online: http://m.chinanews.com/s/gn/2014/08-13/1469090.htm (accessed on 28 August 2014). (In Chinese)

74. Wildau, G. Small Chinese Cities Steer Away from GDP as Measure of Success. Available online: http://www.ft.com/cms/s/0/a0288bd4-22b0-11e4-8dae-00144feabdc0.html#axzz3ANLOtNhh (accessed on 28 August 2014).

75. Kahya, D. Why the World's Biggest Coal Company Has Stopped Extracting Chinese Groundwater and What It Could Mean for China's Coal Use. Available online: http://www.greenpeace.org.uk/newsdesk/energy/news/why-world%E2%80%99s-biggest-coal-company-has-stopped-extracting-chinese-groundwater-and-what-it-could-mean-china%E2%80%99s- (accessed on 28 August 2014).

76. Cooke, B.; Kothari, U. *Participation: The New Tyranny?* Zed Books: London, UK, 2001.

77. Hickey, S.; Mohan, G. *Participation, from Tyranny to Transformation: Exploring New Approaches to Participation in Development*; Zed Books: London, UK, 2004.

Improving China's Environmental Performance through Adaptive Implementation—A Comparative Case Study of Cleaner Production in Hangzhou and Guiyang

Ting Guan, Dieter Grunow and Jianxing Yu

Abstract: This paper examines local policy implementation of Cleaner Production (CP) in China. As the major policy implementer, China's local government plays a crucial role in promoting CP. A better understanding of the factors affecting local government's incentives regarding CP and different strategies available to the local government can help policy makers and implementers improve CP practices and other environmental policy outcomes. This paper uses the cases of Hangzhou and Guiyang to demonstrate that local conditions of policy implementation have a direct impact on the success of CP promotion. Based on 35 in-depth interviews, statistical data and internal government reports, we find that the location-based incentives of local government strongly influence their implementation strategies; and that the choices of different strategies can bring out various policy results. From this study, the identified location-based incentives are affected by energy resource endowment, economic development stage and technological competence. The successful implementation strategies involve using different policy instruments synthetically, regulating CP service organizations by controlling their qualifications, differentiating CP subsidizations, and improving transparency of project progress and outcomes.

Reprinted from *Sustainability*. Cite as: Guan, T.; Grunow, D.; Yu, J. Improving China's Environmental Performance through Adaptive Implementation—A Comparative Case Study of Cleaner Production in Hangzhou and Guiyang. *Sustainability* **2014**, *6*, 8889-8908.

1. Introduction

Cleaner Production (CP) is considered as one of the most important preventive strategies to improve environmental performance of industries [1,2]. Generally, CP practices aim to reduce pollution and increase resource efficiency during industrial activities. As an attractive strategy for pollution prevention and cost-effective production, CP practice has been carried out for more than 20 years in many countries all over the world [3]. During the 1990s, the concept of CP was first introduced into China by the government and CP projects had been run under different programs for more than a decade before it was officially enacted by legislative authority [4]. The Cleaner Production Promotion Law of China (CPPLC) took effect on 1 January 2003 and marked the normative and legal management for CP practice, and it was further revised in March 2012 to update the new developments of CP regulations. This legislation of CP highlights China's green ambition on decoupling economic growth and its energy use from environmental deterioration. Nevertheless, the realization of these ambitions was, and still is, a great challenge for China [5]. How to improve policy implementation at the local level in order to promote CP practices is the key of the challenge.

One popular perspective to address how the CP implementation can be successfully achieved is to analyze the barriers of CP implementation focusing on the need of enterprises. A wide range of barriers have been identified, including the approaches of technology innovation, management strategies [1], market environment and organizational features [6–8]. Technology know-how and CP tools are insufficiently catered to the complexity of industries [9]; information dissemination and communications act as key roles in market promotion [10]; characteristics of firms influence CP outcomes [8]; internal organizational factors, including commitment, leadership, support, communication, staff involvement, and program design, affect CP practices [6,7]. Therefore, the success of CP practices is the result of breaking down the barriers of incentives and capacities of enterprises [8]. Consequently, a lot of CP policy recommendations focus on helping the entrepreneurs to break through the barriers and give detailed advice to public authorities on how to help those entrepreneurs. Nevertheless, many of the barriers-focused analyses are carried out in regions of relatively well-developed CP market environment, where entrepreneurs are the key players, while the government has little ambition or direct control on enforcing CP projects. The pitfall of such analysis is that it fails to consider the policy design and implementation under circumstances in which the CP market is not well-established and government exerts strong influence on CP promotion.

In current social economics of China where a sound free market for CP practices is lacking and government impact is significant, it is not surprising that the roles of state authority and governmental organization are more prominent in promoting CP programs and nurturing the formation of a CP program market [11]. Unlike other countries where CP was introduced as a voluntary participating program, the CP programs in China fall under voluntary and mandatory schemes, in which the latter is forcing designated enterprises to conduct CP projects [12]. Since the central CP policies are only promulgated with certain basic principles and general regulation, local governments become major promoters in enforcing CP implementation. They are given high flexibility in directing local CP policies. As a result, local CP actions turn out to be quite different from one region to another [13]. Specifically, under the same central policy, innovative strategy and active implementation are found in some local governments while passive implementers applying little effort exist in other locations.

Previous studies explored influential factors in local CP policy implementation in China from the approaches of objective barriers and policy "prescriptions". Many barriers exist in China's CP promotion procedure, such as the lack of co-benefit calculation demonstration [14], low awareness and misconception of CP, inadequate institutional framework, constraints in technological facilities and financial support, limited market of CP services, internal conflicts of implementing agencies, and so on [13,15–17]. In order to overcome these barriers, regional government could improve coordination between stakeholders, provide subsidies and increase capacity-building programs [17]. In addition, government could also ensure the continuity and rigorousness of CP projects [13] and encourage enterprises by calculating their project's profit-generating potential [13,18]. These arguments are mostly drawn from approaches of CP implementation barriers and corresponding policy recommendations. However, until now, little has been known about why and how the CP policies are reshaped at the local level, within China's authoritarian regime. Therefore, the purpose of this paper is to examine local CP policy implementation and to explore the reasons why different localities act differently and have varied CP outcomes under the same central policy—by analyzing incentives and actions of local government.

To identify the incentives and strategies of local governments in CP promotions, this paper will analyze China's governmental structure, local government's incentive structure and implementation strategies of CP policies. The question we want to demonstrate is, under similar institutional arrangements, how local governments' incentives and corresponding strategy choices influence local CP practices. We have selected Hangzhou and Guiyang as two comparable, yet quite different cases. They have similar institutional arrangements but different site-specific conditions, which lead to diversified implementing strategies by the local governments. The study is based on 35 interviews in Hangzhou, Guiyang and Beijing and a review of policy documents and statistical data, focusing on local implementing strategies and their outcomes.

The structure of the paper is as follows: after a brief review of the literature, an overview of CP implementation, including related actors, governmental structure and key process of CP policy will be given in Section 2. In Section 3 we will present the key concepts and the theoretical framework guiding the analysis, and in Section 4, we describe the methods and data used. In Sections 5 and 6, we present our analysis of the two city cases in detail. Finally, we will discuss the findings and conclusions in Section 7.

2. Overview of CP Implementation in China

Apart from central government's guidance and scrutiny, successful CP practices at the local level in China rely on the endeavors of an array of actors and stakeholders: local government, enterprises, CP service organizations, and others such as research institutions and media. Local governments are responsible for providing information, announcing regulations/legislations, allocating financial support (optional), and supervising that mandatory enterprises practice CP. Although the duty of CP initiation belongs to the local government, enterprises are the actual executive body for carrying out the various CP projects. As mentioned in Section 1, enterprises are either compulsorily or voluntarily implementing CP. According to Article 11 of CPPLC (2012), those enterprises with high energy consumption, overweight pollution or poisonous discharges are defined as targeted enterprises which must enroll in the mandatory scheme to undertake the CP project [19]. The third actor is known as "CP service organizations", such as energy service companies and intermediary organizations. They are involved because they provide a broad range of energy-saving and pollution-reducing solutions and consultancy, including specific designs and detailed implementation for individual CP projects. Research institutes also play important roles—in the development of CP technologies. In addition, public media have some influence in promoting local CP practices by raising public awareness.

Among all these actors, governments are the policy initiators of CP projects. The governmental structure regarding CP policy implementation involves a variety of government departments at multiple levels (Figure 1). At the national level, the program is supervised by the National Energy Commission (NEC) under the State Council, and the overall implementation of CP policy and its target is in the charge of the National Development and Reform Commission (NDRC) and Ministry of Environmental Protection (MEP). Following the allocated quota and plans from central government, local governments are directly involved in the individual CP project, including project promotion, auditing, and evaluation. Specifically, under the local leading group for Energy Saving and Emission Reduction, the Environmental Protection Bureau and the Economic Development and Trade Commission are mainly responsible for CP policy implementation. They organize and coordinate the overall promotion of CP and the Bureau of Planning, Science and Technology, Construction, Water Resource and Quality and Technical Supervision all work together in the accordance with their own responsibilities [19].

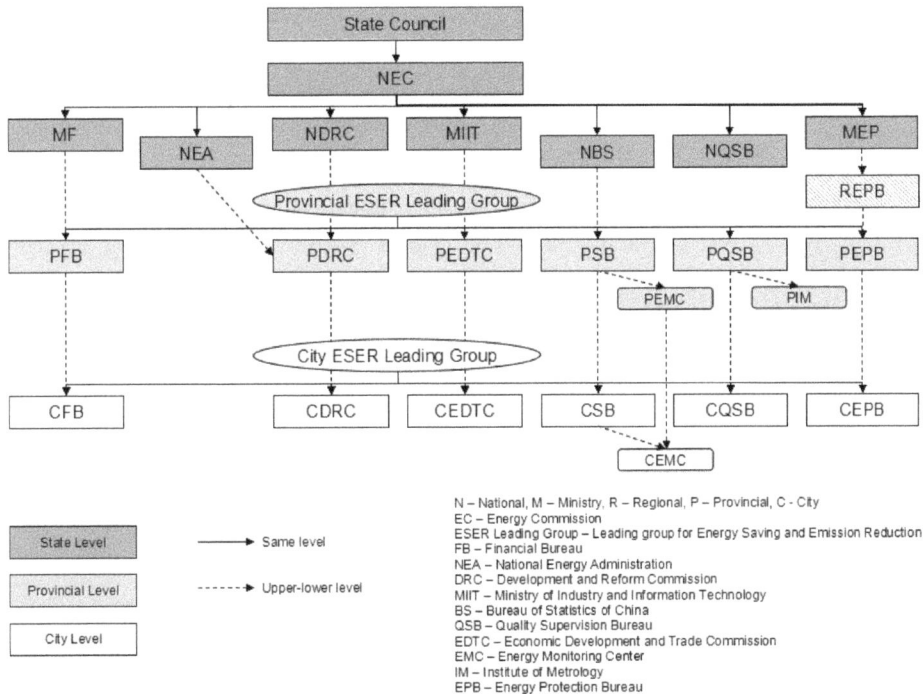

N – National, M – Ministry, R – Regional, P – Provincial, C - City
EC – Energy Commission
ESER Leading Group – Leading group for Energy Saving and Emission Reduction
FB – Financial Bureau
NEA – National Energy Administration
DRC – Development and Reform Commission
MIIT – Ministry of Industry and Information Technology
BS – Bureau of Statistics of China
QSB – Quality Supervision Bureau
EDTC – Economic Development and Trade Commission
EMC – Energy Monitoring Center
IM – Institute of Metrology
EPB – Energy Protection Bureau

Figure 1. Governmental structure of CP (Cleaner Production) policy implementation in China.

The strategy of local CP policy implementation varies in different areas; however, the implementation mode of mandatory CP projects has been formalized by CPPLC, which is composed of at least four stages (Appendix A). First of all, targeted enterprises are identified by the local government on the basis of the three conditions stated in the above paragraphs: those enterprises with high energy consumption, overweight pollution or poisonous discharges will be included. Then the list and basic information of these targeted enterprises is publicized in local newspapers or other influential media, to accept scrutiny from the public. These targeted enterprises, then, must conduct CP projects that are constituted by seven steps: planning and organizing, pre-assessment, assessment, development and screening measures, feasibility analysis, CP option implementation and sustaining CP [13]. Finally, these mandatory CP projects are reviewed and evaluated by an experts committee, which is organized by local governments.

Although this process description seems to indicate a rather directive implementation procedure, there is room for divergent CP promotion by local administration. Even for mandatory CP projects, local governments are given high discretionary powers to supervise the process, let alone the voluntary ones. Therefore, local governments have considerable decision-making power within their jurisdiction regions.

3. Framework for Analyzing Local CP Policy Implementation

Considering the high influence of local governments in CP promotion, city governments and bureaucracies are conceptualized as the "local governmental group", taken as the main study object in this paper. We propose *a priori* that local government's incentive is one of the most influential factors to the formations of local CP strategies; and varied strategies bring out different CP policy impacts and outcomes in various places. Adapted from conceptual framework by Oliver and Ortolano (2006) [17], which was designed for analyzing the implementation of the CP program, we focus the analytical framework on internal incentives and the corresponding implementation strategy of local government. Figure 2 shows our framework for an analysis of CP policy implementation at city level in Hangzhou and Guiyang. The actors of local government and targeted enterprises are involved. Overall, local governments must develop practical strategies for promoting local CPs to meet targets prescribed by higher authorities and by local interests. In order to achieve this, the cadres are expected to conduct policy experiments, create models, and advance policy innovations to promote CP [20].

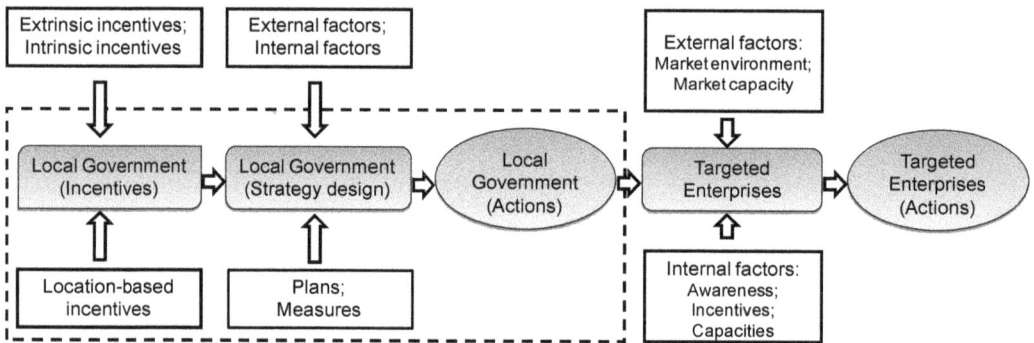

Figure 2. Conceptual framework for analyzing local CP policy implementation.

By referring to the work of Heberer and Schubert about strategic groups [21], the given incentives of developing local strategies can be categorized into extrinsic and intrinsic types: (1) Extrinsic incentives strengthen the motivation of local cadres to improve their performance. One key interest for local leading cadres is promotion within the administrative hierarchy [20]. In addition, social standing and prestige (face "mianzi"), remuneration, bonus, and threat of punishment and demotion all belong to this category [21]; (2) Intrinsic incentives indicate the local cadre's former position, socialization, and personal background. However, neither extrinsic nor intrinsic incentives can explain diverging patterns of local implementation strategy in a sufficient way [21]. Therefore, we define a third type of incentive: location-based incentive. These types of incentives are drawn from local conditions regarding geography, socio-economic development, culture and so on. This study examines which and how those site-specific conditional factors are affecting local CP promotion strategy and actions.

Another important aspect of local government implementation is the content of specific strategies for CP promotions. As mentioned above, central CP policies in China are formulated by enabling adaptive strategies, to allow more flexibility in local implementation. Paul Berman has argued that the local implementing strategy should be designed to match policy situations [22] (p. 2). Based on this concept, adaptive implementation, defined as "a process that allows policy to be modified, specified, and revised—In a word, adaptive—According to the unfolding interaction of the policy with its institutional setting" is considered to be crucial for promising policy outcomes [22] (p. 9). In terms of CP promotions, local implementation and initiative measures are developed and formalized by local cadres who are influenced by various external and internal factors. This investigation demonstrates CP promotion strategies of local government in Hangzhou and Guiyang, and analyzes the relationship between their strategies (*i.e.*, policy impact) and the CP policy outcomes.

The second set of actors in Figure 2 refers to enterprises that are listed in compulsory CP programs. Actions of these enterprises depend on external factors, including market environment and capacity; and internal factors, such as awareness, incentives and capacities within the organizations. As mentioned earlier, these enterprises are strongly influenced by local governments' encouragements and regulations. CP actions of enterprises are measured by their saved energy, reduced waste and pollution, and generated profit.

Based on the framework for analyzing local CP policy implementation in Figure 2, this study concentrates on (1) location-based incentives of local government; and (2), the adapted local strategies (policy impact) for promoting CP practices in Hangzhou and Guiyang, respectively. The aim is to conclude implementation modalities and their effects based on observation and analysis. Further implications for CP policy design in China will be drawn.

4. Methodology

By analyzing the incentives and strategies of CP promotion with a focus on the above dimensions in a case comparison of two cities (Hangzhou and Guiyang), we seek for explanations for the different implementation strategies and their outcomes of CP promotion at the city level.

Hangzhou and Guiyang have been chosen as case cities due to their similar local leaders' support, institutional arrangements, and different local site-specific conditions. Both Hangzhou and Guiyang have been ecological provincial capitals since the early 1990s, thus local leaders in both cities are supportive for environmental policies. Consequently, the extrinsic incentives of local cadres to implement CP policy are similar. Moreover, treating the government officers as a group, we ignore the intrinsic incentives of individuals. Furthermore, similar institutional arrangements are found in these two cities as described in Section 2. The third incentive for local government is related to site-specific conditions, which are different in these two cities: (a) As shown in Table 1, Hangzhou's economy is more advanced than that of Guiyang. Despite a strong economic development in both cities, GDP per capita in Hangzhou is always more than triple that of Guiyang; (b) Resource endowments are fundamentally different. Hangzhou is in an area with scarce energy resources and, therefore, heavily dependent on energy imports, while Guizhou province, where Guiyang is located, is a resource-abundant area. Guizhou produced 170 million tons of coal in 2012 alone and it is also famous for mineral products, such as bauxite and phosphate rocks; (c) The two cities are quite different in terms of technology development. Hangzhou is a national center of technology innovation; it has 24 national research institutes and well-known universities, and many of these research entities are involved in environmental and energy-saving research [23]. In contrast, technology development related to energy and the environment is lacking in Guiyang.

Table 1. GDP per Capita in Hangzhou and Guiyang.

Year		2005	2006	2007	2008	2009	2010	2011	2012
GDP per capita (RMB)	Hangzhou	44,871	51,908	61,315	70,948	74,761	86,691	101,370	111,758
	Guiyang	14,934	17,025	17,732	20,638	22,832	26,209	31,712	37,822

Note: (Resource: Hangzhou Statistic Yearbook 2006–2013; Guiyang Statistic Yearbook 2006–2013).

The incentives of local government and specific strategies to promote CP practices in these two cities are described and analyzed based on our analytical model, shown in Figure 2. The analysis focuses on the incentives and decision-making of the local government group in the dashed line box: (a) the influence of local conditions on government group incentives; (b) adapted strategies and measures in Hangzhou and Guiyang respectively. The overall outcome of CP implementation is measured by the numbers of CP projects, the number of enterprises involved, and the results of these CP projects. Particularly, the "success" and "failure" standard of CP implementations is measured by three aspects: (a) the number of CP projects; (b) the cost-benefits of CP projects; (c) the rate of "passing" (or "failing") rating by the expert-reviews of the CP projects. The analytical model identifies the influential aspects of local conditions and strategies used by local government with regard to these outcome measures.

The data and materials that we collected and analyzed are from government documents, publicly accessible, as well as internal government statistics, local newspaper articles, related documents from associations, and internal governmental reports. In addition, onsite fieldwork was conducted by the author: 35 governmental officers, experts, representatives of industrial companies, research institutes, and CP service organizations in Hangzhou and Guiyang were interviewed between October 2012 and April 2013 (Appendix B). The interviews were semi-structured and lasted for 1–2 h. The list of questions which have been covered in the interviews is shown in Appendix C.

5. Comparison of CP Policy Implementation in Hangzhou and Guiyang

The case comparison starts with a longitudinal overview of the development of implementation initiatives (impact) and results (outcome) in the two cities—followed by a case specific reconstruction of these differences (based on the interviews). CP had been introduced to Hangzhou and Guiyang since 2002 and 2006, respectively. About one decade since its launch, the effects (in terms of impact on local implementation arrangements) of CP promotions in these two cities varied a lot.

From 2002 to 2014, 57,422 projects with more than 2720 enterprises involved in Hangzhou have been conducted and the number of CP enterprises has soared dramatically in recent years [24]. (Some enterprises are calculated more than once as they conduct two or three rounds of CP projects). As shown in Table 2, the number of CP enterprises in Hangzhou has risen from 4 in 2002 to 322–542 per year in the latest three successive years. The total investment of CP projects in Hangzhou is a little short of RMB3 billion. And we are informed by the vice director from EDTC in Hangzhou that the local passing rate is relatively high, though the specific data was not available.

Table 2. Number of enterprises within CP program in Hangzhou and Guiyang.

Year		2002	2004	2007	2008	2009	2010	2011	2012	2013	2014
Number of CP	Hangzhou	4	16	239	264	314	228	184	542	322	356
Enterprises	Guiyang	/	/		20			4		17	

Note: Resources: data from [24,25].

In Guiyang, the first CP practices began in 2006 and only 1453 projects within 41 enterprises finished conducting CP projects until September 2014. Even considering the enterprises with still ongoing projects, there are only 55 more [25]. Moreover, CP projects in 19 enterprises failed to pass the expert-review due to site relocation, technology upgrade or closure [25]. The failing rate is roughly one third, excluding the ongoing CP projects. This high failing rate indicates that a high investment for CP projects is wasted as it produces no economic values. Such waste is a squeeze of the investment, because the effective investment is only roughly two thirds of total investment. The total direct investment is about RMB1.5 billion. Therefore, effective investment is only one billion [25]. Despite the relatively high investment in Guiyang, as the direct investment is about 55% of that in Hangzhou, the annual direct economic benefit in Guiyang is less than 5% of that compared to Hangzhou (Table 3). According to the longitudinal data, the annual economic benefits of RMB2400 million in Hangzhou in contrast to RMB173 million in Guiyang was produced, as summarized in Table 3 [24].

All these numbers indicate that the quantity and quality of CP projects in these two cities varied a lot. This also holds true for saved energy. Annual saved energy equals almost 450,000 tons of standard coal in Hangzhou (with its much higher GDP/capita), while it is not clear of the amount in Guiyang. No specific data was applied due to an unclear calculation method. Thus it is perhaps fair to conclude that CP policy implementation has received much better results (outcome) in Hangzhou compared to that in Guiyang.

5.1. CP Policy Implementation in Hangzhou

The eagerness of the Hangzhou government to promote energy saving actions started from 2003 to 2004, when Hangzhou suffered serious power shortages. During the summer time, when demand peaks (due to the hot weather), this resulted in the establishment of a temporary power quota for enterprises and power cuts for residential areas. In response, the Hangzhou government was forced to put its effort into vigorous measures to tackle such problems as they are detrimental to local economy development. A severe challenge faced by the government is that Hangzhou depends heavily on energy import, which is difficult to grow quickly enough in a short time period to meet soaring demands—since it needs corresponding infrastructure, such as electric grid and power plants. Improving energy efficiency seems to be the only practical choice. As a result of this, Hangzhou initiated CP practices even earlier than the respective decisions taken by the central government.

Table 3. CP policy outcomes in Hangzhou and Guiyang.

City	Hangzhou	Guiyang
Year	2002–2014	2006–2014
Number of Finished CP Projects	57,422 projects with more than 2720 enterprises involved	1453 projects with 41 enterprises involved
Direct Investment (Million RMB)	2852	1577
Annual Economic Benefits of the Project (Million RMB)	2400	173
Annual Saved Energy (Standard Coal Equivalent (Tons))	449,480	Not Identified
Total City Governmental Subsidy (Million RMB)	90	Not Identified

Note: (Resources: data from [24,25]).

Nevertheless, early impacts from CP were not promising, particularly the pilot projects. In 2002, only four projects were conducted in the whole province. During the initiating period, there is almost no CP market, mainly because of three challenges: (1) The technology support is limited because of insufficiency of intermediary organizations: in 2002, there was only one CP service organization under the Institute of Environmental Science Research & Design (ESRD) in the whole Zhejiang Province; (2) The cost of the project was too high: ESRD charged enterprises a minimum of RMB100,000 for each project. Without any government subsidies or direct profit generated from CP projects, the initial investment was too high and risky for enterprises to invest in CP, especially those small ones with limited resources; (3) There were very few examples at the beginning that benefited from successful CP projects in the region. Thus, investment in CP was not seen as rational strategies for the enterprises.

In response to these challenges, many measures were carried out. First, the quantity and quality of CP service organizations were improved by the Hangzhou government step-by-step. The first batch of CP service organizations were mainly persuaded by local government promotion. While more and more enterprises and service organizations got involved in CP practices [26], strict regulations on CP service organizations were issued by the Hangzhou government in order to promote quality of CP policy implementation. One of the basic regulations is that all the CP service organizations are required to sign a mandatory two-year contract with the Zhejiang Government. According to the contract, intermediary organizations must meet many strict qualification requirements [27]. For example, each institution must have more than six full-time national clean production auditors, including at least two auditors who have more than five years of work experience and are familiar with energy and environmental issues, one full-time professional and technical personnel. With these explicit regulations, the development of CP service organizations was getting better gradually.

Besides, additional subsidies for CP projects were provided by the Hangzhou government at an early stage as incentives to encourage enterprises to join in CP practices. Later more detailed subsidy regulations were issued to regulate subsidy allocations more efficiently. From 2002, CP service organizations in the Energy Conservation Association were persuaded by the Hangzhou government to set the benchmark price as RMB50,000 for each project. As a result, in 2004, all of the CP service organizations, including the one under ESRD, reduced the minimum price of a CP project to RMB50,000. Meanwhile, the RMB50,000 subsidy for each CP project is provided by the Hangzhou government.

With successful initiation of CP, the Hangzhou government decided to change the way of subsidy allocation in order to promote the quality of CP policy implementation—starting from 2011. Specifically, the companies that passed the Clean Production Audit are divided into different groups according to the scores (remarks) they get during the audit process, which are allocated by an expert-review system. Projects with scores higher than 95 (included) will be rewarded RMB50,000; those with scores higher than 90 (90–94) will be rewarded RMB40,000; and for those marked from 85 to 90 will get RMB30,000. No rewards will be given if scores of projects are below 85. Consequently, the quality of CP projects has greatly improved ever since.

As a result of the measures above, since 2004, the enterprises were able to carry out CP projects under almost no cost because of the regulated price and government subsidies were the same, *i.e.*, RMB50,000. Enterprises started to be attracted to participate in CP practices, since the possibility of energy savings and waste recycling during the production process essentially lowers the production cost, thus increases the profit. As more and more enterprises involved into CP and got benefits from those projects, it became obvious that CP was economically a good practice for enterprises. Owing to these adaptive implementing measures of the Hangzhou's government, more and more enterprises got enrolled into this program. Even after 2011, when government subsidies became more restricted, the rising trend of enterprises that are willing to join in CP continues, indicating the popularity of CP practices among entrepreneurs. Until 2012, the number of enterprises soared to 542, ten times higher than 10 years ago, when CP was first initiated.

Based on our field work, two key aspects are prominent during CP implementation in Hangzhou: (1) the high internal incentives of CP promotion from the Hangzhou government and (2) well-adaptive implementing strategy executed by the government to promote CP projects. In spite of the exceptional results from CP practices, however, during our interviews, it was mentioned that many other problems still exist. For example, some interviewees reflect that CP service organizations are still behaving quite differently: some companies finish their projects carefully while others tend to temporize. In addition, fake companies that target the government's subsidies by fabricating CP projects with enterprises still exist.

5.2. CP Policy Implementation in Guiyang

The internal incentives of the Guiyang government to promote CP projects are vague at best. Due to easy access to energy and other natural resources, combined with a laggard economy, there is neither need nor the availability of financial resources for the Guiyang government to support CP. Consequently, the awareness of CP promotion among local government is low. Besides very few "Green-Promotion" leaders, most of the other cadres are much more in favor of "Economic-Promotion". During our local interviews, we found that the Guiyang officers were, in a way, lacking incentives to implement CP policies. "We are already the poorest province in China and we have fallen extremely behind the national average level", said a senior leader in DRC, "Honestly, we are desperately hungry for GDP, because our current (economic) condition is quite different from east (China)". When we mentioned that CP can also help the enterprises have profitable and efficient development patterns, an officer from the Financial Bureau told us: "These kinds of programs are all long-term work. Because of current cadre turnover and cadre evaluation system, it is crucial for local leaders to have deliverable political accomplishments as soon as possible". Similar views were recorded repeatedly during our fieldwork.

Nevertheless, with the law and policies enforced by the central government, CP promotion has to be tucked into local government agendas, regardless of the attitude of the Guiyang government. During the implementation of CP policy, Guiyang faces similar challenges as Hangzhou. First, Guiyang lacks sophisticated CP service organizations. The current situation is: (a) there are few existing qualified research institutes in Guiyang for CP; (b) newly-founded high-tech service organizations are facing many obstacles, such as insufficient government financial support, limited land, and other facilities; however, little willingness was shown by the Guiyang government to promote such service organizations because of it being such a daunting task that delivers no direct economic accomplishment. So far, neither specific local regulations nor provisional ones have been applied to tackle this issue. In fact, most of the completed CP projects in Guiyang are supported by institutes directly appointed by the local government or by the research and development centers of the enterprises themselves. For example, for one of the top resources-consuming enterprises, China Aluminum Corporation, their manufacturer only trusts and relies on its own research institute for CP technology innovation instead of service centers in Guiyang. As a local officer mentioned, "there are not many CP service organizations and most of them are of quite low quality. So they are hardly trusted by local enterprises. These enterprises also fear the leaking of their energy consumption data when taking part in CP, which are widely considered as commercial sensitive information. The lack of trust in the system is a fundamental problem".

Second, active participation in the CP of local enterprises is rare because of the high cost of the CP project. The deadlock of new CP promotion at the initial stage was broken by strong government willingness and measures in Hangzhou. In contrast in Guiyang, hardly any effort was made to lower the cost of CP projects or provide strong financial support for CPP projects. Based on our review, Guiyang had no special financial support for CP until 2012. Enterprises can only conduct the CP projects by applying for subsidies under other categories, such as the technological upgrading subsidy and the technical innovation subsidy, the special subsidy for small enterprise, tax relief and so on. These subsidies are not easy to apply for because they target a wide spectrum of projects and the overall budgets are relatively limited, which means it is very competitive to get funded. Furthermore, applicants can hardly get the approved subsidies on time. "Part of the problem is that enterprises themselves find it is hard to apply for government funds". Even if they get the fund, they cannot get money on time for various reasons, especially for local special funds, "for example, some of our energy saving projects that were approved last year (2010), has not been fully funded until now and it is November 2011 already", said an officer in district-level EDTC.

6. Summary of Case Comparison: Location-Based Incentives and Implementing Strategies

The outcomes of CP policy implementations in Hangzhou and Guiyang are quite different: the former enjoys the continuing boost of CP participators, while the latter is still struggling with limited pilot projects. Many initiatives and active measures were found in Hangzhou, but Guiyang implemented the policy passively and reluctantly. The reasons for these differences can be interpreted from two aspects: (1) incentives for CP promotions are much higher in Hangzhou, partly due to Hangzhou's site conditions; (2) local implementing strategies are better in adjusting central policies into local conditions, in one word, being more "adaptive", in Hangzhou than that in Guiyang.

With regard to incentives of local government for implementing CP, as mentioned in Section 3, the extrinsic incentives are almost the same for Hangzhou and Guiyang, and the intrinsic incentives can be minimized due conceptualizing local cadres as a group. The biggest difference between Hangzhou and Guiyang are location-based incentives, affected by varying site conditions. Particularly, based on the analysis of Section 5, the incentives of the local governments are affected by the following conditional factors: resource endowment, economic stage, and technology development.

First, resource endowment appears to have a direct impact on the priority of CP promotion on the local government agenda. After experiencing the severe power cut and limitation in 2003, the Hangzhou government had sufficiently realized the significance and urgency to deal with the power shortage in order to secure economic development. Promoting energy saving by CP is, therefore, a feasible choice for local policy makers. The scarce resource endowment became an internal driving force for the local government to actively promote CP practices. In contrast, richness in energy resources allows high energy consuming industries to grow fast continuously in Guiyang. Promoting CP may potentially impede the expansion of the energy-dependent economic sector in the short term, and slow down the economic growth. Therefore, it is comprehensible that, to some extent, resource abundant regions possibly lack incentives for CP promotion.

Second, the economic stage and government fiscal condition limit the options of local governments to subsidize CP projects, which further affect their incentive to promote these projects. A detailed discussion of the fiscal structure of these two cities is beyond the scope of this paper. Nevertheless, a close look at the overall governmental fiscal situation shows that Hangzhou has a much better fiscal position to support environmental policies compared to Guiyang. In 2012, the Hangzhou government (including city and county levels) had revenue of RMB86 billion against the expenditure of RMB78.63 billion, enjoying RMB8 billion surpluses [28]. In the same fiscal year, the Guiyang government revenue was RMB24.1 billion, while the expenditure was RMB35.1 billion, which indicates RMB9 billion deficits [29]. Therefore, allocating an extra budget for CP is an easy decision for the Hangzhou government because of the sufficient fiscal budget. While in Guiyang, increasing the financial support for CP projects means either budget cuts for projects of other purposes or more deficits for the local government; both options are not attractive to Guiyang government. As is shown in Table 3, the Hangzhou government allocates 90 million for CP promotion while almost no special CP fund was found in Guiyang. Financial support for CP in Guiyang is all subsumed under "other fund" title and it is very hard to specify its amount.

Third, the development of technology and service organizations is also a key element to secure the success of CP. The high reputational group of people with CP expertise in Hangzhou is the backbone of CP service organizations and the expert committee. They are not only facilitating the accomplishment of an individual CP project but also ensuring high quality of the project. The best-practice cases, therefore, are more likely to gain a good reputation among entrepreneurs by generating economic values and drawing in further participants. The lack of technology development in Guiyang results in an inability to carry out high quality CP projects and this probably is part of the reason for the high failing rate and low economic benefits of CP projects. Consequently, CP lost government trust as well as entrepreneur's support, which hinders its promotion.

Besides these location-based incentives of local government, the implementing strategies that are used at the local level are also fundamentally important for CP promotion. This study points to the conclusion that initiatives to CP promotion in Hangzhou and Guiyang are both processes of adaptive implementation. However, Hangzhou appears to have done a better job in policy redesigning and implementation, better customized central policy in local context, therefore leading to a better policy outcome, while Guiyang tends to implement passively, with little deliberateness in appropriating CP decisions into local conditions. The strategy of well-adaptive implementation drawn from the case of Hangzhou is summarized below.

A typical strategy of the Hangzhou government is to utilize a bunch of policy instruments, such as the command and control instrument, economic instrument, and communication instrument, in CP promotion. As mentioned in Section 2, CP is mainly promoted by China's government through a hierarchal system. Therefore, in Hangzhou, annual CP targets were set and contracts were signed between government and targeted enterprises. Since these contracts were placed into the responsibility of enterprises' leaders and local cadres, they have to commit themselves to achieve CP targets, otherwise their promotions will be affected. Along with these "sticks," in order to kick off this new program, Hangzhou also provides a full subsidy for targeted enterprises. In addition, energy service organizations were persuaded to join the CP program. These multi-pronged efforts have successfully raised local CP awareness and stimulated CP practices.

Furthermore, recognizing many constrains in the CP market, detailed regulations for CP service organizations were issued to overcome CP market barriers. Since 2008, a two-year contract needs to be signed by all the CP service organizations with the Hangzhou government. The contract stipulates that all the CP service organizations must accept expert review and meet CP service requirements. Only with this contract can CP service organizations participate in a CP project; however, no regulations like this were identified in Guiyang. Although it also suffered from insufficiency and low-competence of CP service organizations, rare progress was made by the local government. The relative low competence of CP service organizations in Guiyang damages the reputation of CP among enterprises.

With more and more enterprises involved, differentiated CP subsidizations were introduced to Hangzhou's CP subsidy system. Since 2011, stricter regulations were issued and different subsidies were provided for CP enterprises, based on the scores given by expert reviews of conducted projects.

In order to get higher subsidies from the government, targeted enterprises are encouraged to improve the quality of their projects. By introducing an expert review system, a fair and open competitive platform was provided for enterprises. Varied subsidization further promoted enterprises' enthusiasm for conducting CP projects.

Another pillar for improving CP practices is increasing transparency of project information. Although CP information publication has been claimed by the central government, local implementation seems to be quite different. For example, most CP information can be easily found on the official website in Hangzhou and subjected to public scrutiny, while much less information can be observed in Guiyang via similar channels. Some CP news in Guiyang was released, yet in a rather random and unsystematic way. Thus, it is perhaps difficult to inform and stimulate local enterprises' actions in Guiyang. Lack of transparency may create additional barriers for local enterprises to participate in CP. In Hangzhou, a fair and open process to distribute the funding gives more motivation to enterprises to ensure a high quality project. Contrastingly, the black box funding distribution in Guiyang hampered the enthusiasm of local enterprises.

7. Conclusions and Discussions

CP has been promoted in many countries and deemed as a powerful strategy to improve environmental performance. Unlike the voluntary-based CP program in western countries, China's government is the most important initiator and promoter for CP practices. Because of heavy intervention of governments, decision-making and execution of China's government is crucial for CP promotions. As local governments have enough flexibility to implement CP policies, they are apparently playing pivotal roles in initiating local CP practices. In order to uncover the influence of local government to CP implementations, this study takes the incentives and strategies of local governments as the main objectives and develops a conceptual framework for analyzing local CP policy implementation in China.

The results from this study suggest that unbalanced implementation of CP at the local level remains a challenge for policy makers. The differences of policy impacts and outcomes are highly dependent on local governments' motivations and implementing strategies. As mentioned previously, the incentives of local cadres include three types: intrinsic incentives, extrinsic incentives and location-based incentives. Due to existing similarities between Hangzhou and Guiyang, this study allows us to focus on the location-based incentives, and we find that this type of incentive is mainly affected by three conditional factors that are different from one region to another: resource endowment, economic stage, and technology development. Local government is more likely to be incentivized to implement CP policies innovatively under the following conditions: (a) local resource endowment is scarce; (b) the economy is relatively well-developed and the local government financial situation is strong; and (c) technology providers in the CP- related environment and energy domain are technologically competent.

With regard to CP policy implementation, it is fair to conclude that CP projects in China are mostly promoted in a command and control way. In both Hangzhou and Guiyang, the progress of CP implementation, whether good or not, is under the relevant control of local government as they involve virtually all aspects of CP from setting target enterprises to evaluating the results. As an encouraging progress, market-base mechanisms are increasingly used in CP practices, but still in a rather controlled manner. Therefore, it is not surprising that this study concludes the different implementing strategies of local government tend to produce completely different results. This investigation has revealed a number of strategies that contributed to the success of local CP promotion: (a) different policy instruments are used in a combined way for promoting CP practices; (b) detailed regulations of local CP service organizations are required for building CP market capacity; (c) differentiated CP subsidizations, based on the results of expert reviews, are more likely to promote high quality CP practices; and (d) high transparency of project information may promote activities of targeted enterprises towards CP.

Although some insight into CP implementation is drawn based on the two cases of Hangzhou and Guiyang, it is worthy to emphasize that there is no universally best way to implement the CP policy: implementation can only be effective if the policies can be and were adjusted to suit local conditions [22]. In China, under the same institutional arrangement, central policies are implemented in all 2862 Chinese counties with diversified local site conditions. As a result, to adapt central policies into local conditions, proper implementing strategies of modification, specification and revision are indispensable. In order to improve CP practices effectively, well-adaptive implementation at the local level is necessary.

As is known that a top-down approach is frequently adopted in China's context: some experts and policy makers have sought to improve implementation quality by passing cleaner and stricter central policies [30]; however, the practice of CP promotion in China shows a different scenario: guided by central CP policies, local governments are given space and flexibility to develop local strategies that may meet both the central government's target and local interests. It has been noticed that as legislation gets stricter and/or resources invested into innovative strategies are very limited and conditioned, the implementers find it more difficult to adapt it to the different local interests involved [30]. Therefore, certain flexibility should be incorporated into policy design on the central level for more effective policy implementation, even if that means missing the policy targets in areas where the policy doesn't fit local interests. As Berman and McLaughlin conclude, excessive control can lead deliverers to follow guidelines only symbolically [31]. For future studies, it would be interesting to assess the needed extent of the "strictness" or "discretion" of CP central policies, in order to ensure both "discretion" and "fidelity" of local adaptive implementation. Another remaining aspect is how to build CP capacity in enterprises, service organizations and research centers in China. As reviewed at the beginning, capacity building of these actors is considered as an important requirement for CP promotion [1,8]. The factors which contribute to the capacity development of these actors should be addressed by future research.

Acknowledgments

The authors would like to thank the German Federal Ministry of Education and Research (BMBF) for funding Ting Guan's research visit in Germany under Green Talent Program as an initiative of this study. The research is funded by the School of Public Affairs of Zhejiang University and Zhejiang Province Institute of metrology (Project No. 2013C25105). For invaluable comments and suggestions, we are indebted to Jørgen Delman, Thomas Heberer, Sujian Guo and Xiang Gao. We are also thankful for feedbacks and comments during the Association of Asian Studies Annul Conference, San Diego, CA, USA, 2013 and the conference of Asian Dynamics Initiative, Copenhagen, Denmark, 2013.

Author Contributions

All authors contributed to the research design. Jianxing Yu provided network support for conducting interviews; Ting Guan and Jianxing Yu performed interviews and document collections in Hangzhou and Guiyang. With the advices of Dieter Grunow and Jianxing Yu, exploratory analysis was conducted by Ting Guan. All authors conceptualized the line of argumentation in the manuscript. Ting Guan wrote the manuscript. All authors have read and approved the final manuscript.

Appendix A

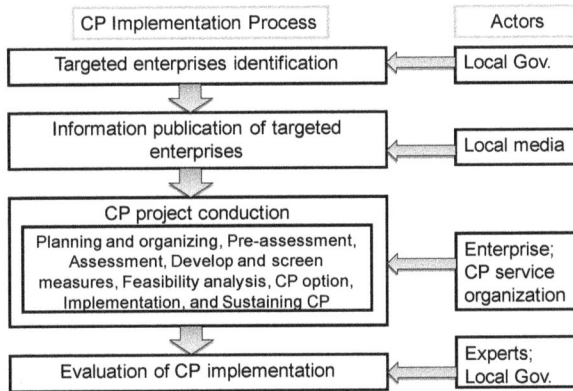

Figure A1. Local mandatory CP policy implementation and related actors. Note: (Resource from [13,19]).

Appendix B

Table A1. Interview List.

No.	Interviewee	Type of Organization	Title	Date
iH01	C	Research Institute (Zhejiang Province Institute of Metrology)	Vice president	21 September 2012
iH02	W	Research institute (Institute for Thermal Power Engineering of Zhejiang University)	Researcher	18 September 2012
iH03	Q	Hangzhou Government (Development and Reform Commission)	Vice Director	24 September 2012
iH04	L	Hangzhou Government (Economic Development and Trade Commission)	Vice Director	26 September 2012
iH05	S	Research institute (Institute for Thermal Power Engineering of Zhejiang University)	Researcher	29 September 2012

Table A1. *Cont.*

No.	Interviewee	Type of Organization	Title	Date
iH06	S	Energy service institute (Energy Saving Association)	Professor; Energy expert	15 October 2012
iH07	S	Energy service institute (Energy Saving Association)	President	16 October 2012
iH08	L	Hangzhou Government (Economic Development and Trade Commission)	Vice Director	22 October 2012
iH09	X	Zhejiang Provincial Government (Energy Auditing Team)	Chief Engineer	25 October 2012
iH10	H	Energy service institute (CP Energy Evaluation Centre)	Researcher	29 October 2012
iH11	L	Hangzhou Energy service institute (Energy Saving Association)	Former president	5 November 2012
iH12	H	Hangzhou Energy service institute (Zheda CP center)	President	14 November 2012
iH13	H	Hangzhou Energy service institute (Zheda CP center)	Consultant	15 November 2012
iH14	X	Company (Zhejiang Diyuannengyuanhuanjing company)	Manager	25 February 2013
iH15	S	Company (Fenghuangcheng Company)	Manager	25 February 2013
iH16	W	University (Zhejiang Unviersity)	Professor	20 March 2013
iG01	W	Guizhou Provinvial Government (Development and Reform Commission)	Vice Director	31 October 2012
iG02	Z	Guiyang Government (Development and Reform Commission: Resource Conservation and Environmental Protection Bureau)	Director	31 October 2012
iG03	H	Univeristy (Guizhou Party School)	Professor	1 November 2012
iG04	X	Guiyang Government (Industry and Information Technology Commission: Energy Saving Bureau)	Officer	1 November 2012
iG05	Z	Guiyang Government (Industry and Information Technology Commission: Energy Saving Bureau)	Director	1 November 2012
iG06	H	Guiyang Government (Comprehensive demonstration office for energy saving and emission reduction)	Director	2 November 2012
iG07	L	Enterprise (Guiyang aluminium Enterprise)	Manager	2 November 2012
iG08	W	Enterprise (Guiyang aluminium Enterprise)	Manager	2 November 2012
iB01	L	Central government (National Development and Reform Commission: Resource Conservation and Environmental Protection Department)	Vice Director	3 September 2012
iB02	W	Central government (Ministry of Industry and Information Technology: Energy Saving Department)	Director	8 October 2012

Table A1. *Cont.*

No.	Interviewee	Type of Organization	Title	Date
iB03	L	Central government (National Development and Reform Commission: Resource Conservation and Environmental Protection Department)	Director	8 October 2012
iB04	X	Central government (National Energy Saving Center)	Chairman	8 October 2012
iB05	Z	Central government (National Energy Saving Center: Energy Saving Department)	Vice Director	8 October 2012
iB06	S	Central government (National Energy Saving Center: General Office)	Vice Director	8 October 2012
iB07	Z	Research institute(National Energy Research Institute)	Researcher	9 October 2012
iB08	D	Energy service company (Energy Efficiency Project Investment Company)	Chairman and CEO	10 October 2012
iB09	H	Central government (State Council)	Department Chief	5 April 2013
iB10	L	Central government (State Council; Energy Saving Department)	Vice Director	5 April 2013
iB11	C	China National School of Administration	Professor	10 April 2013

Appendix C. Questionnaire Outline

Questionnaires are designed to be different for varied individuals, depending on the background of interviewees. Most questions are addressed in the following way:

(1) Background of the interviewees, such as positions, educational/working experiences and duties.
(2) The significance of involvement of their work in local CP promotion: How important is their duties in local CP implementation?
(3) The organizational arrangement: Which organization is responsible for monitoring all of the programs/projects? How is it done?
(4) Challenges and opportunities of the organizations in which the interviewees worked.
(5) The positive/negative factors that affect the outcome of implementation of CP programs.
(6) Specific measures that are adopted in local CP promotion: What is the status of local CP implementation and what are the results of the CP implementation so far?
(7) The evaluation of local CP implementation: are the interviewees satisfied with the way they have been implemented?
(8) Implications of CP policy implementation: What are the major lessons learnt so far from implementing?
(9) Implications of CP policy implementation: What are the challenges of opportunities of local CP policy implementation in the future?

Conflicts of Interest

The authors declare no conflict of interest.

References

1. Baas, L. To make zero emissions technologies and strategies become a reality, the lessons learned of cleaner production dissemination have to be known. *J. Clean. Product.* **2007**, *15*, 1205–1216.

2. Van Hoof, B. Organizational learning in cleaner production among Mexican supply networks. *J. Clean. Product.* **2014**, *64*, 115–124.

3. Kjaerheim, G. Cleaner production and sustainability. *J. Clean. Product.* **2005**, *13*, 329–339.

4. Ning, D.; Yan-ying, B.; Xiu-ling, Y.; Jie, Y.; Dan-na, S. Analysis on Cleaner Production policy and its results in China. In Proceedings of the 2nd International Workshop Advances in Cleaner Production, São Paulo, Brazil, 20–22 May 2009.

5. Xue, B.; Mitchell, B.; Geng, Y.; Ren, W.; Müller, K.; Ma, Z.; Puppim de Oliveira, J.A.; Fujita, T.; Tobias, M. A review on China's pollutant emissions reduction assessment. *Ecol. Indic.* **2014**, *38*, 272–278.

6. Stone, L.J. Limitations of cleaner production programmes as organisational change agents. I. Achieving commitment and on-going improvement. *J. Clean. Product.* **2006**, *14*, 1–14.

7. Stone, L.J. Limitations of cleaner production programmes as organisational change agents. II. Leadership, support, communication, involvement and programme design. *J. Clean. Product.* **2006**, *14*, 15–30.

8. Van Hoof, B.; Lyon, T.P. Cleaner production in small firms taking part in Mexico's Sustainable Supplier Program. *J. Clean. Product.* **2013**, *41*, 270–282.

9. Van Berkel, R. Cleaner production and eco-efficiency initiatives in Western Australia 1996–2004. *J. Clean. Product.* **2007**, *15*, 741–755.

10. Rowley, J. Promotion and marketing communications in the information marketplace. *Libr. Rev.* **1998**, *47*, 383–387.

11. Mol, A.P. Environment and modernity in transitional China: frontiers of ecological modernization. *Dev. Chang.* **2006**, *37*, 29–56.

12. Dan, Z.; Yu, X.; Yin, J.; Bai, Y.; Song, D.; Duan, N. An analysis of the original driving forces behind the promotion of compulsory cleaner production assessment in key enterprises of China. *J. Clean. Product.* **2013**, *46*, 8–14.

13. Hicks, C.; Dietmar, R. Improving cleaner production through the application of environmental management tools in China. *J. Clean. Product.* **2007**, *15*, 395–408.

14. Mestl, H.E.; Aunan, K.; Fang, J.; Seip, H.M.; Skjelvik, J.M.; Vennemo, H. Cleaner production as climate investment—Integrated assessment in Taiyuan City, China. *J. Clean. Product.* **2005**, *13*, 57–70.

15. Zhang, T.Z. Policy mechanisms to promote cleaner production in China. *J. Environ. Sci. Health Part A* **2000**, *35*, 1989–1994.

16. Fang, Y.; Côté, R.P. Towards sustainability: Objectives, strategies and barriers for cleaner production in China. *Int. J. Sustain. Dev. World Ecol.* **2005**, *12*, 443–460.
17. He Oliver, H.; Ortolano, L. Implementing cleaner production programmes in Changzhou and Nantong, Jiangsu province. *Dev. Chang.* **2006**, *37*, 99–120.
18. Van der Tak, C. The Need for Creating Framework Conditions for Cleaner Production. In Proceedings of the International Conference on Cleaner Production, Beijing, China, September 2001.
19. National People's Congress of China (NPC), Cleaner Production Promotion Law, 2012. Available online: http://www.china.com.cn/policy/txt/2012-03/01/content_24769874.htm (accessed on 19 November 2014).
20. Heberer, T.; Senz, A. Streamlining local behaviour through communication, incentives and control: A case study of local environmental policies in China. *J. Curr. Chin. Aff.* **2011**, *40*, 77–112.
21. Heberer, T.; Schubert, G. County and Township Cadres as a Strategic Group. A New Approach to Political Agency in China's Local State. *J. Chin. Polit. Sci.* **2012**, *17*, 221–249.
22. Berman, P. *Designing Implementation to Match Policy Situation: A Contingency Analysis of Programmed and Adaptive Implementation*; Rand Corporation: Santa Monica, CA, USA, 1978.
23. Wang, J.J. 24 National Research Institutes in Hangzhou. Available online: http://news.sina.com.cn/c/2009-11-03/045416542232s.shtml (accessed on 26 November 2014).
24. Hangzhou Government. Most Data Can also Be Retreived from Official Website of Hangzhou Economic and Information Commission. Available online: http://www.hzjxw.gov.cn/hz/web/Info.asp?TypeInfoID=004&TypeID=4&FileID=100 (accessed on 19 November 2014).
25. Guiyang Government. Some Data Can also Be Retreived from Local News. Avalibale online: http://www.ghb.gov.cn/doc/201465/997424092.html (accessed on 24 November 2014).
26. Note of the Organizing Enterprise Reporting Cleaner Production Audits and Electrical Balance Test Grant Funding. Available online: http://www.hedajfj.gov.cn/cms/list!detail.action?nav2.id=50&info.id=1755 (accessed on 19 November 2014).
27. Zhejiang Provincial Government. *Clean Production Audit Interim Measures of Zhejiang*; Zhejiang Provincial Government: Zhejiang, China, 2003.
28. Bao, Y. Hangzhou 2012 Fiscal Expenditure is RMB78.628 Billion Yuan with Livelihood Expenditures Accounted for More Than 70%. Available online: http://hangzhou.zjol.com.cn/hangzhou/system/2013/01/28/019118613.shtml (accessed on 19 November 2014).
29. Guiyang Government of Financial Bureau. *Guiyang City Budget Implementation in 2012 and Executive Budget for 2013*; Guiyang Government of Financial Bureau: Guiyang, China, 2013. Available online: http://www.gygov.gov.cn/art/2013/3/28/art_10753_417668.html (accessed on 28 November 2014). (In Chinese)
30. Van Rooij, B. Implementation of Chinese environmental law: Regular enforcement and political campaigns. *Dev. Chang.* **2006**, *37*, 57–74.
31. Berman, P.; McLaughlin, M.W. *Federal Programs Supporting Educational Change: Implementing and Sustaining Innovations*; Rand Corporation: Santa Monica, CA, USA, 1978.

Environmental Legislation in China: Achievements, Challenges and Trends

Zhilin Mu, Shuchun Bu and Bing Xue

Abstract: Compared to the environmental legislation of many developed countries, China's environmental legislation was initiated late, beginning in 1979, but nevertheless has obtained considerable achievements. As many as thirty environmental laws have provided rules regarding prevention and control of pollution, resource utilization, and ecological protection in China. However, China's environmental legislation still faces a series of challenges and problems, including that the sustainable development concept has not yet been fully implemented, as well as presence of gaps and non-coordination phenomena between laws and regulations, unclear responsibility, imperfect system design, imbalance between rights and obligations, higher impacts resulted from the GDP-centralized economy, lack of operability and instruments in the legal content, as well as difficulty of public participation. In contrast, China's environmental legislation has improved, as a result of learning from experience in developed countries and introducing innovations stimulated by domestic environmental pressure. Looking into the future, increased attention to environmental protection and ecological consciousness paid by China's new leaders will bring a valuable opportunity to China's further development concerning environmental legislation. In the future, there are prospects for the gradual improvement of legal approaches, continuous improvements of legislation to mitigate environmental problems, and more opportunities to strengthen public participation can be predicted.

Reprinted from *Sustainability*. Cite as: Mu, Z.; Bu, S.; Xue, B. Environmental Legislation in China: Achievements, Challenges and Trends. *Sustainability* **2014**, *6*, 8967-8979.

1. Introduction

Environmental legislation has become an important component of domestic and international legislation throughout the world. From a global perspective, environmental legislation dates back to the 18th Century, or even earlier. European countries, such as Finland, France and Germany, had passed acts with the purpose of improving natural resources conservation [1] (pp. 9–10); [2] (pp. 40–41). In the 1870s, the rise of nature conservation ideology prompted the creation of national park and forest protection laws in the United States [3] (pp. 10–18). Since the 1960s, environmental pollution and related incidents gradually increased in many nations, along with rapid industrialization and urbanization development, requiring national and state governments to deal with environmental pollution. Therefore, environmental legislation emerged largely in Europe and other developed countries to control pollution of air, water and land. Typical examples are the *National Environmental Policy Act* of the United States in 1970 and the *Basic Law for Environmental Pollution* of Japan in 1967 [1,3,4]. With the proposed concept of sustainable development, being popularized in the late 1980s, environmental legislation in developed countries began to shift, and has changed from focusing on pollution and treatment to prevention and a holistic approach to the whole process of management of natural resources development and utilization. Moreover, an accompanying shift has also been to focus on national legislation to address deals with the international common environmental issues through legislation [5] (p. 199); [6] (pp. 75–77). Today, legislation has become sophisticated and comprehensive, even with acknowledged limitations in applying it effectively. At the same time, environmental legislation in developing countries has also been improving step by step [7].

In China, the start of environmental legislation came late compared to many other nations. The first special environmental law was passed in 1979. With the rapid economic development after the reform and opening up in China beginning in the early 1970s, environmental pollution problems became prominent. Consequently, initial legislation focused upon environmental pollution control reaching a climax in the 1990s. Entering the new century, the bottlenecks and contradictions between resources development and environmental problems were apparent, and related to the long-term, accumulated development in China. Along with the concept of the Scientific Outlook on Development, the legislation on resource protection and management has been steadily expanded and strengthened, but unfortunately the legal quantity was not as satisfactory as the increasing quality, which means, legal enforcement effectiveness needs to be improved continuously.

2. Retrospect of China's Environmental Legislation

2.1. Achievements

As already noted, China's first environmental legislation was passed in 1979. The first statute was the *Environmental Protection Law (Trial)*, which was formulated as a landmark symbol for China's environmental legislation [8]. During the development process of China's environmental legislation, a key internal motivation was the reality of increasing serious environmental pollution problems. On the other hand, growing international attention about environment and development issues in the world and increasing introduction by developed countries of environmental legislation influenced leaders in China, resulting in two peaks of environmental legislation in China. The first peak was influenced by the United Nations Conference on the Human Environment in 1972 with its interest in creating legal systems for environmental protection. It was the trigger that led to the Environmental Protection Law in 1979; the second peak was affected by the United Nations Conference on Environment and Development in 1992, for which a main target was to fill legislative gaps and improve the existing legal systems. From 1993 onwards, the National People's Congress, the highest legislative institution in China, not only adopted new environmental protection laws but also revised many existing environmental laws [9]. After thirty years of unremitting efforts, China's environmental legislation has developed from a blank space into one of the most active legal fields, as well as has been playing an important role in the Chinese legal system [10,11]. Until the end of August 2014, the Standing Committee of the National People's Congress had approved thirty laws about environmental protection and resources conservation, which include five comprehensive laws, five pollution prevention and treatment laws, eleven resources conservation and utilization laws, four energy laws, and five other various types of laws. Thus, China's environmental legal framework has been basically established. The various elements of the environment have been basically covered, and there are basic laws to relative to the main areas of environmental protection (Table 1).

Table 1. Existing environmental and resources laws in China.

Note	Name	Adopted	Went into Effect	Revised	Went into Effect
1	Environmental Protection Law	1979-09-13	1979-09-13	1989-12-26 / 2014-04-24	1989-12-26 / 2015-01-01
2	Marine Environment Protection Law	1982-08-23	1983-03-01	1999-12-25	2000-04-01
3	Law on Prevention and Control of Water Pollution	1984-05-11	1984-11-01	1996-05-15 / 2008-02-28	1984-11-01 / 2008-06-01
4	Forestry Law	1984-09-20	1985-01-01	1998-04-29	1998-07-01
5	Grassland Law	1985-06-18	1985-10-01	2002-12-28	2003-03-01
6	Fisheries Law	1986-01-20	1986-07-01	2000-10-31	2000-12-01
7	Mineral Resources Law	1986-03-19	1986-10-01	1996-08-29	1997-01-01
8	Land Administration Law	1986-06-25	1987-01-01	1988-12-29 / 1998-08-29 / 2004-08-28	1988-12-29 / 1999-01-01 / 2004-08-28
9	Law on Prevention and Control of Atmospheric Pollution	1987-09-05	1988-06-01	1995-08-29 / 2000-04-29	1995-08-29 / 2000-09-01
10	Water Law	1988-01-21	1988-07-01	2002-08-29	2002-10-01
11	Law on the Protection of Wildlife	1988-11-08	1989-03-01	-	-
12	Law on Urban and Rural Planning	1989-12-26	1990-04-01	2007-10-28	2008-01-01
13	Law on Water and Soil Conservation	1991-06-29	1991-06-29	2010-12-25	2011-03-01
14	Surveying and Mapping Law	1992-12-28	1993-07-01	2002-08-29	2002-12-01
15	Law on Prevention and Control of Environmental Pollution by Solid Waste	1995-10-30	1996-04-01	2004-12-29	2005-04-01
16	Electric Power Law	1995-12-28	1996-04-01	-	-
17	Law on the Coal Industry	1996-08-29	1996-12-01	2011-04-22	2011-07-01
18	Law on Prevention and Control of Environmental Noise Pollution	1996-10-29	1997-03-01	-	-
19	Flood Control Law	1997-08-29	1998-01-01	-	-
20	Law on Energy Conservation	1997-11-01	1998-01-01	2007-10-28	2008-04-01
21	Law on Protecting Against and Mitigating Earthquake Disasters	1997-12-29	1998-03-01	2008-12-27	2009-05-01
22	Meteorology Law	1999-10-31	2000-01-01	-	-
23	Law on Prevention and Control of Desertification	2001-08-31	2002-01-01	-	-
24	Law on the Administration of the Use of Sea Areas	2001-10-27	2002-01-01	-	-
25	Law on Promotion of Cleaner Production	2002-06-29	2003-01-01	2012-02-29	2012-07-01
26	Law on Evaluation of Environmental Effects	2002-10-28	2003-09-01	-	-
27	Law on Prevention and Control of Radioactive Pollution	2003-06-28	2003-10-01	-	-
28	Renewable Energy Law	2005-02-28	2006-01-01	2009-12-26	2010-04-01
29	Law on Promotion of Circular Economy	2008-08-29	2009-01-01	-	-
30	Law on the Protection of Offshore Islands	2009-12-26	2010-03-01	-	-

Note: In order of adopted date; data current to 31 August 2014.

In addition, some other relevant laws, such as the Criminal Law, Cultural Relics Protection Law, Standardization Law, Administrative Punishment Law, Administrative Procedure Law, and Agriculture Law, also contain environmental terms or content in their articles. In order to implement these laws, the State Council of China has formulated twenty-five administrative regulations about environmental protection under the authority of the Constitution and other relevant laws, and the relevant ministries of the State Council and local governments have also developed hundreds of regulations or departmental rules regarding environmental protection. At the same time, the Supreme People's Court and the Supreme People's Procuratorate of the People's Republic of China have provided relevant judicial interpretations to punish environmental crimes. These rules, regulations and normative documents have covered most aspects of environmental protection, and are also constituent parts of China's environmental legal system.

Overall, China's environmental legislation has developed over 35 years from nothing to providing comprehensive coverage of various environments and natural resources, and of activities related to their exploration, development, extraction, processing, use and disposal. It is thought that China's environmental legislation is playing a positive role for the environmental protection and resources conservation progress in China.

2.2. Challenges

Although the formulation and implementation of China's environmental laws have played an important role in China's environmental protection work, there are still many defects and deficiencies in China's environmental legislation. The reason why China's current environmental pollution controls are ineffective, which shows the trend of partial improvement but deterioration from the overall view, the enforcement of environmental law is certainly an important reason, at the same time the existing defects and also the problems with environmental legislation cannot be ignored.

A mass of viewpoints and discourses about the shortcomings of China's environmental legislation has been discussed. Professor Zhou Ke believes that the numerous number of environmental laws did not mean good quality equally; problems with the system design of environmental laws are the root cause of poor effects about environmental protection work; people have no qualification to talk about the completeness of China's environmental legislation because there are still too many gaps inn environmental legislation between China and Western countries, whether from the view of legislative ideas, legislative techniques or legislative approaches [12]. Professor Wang Jin has also pointed out bluntly that China's environmental legislation has "no big mistakes but also no obvious effects", "everything looks perfect and extensive, but it's difficult to find a specific solution to the problem in the law provisions when facing real problems" [8]. In addition, there are also some opinions to place the problems with China's environmental legislation as the malposition of legislative concept, the lag of legislative content, and non-coordination of relevant regulations from other scholars [13]. Based on the multiple standpoints of scholars and the legislative implementation in practice, this article regards the five following problems of China's environmental legislation as the main challenges.

222

Firstly, the concept of sustainable development has not yet been fully implemented in China's environmental legislation. The concept of sustainable development is an important guiding principle which environmental legislation should follow. However, among the current existing thirty environmental laws, only a few new laws revised or adopted in recent years reflect the sustainable development requirements in legislative purpose and in specific contents [14]. But in many other laws, especially the resources laws, including the *Mineral Resources Law, Law on the Coal Industry, Forestry Law* and others, the concept of sustainable development has not been fully reflected, neither in the legislative purpose nor some specific contents [14]. The thinking of "Economic Interests Above All" is still a natural guiding ideology for many legal systems designs. The principle of "Prevention priority, combining prevention with control" should be emphasized, both "prevention in the first" and "treatment in the end", however, the existing legal system is more often confined to "treatment in the end", which means the pollutant emission control at the end of pipe, while the ideology of "source control" is rarely embodied.

Secondly, there are still many gaps in China's environmental legislation, and the non-coordination phenomenon between laws and regulations stands out. To check from the integrity of the environmental legal system, many legislative gaps exist at a number of important environmental protection areas. For example, there are still no specific laws in the fields of soil pollution control, toxic chemicals management, nuclear safety, bio-security, nature conservation, environmental damage compensation, and some environmental technical specifications and standards are also lacking. Especially in regard to soil pollution, it has been one of the most severe environmental problems in China, but, unfortunately, it is still not well addressed by current laws and regulations. Furthermore, some laws are often difficult to be implemented because many specific relevant regulations required by the laws were not finished in a timely manner. It has been shown from some statistical data that, in the current environmental laws that have published authorization, there are in total more than 140 regulations that should be formulated by the State Council and other ministries, but only less than 100 were finished up to now, which means the average completion rate is less than 70%. And many supporting rules and regulations were completed too slowly after the law enforcement, waiting a long time instead of simultaneously being implemented with the law, which was obviously not conducive to its functioning well [15]. Besides, problems of repetition and non-coordination between the environmental laws and regulations also exist. Take for example the "Environmental Protection Law", before its revision this time there were 31 articles out of the total 47 articles that were repeated in other environmental pollution control laws, which means the repetition rate was up to 66%; meanwhile, rules were also inconsistent in fundamental principles, basic procedures, applicable conditions, management subjects and other aspects [11] (p. 31).

Thirdly, there are problems of unclear responsibility, unreasonable system design, imbalance of rights and obligations, and lack of operability in the legal content. Many environmental laws state clearly that the local government should take responsibility for local environmental quality, but it is a pity that this rule usually exists in name only and could not be implemented very well because of the lack of investigation and affixation of legal liability mechanisms [16], the Law on Prevention and Control of Water Pollution and Law on Prevention and Control of Atmospheric Pollution are typical examples. It is known that one of the fundamental characteristics of the law is the consistency of rights and obligations. However, at present many of China's environmental laws set too much power, but too little obligations, to the administrative department, and too few rights, but too many obligations, to the administrative relative counterpart, so that the balance of rights and obligations in environmental legislation is broken [17] (pp. 20–21). It is also known that a good law should be implemented with a strong operability, but many articles of China's environmental laws were expressed vaguely without relevant supporting procedural requirements, or did not comply with Chinese national conditions, which made the laws difficult to implement effectively [16]. In addition, the tendency of departmental interests during the environmental legislative process increased the difficulty of implementing environmental laws [17] (pp. 23–24).

Fourthly, impacted by the planned economy for a long time, China's environmental legislation paid more attention to administrative control commands instead of using market economic instruments. From the aspects of guiding principles, system design and construction, approach selection and law enforcement, China's environmental laws used to emphasize the role of government and provided a lot of administrative enforcement mechanisms, but lacked experiences about rational use of market mechanisms [9]. According to the division of China's legal system, most environmental laws belong to the administrative law department. Therefore, the main principles and legal systems largely based on the establishment of administrative control, even some measures of economic encouragement system, also depend on the administrative order [18]. "Legal articles run through administrative arrangements, while the market supply rarely enters the legal system" [9]. Nowadays, China has entered the stage of socialist market economy, which made the old method of administrative control unsuitable for the new situation. It is inevitably going to cause local protectionism on environmental protection work if the fact that legislative approaches are too dependent on administration does not change.

Fifth, the phenomenon of being administration-led stands out during legislation, and it is difficult for the public to participate in the legislative process. China's current legislative work was led seriously by the administrative departments. The data shows that among the laws adopted by National People's Congress in the last 20 years, the bills which proposed by the relevant administrative departments of the State Council accounted for 75%–85% of the total. Besides, a popular behavior was to authorize the State Council to make regulations separately when some difficult or sensitive issues appeared in legislation [19]. In addition, it is also very common to authorize the administrative department to draft the legal text in environmental legislation practice, which contributes directly to the phenomenon of administration-leadership and then increases departmental interests. At the same time, the public and other common people have a very difficult time getting the opportunity to participate in the legislation. According to Article 35 of the existing *Legislation Law*, only some important legal texts could be open for public comment, with the decision by chairman's meeting after the first deliberation [20]. The lack of public participation procedures in some legislative preparation stages, including legislative program planning, legal text drafting and some important articles hearings, lead to the absence of public participation, which has brought tremendous obstacles for the implementation of environmental laws, the protection of environmental rights and interests of citizens.

3. Analysis of Legislative Cases in Recent Years

Things are always changing with the situation. Although there are still many problems with China's environmental legislation, some positive changes have appeared in some legislative cases of recent years. By analyzing the drafting and amendment of these laws, we can conclude that China's environmental legislation is progressing gradually to integrate with international trends during the evolution of development stage.

3.1. Renewable Energy Law: Favorable Inspiration from International Legislative Experiences

Renewable Energy Law of China was adopted on 28 February 2005, and has been enforced from 1 January 2006. The law gave a definite legal status to renewable energy, as well as the priority of its development. There are both domestic and international backgrounds for this legislation. To analyze from the domestic side, China's energy requirements increased rapidly with the high-speed economic growth. The gap between the energy supply and requirements became wider and wider, which brought huge pressure to look for substitute energy [21]. From the international side, the international society has recognized the importance of climate change issues and attached more and more attention from the 1990s onward. In the aim of dealing with this phenomenon, the *United Nations Framework Convention on Climate Change* and the *Kyoto Protocol* were adopted [22]. The principle of "Common but Differentiated Responsibilities" was confirmed by these two international legal documents. Although China has no forced emission reduction responsibility, the fact that China has the highest GHG emissions of any country brought more and more pressure. For these reasons, the development of renewable energy became one of the major measures, and the legislation on renewable energy was soon put on the legislative agenda in China.

The main legal approaches of the *Renewable Energy Law* include the total volume objective, full-acquisition, fixed price, different price sharing and renewable energy development funds approaches. A significant specialty of this legislation is the combination of domestic practice and international advanced experiences. In the basis of China's renewable energy situation, the legislative modes and contents of Germany, Denmark, USA, UK and other countries were referenced and compared to ensure the suitable legal system for China. Here into, the progressive experience of Germany was used for reference. A brief comparison can be found from Table 2.

Table 2. Brief comparison of renewable energy laws between Germany and China [21,23].

Renewable Energy Law of Germany	Renewable Energy Law of China
Legislative mode and goal	Legislative mode and goal
RE percentage (30%, 2020)	Total volume objective system
Duty of preemption	Full-acquiring system
Feed-in Tariffs	Fixed price system
Price balance and sharing	Different price sharing
Payment obligations	RE development funds system

In 2009, the *Renewable Energy Law* was amended to solve some new issues according to the situation of rapid development of the renewable energy industry. It is comforting that the percentage of renewable energy in all of the energy sources has growing from 5.5% in 2005 to 9.3% in 2013 [24]. Therefore, we can draw a conclusion from the implementation effect of the *Renewable Energy Law* that the renewable energy legislation improved successfully and strongly the renewable energy industry in China.

3.2. Revision of Environmental Protection Law: Multiple Innovations of Legislative Concepts, Content and Model

As the newest progress of China's environmental legislation, the *Environmental Protection Law* has been revised and adopted on 24 April 2014, and will be enforced from 1 January 2015. This law, which went through four rounds of deliberations within two years, is considered the most strict environmental protection law in China [25]. Different from past environmental legislation, this revised law has two sides to its background. One side was that air pollution, especially the haze that occurs frequently and widely provoked the public's consciousness on environmental protection, made the law-amendment stronger and stronger [26]. The other reason was that China's new leaders paid unprecedented attention to the environmental protection issues, which helped to break up the rigidities in traditional legislation and produced many new bright spots in the new law. Specifically speaking, there are three kinds of innovations, including legislative concept, administrative mechanism and legal approaches.

About the innovation of legislative concepts, the new law is located as the fundamental and comprehensive law in the environmental protection field in order to make explicit differences for the relationship with other specific laws [27]. Sustainable development concepts and ecologically conscious construction are clearly regulated as the guiding values in the new law. Meanwhile, environmental protection is also given the status of basic national policy. The new law also gave the principle of "Priority of Environmental Protection", which requires that economic development should coordinate with environmental protection [28].

About the innovation of administrative mechanisms, the new law emphasizes the multiple-governance, which means not only the government, but also enterprises and citizens, should share the environmental protection obligation burden together. At the same time, the public has been granted the rights of environmental information knowledge, participation, and supervision. Based on the principle of correspondence, the powers and responsibilities on environmental protection issues, which belong to different ministries, have also been regulated definitively and improved obviously [28]. In addition, the non-corrupt judiciary with power to imprison and impose substantial fines and other enforcement mechanisms has been also improved in this new law.

About the innovation of legal approaches, the total quantity control of pollutants, the permission of pollutant discharge, public interest litigation, the coalition prevention for cross-administrative district pollution and other important approaches are added, and the existing ones are also completed in the new law. Specifically, the government responsibility of environmental protection has been emphasized, and it will be supervised and evaluated at regular intervals [28].

One specific point that should be underlined is the innovation of legislative mode during the revised progress of *Environmental Protection Law*. As has been mentioned before, environmental quality has a direct bearing on public interest, so that this revision paid more attention to seeking

226

opinions from all society. As a result, the revising plan changed from the partial amendment in the beginning to the full-scale revision, and the draft passed a process seeking twice as many opinions and four times the deliberation. Just by progress like this, public participation rights in the legislation have been strengthened, and reasonable opinions and suggestions can be adopted easier than before. Some information can be found from Table 3.

Table 3. Revised process of Environmental Protection Law [29].

Time	Name of the Draft	Revise Model	Main Contents of Revise	Opinions Feedback
August 2012	Decision on Amending the EP Law	Partial Amend	Add Government Responsibility, Strengthen Legal Liability	1th Consultation on-line, 11,748 Opinions Feedback
June 2013	Decision on Amending the EP Law	Partial Amend	Add Public Interest Litigation, Add Daily Fine System, Strengthen Legal Principles	2th Consultation on-line, 2434 Opinions Feedback
October 2013	Revised Draft of EP Law	Full-scale Revise	Expand Public Interest Litigation Subject Scale, Strengthen Target-oriented Responsibility	Plenary Deliberation
April 2014	Revised Draft of EP Law	Full-scale Revise	Add Public Pre-warning System, Strengthen Environment Publicity, Detailing Related Rules	Adopted after 129 Members Spoke

4. Prospects of Environmental Legislation in China

In 2013, the new Chinese leaders put forward a new ruling idea of constructing an ecologically sound civilization, which indicated a high degree of importance on sustainable development. It also means that the road to sustainable development in China is working toward a deeper, stronger, more all-around way, from the publication of *China's Agenda of 21st Century* to the Scientific Outlook on Development and then to the construction of the ecological civilization. However, along with the rapid economic development and the large economic aggregation, China is also facing increasingly severe environmental and resource problems. The unsustainable aspects of development are also increasingly prominent. China did not hope to repeat the mistake of "treatment after pollution" during the developing progress of some Western countries, but it is really a big challenge. Therefore, in addition to the reliance on technology innovation and economic and industrial restructuring, it is indispensable to have legal safeguards for the future of China's sustainable development.

By reviewing the development process of China's environmental legislation, it can be found that one major feature of China's environmental legislation is it is deeply influenced by national development strategies and the Chinese leadership governing philosophy. Therefore, based on the above background, China's environmental legislation is expected to open up a new period of development. Some tendencies of China's environmental legislation can be foreseen as follows: Firstly, the future environmental legislation will fully implement the concept of sustainable development, and the economic development will be accelerated to transfer from extensive mode to sustainable mode supervised with clearly defined responsibilities and legal force; secondly, the future environmental legislation will promote the construction of the ecological civilization by paying more attention to the legal practice of the concept of prevention and the whole process of managing in law; thirdly, the future environmental legislation will also encourage more and more public citizen participation as it can enhance the supervision of legal drafting and implementation by strengthening public citizens' rights, and it can also urge achieving environmental quality objectives by intensifying public citizens' duties.

From the view of specific legislative fields, the direction has been shown clearly by the Legislative Planning of the 12th Standing Committee of National People's Congress. During the next five years, China's Environmental Legislation will focus on revising and improving the existing pollution control laws and natural resources laws, such as the *Law on Prevention and Control of Water Pollution, Law on Prevention and Control of Atmospheric Pollution, Law on the Protection of Wildlife, Forestry Law,* and *Mineral Resources Law.* On the other hand, the legislative work will also attach importance to some new areas, like *Law on Prevention and Control of Soil Pollution, Nuclear Safety Law, Law on Resources of Deep Sea* and so on, to fill in the gaps of China's environmental legal system [30]. What should be emphasized is that, looking ahead for the future, China's environmental legislation still needs to draw on global experiences, not only from the Western developed countries, but also other developing countries and emerging countries, which is being reflected in the legislative research on tackling climate change in China.

To look forward to the far future, without a doubt, the shortage of natural resources and energy scarcity are two of the most critical issues for China to achieve a sustainable future. That means, it is so important and urgent to enhance the legislation of resources recycling and low-carbon development to achieve the rational use of resources and security of energy supply. Fortunately, the international community has had good experience in legislation of both areas, and there are also productive practices and explorations within the national conditions in China. Therefore, it will be significant to bridge global experiences and local action by the way of legislation, not only to mitigate and adapt to global climate change, but also to establish the foundation and provide protection for the true sense of sustainable development in the largest developing countries of the world.

5. Conclusions

By reviewing the progress of China's environmental legislation for more than thirty years and the analysis of some new environmental legislative cases in recent years, this paper considers that China's environmental legislation will go into a brand new stage of development along with the new national development strategy of the construction of the ecological civilization raised by China's new leaders. The increasing problems of resource shortage and environmental pollution are also another reason. A large number of resource and environmental laws will appear in the next five-to-ten years. The legislative model will be in the forms of both existing law revisions and new legislation for empty fields. The legislative content will probably be the legal system improvement to achieve sustainable development by balancing economic interest and environmental interest. And the legislative procedure could be imagined to strengthen public participation to realize scientific and democratic legislation for the strategy of "Ecological Civilization".

Acknowledgments

This research is supported by Natural Science Foundation of China (41101126, 41471116), Ministry of Science and Technology of China (2011DFA91810), China Postdoctoral Science Foundation (2014M551142), International Postdoctoral Exchange Fellowship Program under China Postdoctoral Council (20140050); and the IKS program of the Alexander von Humboldt Foundation of Germany. Special thanks go to Martin Eiffert of Humboldt University of Berlin, Timo Koivurova and Paolo Davide Farah for their valuable comments.

Author Contributions

Zhilin Mu and Bing Xue designed research, performed research and analyzed the data; Zhilin Mu and Shuchun Bu wrote the paper, and all authors read and approved the final manuscript.

Conflicts of Interest

The authors declare no conflict of interest.

References

1. Kiss, A.; Shelton, D. *Manual of European Environmental Law*; Cambridge University Press: Cambridge, UK, 1994.
2. Wang, J. *A Study of the Objectives of Environmental Law*; Law Press China: Beijing, China, 1999; pp. 40–41.
3. Campbel, M.; Breen, F. *Sustainable Environmental Law*; West Publishing Company: St. Paul, MN, USA, 1993.
4. Wang, J. *Introduction of Environmental Law in Japan*; Wuhan University Press: Wuhan, China, 1994.
5. Mugong, G.Y. *Introduction of Public Hazard*; Youfei Press Japan: Tokyo, Japan, 1974.
6. Ye, J.R. *Environmental Policies and Laws*; Yuanzhao Press Company: Taibei, Taiwan, 1993.
7. Yecun, H.H. *Environmental Law in Developing Countries*; Asian Economic Institute: Tokyo, Japan, 1993.
8. Wang, J. Thirty Years Rule of Environmental Law in China: Retrospect and Reassessment. *J. China Univ. Geo-Sci. (Soc. Sci. Ed.)* **2009**, *9*, 3–9.
9. Li, Q.J. Chinese Environmental Legislation Assessment: Sustainable Development and Innovation. *China Popul. Resour. Environ.* **2001**, *11*, 23–28.
10. Wang, C.F. Glory Challenges and Prospects of Environmental Law. *Trib. Polit. Sci. Law* **2010**, *28*, 106–107.
11. Mu, Z.L. Research on the Interests of Environmental Legislation. Ph.D. Thesis, Renmin University of China, Beijing, China, 2014.

12. Zhou, K. Environmental System Change is the Necessary Condition of Inflection Point. *J. Chin. Youth* **2013**, *2*, 6–7.
13. Liu, A.J. Ecological Civilization and China's Environmental Legislation. *China Popul. Resour. Environ.* **2004**, *14*, 36–39.
14. National People's Congress. Searching System of China's Laws and Regulations. Available online: http://law.npc.gov.cn:87/page/browseotherlaw.cbs?rid=bj&bs=270682&anchor=0#go0 (accessed on 18 December 2013).
15. Sun, Y.H. Basic Experiences and Problems of China's Environmental Legislation from Reform and Opening-up. *J. China Univ. Geosci.* **2008**, *8*, 41–49.
16. Sun, Y.H. Research on Improving the Quality of Environmental Legislation. *J. Environ. Prot.* **2004**, *8*, 3–11.
17. Li, D. *Interests Analysis of Environmental Legislation*; Intellectual Property Press: Beijing, China, 2009.
18. Wang, C.F.; Yu, W.X.; Li, D.; Li, J.H. Dilemma and Outlet of China's Environmental Legislation: From the View of Songhuajiang River Pollution Incident. *Acad. J. Zhongzhou* **2007**, *1*, 91–97.
19. Ma, S.M.; Cao, K.Q. Impulse and Contain of Department Interests Legislation. *J. People's Court* **2008**, *7*, 1–2.
20. Xin, C.Y. Legislative Law and Legislative Work of National People's Congress. Available online: http://www.npc.gov.cn/npc/xinwen/2008-05/30/content_1466382.htm (accessed on 15 February 2014).
21. Shi, J.L.; Li, J.F. The Background and Basic Thinking of Renewable Energy Legislation. *J. Shanghai Electr. Power* **2005**, *6*, 551–554.
22. Han, W.K. Some Issues about the Renewable Energy Strategy of China. Available online: http://www.npc.gov.cn/npc/xinwen/2009-10/31/content_1525123.htm (accessed on 16 February 2014).
23. Chen, H.J. Germany's Renewable Energy Law and It's Experience for Learning. *J. Environ. Sci. Manag.* **2006**, *31*, 32–34.
24. China National Renewable Energy Centre. *The Renewable Energy Industry Development Report (2013)*; China National Renewable Energy Centre: Beijing, China, 2014.
25. Wang, J. Another Milestone in the History of Environmental Legislation. Available online: http://opinion.people.com.cn/n/2014/0425/c1003-24940245.html (accessed on 26 May 2014).
26. Mu, Z.L. Legal Thinking about the Coping with Haze Pollution. *J. Environ. Sustain. Dev.* **2014**, *1*, 52–55.
27. Zhang, D.J. The New Environmental Protection Law Should be Publicized and Implemented Well. Available online: http://politics.people.com.cn/n/2014/0425/c1024-24940199.html (accessed on 26 May 2014).
28. National People's Congress. Environmental Protection Law. Available online: http://npc.people.com.cn/n/2014/0425/c14576-24944726.html (accessed on 26 April 2014).

29. National People's Congress. Asking for Opinions on the Environmental Protection Law. Available online: http://www.npc.gov.cn/npc/xinwen/lfgz/flca/2013-07/17/content_1801189.htm (accessed on 17 June 2014).

30. National People's Congress. The Legislative Plan of the 12th Standing Committee of National People's Congress. Available online: http://www.npc.gov.cn/npc/lfzt/lfgzhy/node_21894.htm (accessed on 17 August 2014).

An Entropy-Perspective Study on the Sustainable Development Potential of Tourism Destination Ecosystem in Dunhuang, China

Huihui Feng, Xingpeng Chen, Peter Heck and Hong Miao

Abstract: This paper analyzed the characteristic of the tourism destination ecosystem from perspective of entropy in Dunhuang City. Given these circumstances, an evaluation index system that considers the potential of sustainable development was formed based on dissipative structure and entropy change for the tourism destination ecosystem. The sustainable development potential evaluation model for tourism destination ecosystem was built up based on information entropy. Then, we analyzed each indicator impact for the sustainable development potential and proposed some measures for the tourism destination ecosystem. The conclusions include: (a) the requirements of Dunhuang tourism destination ecosystem on the natural ecosystem continuously grew between 2000 and 2012; (b) The sustainable development potential of the Dunhuang tourism destination ecosystem was on an oscillation upward trend during the study period, which is dependent on government attention, and pollution problems were improved.

Reprinted from *Sustainability*. Cite as: Feng, H.; Chen, X.; Heck, P.; Miao, H. An Entropy-Perspective Study on the Sustainable Development Potential of Tourism Destination Ecosystem in Dunhuang, China. *Sustainability* **2014**, *6*, 8980-9006.

1. Introduction

Sustainable tourism (ST) is an important part of global sustainable development (SD) [1]. The rapid development of tourism has brought great benefits to tourism destinations, while a variety of other problems are emerging, such as, resources and environmental issues and poor management of the tourism industry. Generally, as a reception carrier, the tourism destination of tours concentrates all elements of tourism on an effective framework, which is the most vital part for examining the impact of tourism. Hence, research for sustainable development of tourism destinations could improve the overall efficiency of tourism and optimize the ecological services related to the tourism. Until now, a number of organizations and academics have paid attention to this topic as well as achieved great progress in research methods and practices.

Particularly, their research focus on the following contents [2–12]: (a) The concept of ST [13–22], for example, Hunter suggested that the concept of ST be redefined in terms of an over-arching paradigm which incorporates a range of approaches to the tourism/environment system within destination areas [13]. Swarbrooke [14] and Aall [17] noted that ST is not just about protecting the environment; it is also concerned with economic viability and social justice, and a suitable balance must be established between these three dimensions to guarantee its long-term sustainability. Sharpley point that there has been significant differences between the concepts of ST and SD, the principles and objectives of SD cannot be transposed onto the specific context of tourism [15].

Hardy argued that ST has traditionally given more focus to aspects related to the environment and economic development, which should be more focused on community involvement [4]. Gianna noted the need to very clearly distinguish between the concept of ST and the idea of tourism as one possible tool to support sustainability at multiple levels [16]. Saarinen concluded that perspectives of the resource-based tradition and the community-based tradition have their advantages in different use contexts and they can complement each other, but in respect to the idea of sustainability and the future challenges of humanity, they all share the same major limitation, which is the strong focus on the local scale [18]. Moyle and McLennan noted that the frequency of occurrence of sustainability as a concept has slightly increased in strategies over the past decade. At the same time, there has been a shift in the conceptualisation of sustainability, with thinking evolving from nature-based, social and triple bottom line concepts toward a focus on climate change, responsibility, adaption and transformation [20]; (b) The indicators of ST [23–41], for example, McElroy constructed a "Composite Tourism Penetration Index" from per capita visitor spending, daily visitor densities per 1000 population and hotel rooms per square kilometer. They tested it on 20 small Caribbean islands and yielded three levels of increasing penetration [26]. McCool and Moisey provided a tourism industry perspective on what items could be sustained and what indicators should be used to monitor for sustainability policies [27]. Wang analyzed the principle of indicators of ST, constructed the indicators of ST, the indicator weight and selected the comprehensive evaluation method [40]. Ward and Butler investigated how to monitor sustainable tourism development (STD) in Samoa. It described some of the methodological considerations and processes involved in the development of STD indicators and particularly highlighted the importance of formulating clear objectives before trying to identify indicators, the value of establishing a multi-disciplinary advisory panel, and the necessity of designing an effective and flexible implementation framework for converting indicator results into management action [31]. Ko proposed that the "Barometer of tourism sustainability" (BTS) model represents the comprehensive level of tourism sustainability in a given destination, combining human and natural indicators into an index of sustainable tourism development, without trading one off against the other. The "AMOEBA of tourism sustainability indicators" (ATSI) model was introduced to complement the BTS analysis and to illustrate individual levels of sustainability of tourism indicators [24]. Chris and Sirakaya employed a modified Delphi technique to constructed indicators from political, social, ecological, economic, technological and cultural dimensions for community tourism development (CTD) [25]. Schianetz and Kavanagh proposed the methodological framework for the selection and evaluation of sustainability indicators for tourism destinations, the systemic indicator system (SIS) this framework takes the interrelatedness of sociocultural, economic and environmental issues into account [33]. Reddy engaged in the identification, selection and evaluation of sustainability indicators for rapid assessment of tourism development in Andaman and Nicobar Islands of India (ANI). These indicators are developed and assessed mainly for developed countries and evaluated a feasible bottom-up approach based on local knowledge [38]. Blancas and Gonzalez introduced an indicator system to evaluate sustainability in established coastal tourism destinations, as well as developing a new synthetic indicator to simplify the measurement of sustainability and facilitate the comparative analysis of destination ranking [30]. Buckley suggested that the indicators of ST should include: population, peace, prosperity, pollution, protection [6]. Oyola and Blancas presented an indicator system to evaluate ST at cultural destinations. Also, they suggested a method based on goal programming to construct composite indicators. Then, they proposed three basic practical uses for these indicators: the formulation of general action plans at a regional level, the definition of short-term strategies for destinations and the establishment of destination benchmarking practices [36]. Delgado and Saarinenc examined the significant role of indicators based on literature review in tourism planning and management. The indicator type (set or index) needs to be carefully selected depending on the situation under analysis and the purpose underpinning the study. However, indicator effectiveness to achieve the ideals of sustainable tourism development is affected by the ambiguity in the definition of the concept of ST and problems associated with data availability and baseline knowledge. The main challenge is to overcome strategic guidelines and political and theoretical proposals of indicators and achieve practical applications for the sustainable development of tourism [23]; (c) Ecological security and environment carrying capacity for tourism [42–54], for instance, Ahn used the limits of acceptable change (LAC) framework as a guide to examine and inform the process of ST on a regional scale. Also, he examined resident attitudes toward tourism development in general, toward desirable types of tourism services, toward local conditions and finally, toward perceptions about if and how conditions might change due to tourism [42]. Gössling provided a methodological framework for the calculation of ecological footprints (EF) related to leisure tourism. Based on the example of the Seychelles, it reveals the statistical obstacles that have to be overcome in the calculation process and discusses the strengths and weaknesses of such an approach [44]. Hunter attempted to connect, conceptually, the realms of ST and EF thinking. It is argued that primary research should focus on calculating the TEF associated with individual tourism products, throughout the product's life-cycle. As well as bringing another dimension to our understanding of tourism's actual ecological demand, it is also argued that the concept of the TEF may be used to clarify theoretical aspects of the sustainable tourism debate, helping to rejuvenate this debate in the process [47]. Cui put out the tourism bearing capacity index and its arithmetic model of operation. He defined the tourism environmental bearing capacity as the bearing intensity of tourism destinations during a period which does not do harm to the present and future people in its current state and which can be accepted by the residents. The bearing intensity of tourism destinations mainly includes the tourist density, the tourism land use intensity and the tourism income value [50]. Xiao constructed the models of general ecological security coefficient (GESC) of island tourism destination and special ecological security coefficient (SESC) of island tourism destinations, and then the assessment framework and judgment criterion were proposed on the ecological security of island tourist destination (ESITD) and island tourist sustainable development (ITSD). Furthermore, the models of island tourist ecological footprints were established based on the idea of EF and an empirical analysis of Zhoushan Islands, China was conducted [51]. Salerno and Viviano describes how the concept of Tourism Carrying Capacity (TCC) has shifted from a uni-dimensional approach to incorporating environmental, social and political aspects. Then, an empirical analysis of internationally popular protected area used by trekkers, the Mt. Everest Region, was conducted [52]. Zhong examined the applicability of the model to China's Zhangjiajie National Forest Park. At the same time, both external and internal factors affecting the park's tourism development as well as the

234

environmental, social, and economic changes of the area are also discussed [53]; (d) The development
pattern of ST [55–60] was examined by Rodríguez along with an analysis of the life cycle of
Tenerife (Canary Islands, Spain), and two types of strategic decisions are considered: the political–
legal decisions of the regional government to regulate tourism activity and the decisions to regrade
supply, developed by the administrative institutions related to tourism activity in this destination
[55]. Keitumetse devised a Community-Based Cultural Heritage Resources Management
(COBACHREM) model that merges the technical and academic approaches to illustrate a symbiosis
between cultural and natural resources for sustainable resources conservation at community levels [57].
Rizio explored a forest ecosystem and identified its potential flows of utility, addressing those
which best satisfy tourism activities and recreational purposes; to identify the most appropriate
tools to manage the flows of utility based on sustainable principles which integrate tourism
activities [58]; (e) With regards to perception of residents and visitors [61–75], for example, Xuan
studied residents' perceptions of the economic, socio-culture, environment impacts of tourism and
residents' attitudes to tourism development in Hainan and Sanya, China. It was concluded that the
residents are more aware of positive tourism impacts than negative impacts, and they support
tourism development with some reasonable attitudes [61]. Choi and Sirakaya too developed and
validated a scale assessing residents' attitudes toward sustainable tourism (SUS-TAS). Then, they
administered a 51-item scale of resident attitudes toward sustainable community tourism and 800
households in a small tourism community in Texas. Psychometric properties of SUSTAS along
with its practical and theoretical implications are discussed within the framework of sustainable
tourism development [65]. Nicholas and Thapa examined visitors' perspectives and support for
sustainable tourism development in the Pitons Management Area (PMA) in St. Lucia. Specifically,
the focus was on visitors' environmental, economic, and social attitudes based on a sustainable
tourism development framework and the effect and best predictive validity of attitudes on support
for sustainable tourism development were explored [68]. Bimonte and Punzo analysed how distinct
groups of residents, characterised by different levels of involvement in tourism-related activities,
perceived the tourism phenomenon, and to check whether there exists a latent or potential ground
for conflicts between groups of residents [75]. Cottrell and Vaske examined the relative influence of
four sustainability dimensions (environmental, economic, socio-cultural, and institutional) in predicting
resident satisfaction with sustainable tourism development in Frankenwald Nature Park, Germany.
Structural equation modeling supported the hypothesis that all four dimensions were significant
predictors of satisfaction. The economic dimension was the strongest predictor, followed by the
institutional, social, and environmental dimensions. Findings indicate that all four dimensions
should be included for a holistic approach to planning and monitoring sustainable tourism
development [69]. Sörensson and Friedrichs used importance-performance analysis (IPA) to
examine the performance of one particular tourist destination with regard to social and environmental
sustainability, and to establish whether international tourists and national tourists differ in the
sustainability factors they consider important [70]. Dorcheh and Mohamed reviewed existing
literature on local perceptions of tourism development and its process. It also discusses influential
theories and explains the social exchange theory as an effective framework for sustainable cultural
tourism [71]. Miller and Merrilees examined tourists' pro-environmental behaviours in four major

categories: recycling; green transport use; sustainable energy/material use (lighting/water usage), and green food consumption. They explored five major antecedents to those categories: habitual behaviour, environmental attitudes, facilities available, a need to take a break from environmental duties, and sense of tourist social responsibility. Also, the poorly understood belief that pro-environmental behaviour weakens when residents become tourists was examined [73]; (f) With regards to research for stakeholders [76–84], Hardy and Beeton argued that the nexus involves an understanding of stakeholder perceptions, and applies this to the Daintree region of Far North Queensland, Australia, to determine whether tourism in the region is operating in a sustainable or maintainable manner. In order to do this, an iterative approach was taken and local people, operators, regulators and tourists were interviewed, and content analysis applied to management and strategic documents for the region. The results illustrate the importance of understanding stakeholder perceptions in facilitating sustainable tourism [83]. Timur and Getz examined the concept of sustainable tourism development in urban destinations. Both qualitative and quantitative data were employed, from interviews and questionnaires undertaken in Victoria and Calgary, Canada, and San Francisco, USA. Respondents representing the three clusters of the tourism industry, local government and the host environment were examined on their interpretation of "sustainable tourism", sustainability goals and barriers to achieving sustainable tourism in urban destinations. Results revealed important similarities and differences among key stakeholders, and particularly a lack of appreciation for a triple bottom line approach among the tourism industry respondents [81]. Vellecco and Mancino demonstrate that in lacking shared responsibility, conflicts and tensions inside the local community paralyse innovative environmental behaviors when they ought really to be turned into opportunities for debate so that shared strategies and solutions may be identified in three Italian areas [78]. Holden found that although stakeholders shared positive perceptions of the economic benefits of tourism, its continued use for sustainable development is uncertain. Key challenges include a lack of confidence in the economic certainty of tourism and its use for out-migration, a maturing tourism market, and challenges to the local control of natural resources with external hegemonic forces [80]. Dabphet and Scott explored the diffusion of the sustainable tourism development concept among stakeholders in the tourism destination of Kret Island, Thailand. It is argued that both interpersonal and media communication and the identification of key actors in the community are needed to effectively diffuse sustainable tourism ideas among destination stakeholders. The results validate the use of diffusion theory as a means to understand the transfer of the sustainable tourism development concept among stakeholders, and they also provide information useful for the design of information dissemination programs [82]; (g) In regard to relevant policy for ST [85–94], for instance, Farsari and Butler explored policies for sustainable tourism development and potential interrelationships between policy considerations. Such policies have been characterized as *ad hoc* and incremental, lacking a clear orientation towards sustainable development, and the complex relationships underpinning them have rarely been considered in decision-making for sustainable tourism [87]. Dodds and Butler found that although respondents were aware of sustainable tourism, the individual advantage from exploiting shared pooled or shared resources is often perceived as being greater than the potential long-term shared losses that result from the deterioration of such resources, which means that there is little motivation for individual actors (whether governments, elected officials, or individual operators), to invest or engage in protection or conservation for more sustainable tourism [86]. Yasarataa and Altinay noted key political actors' interests and how to mitigate personal interests to facilitate and maintain sustainable tourism development in small states North Cyprus, Turkish [92]. Whitford and Ruhanen recommends that there cannot be a "one size fits all" framework for indigenous tourism development to suit all circumstances. Policies need to draw upon indigenous diversity and, in a consistent, collaborative, coordinated and integrated manner, provide the mechanisms and capacity-building to facilitate long-term sustainable indigenous tourism [93]. Solstrand suggest that the environmental and socio-cultural sustainability of marine angling tourism (MAT) in Norway and Iceland requires a complex socio-ecological systems perspective, with interactive governance strategies leading management policies. Sustainability requires that a management strategy not only focuses on the economic aspects; priority must also be given to minimizing multi-stakeholder conflicts and providing sufficient resource data to protect vulnerable fish stocks [89]. Xu and Sofield examined the situation that little guidance is provided to promote sustainability principles in tourism development strategies in China. In the future, a pro-active sustainability approach should be integrated with environmental concerns to allow tourism to participate constructively in the national transformation to a sustainable society [90].

Due to tourism research involved in geography, ecology, environmental science, sociology and so on, also combined with different scales covering the micro to the macro [2], the research focus on the multi-index comprehensive evaluation method (MICEM), tourism carrying capacity (TCC), tourism environment impacts assess (TEIA), ecological footprint analysis of tourism (EFT), life cycle assessment (LCA), limits of acceptable change (LAC) and Geographic Information Systems (GIS) [95–97]. The MICEM could quantify the level of tourism sustainable development, which employed AHP and Delphi method by the level of sustainable development and other potential targets. However, selection indicator and its weights usually by personal decision-making [33,39,98–101]. TCC could comprehensively measure the carrying capacity of tourism destinations such as ecology, resources, psychological and space, which employed remote sensing (RS), field measurements, questionnaire and Delphi method, *etc.* However, it has a characteristic of randomness and subjectivity when assessing environmental carrying capacity [102–104]. TEIA is an effectively method to analysis the effect of tourism on ecological environments by mathematical statistical analysis methods form microscopic view, which construct assessment index system and select assessment model based on environmental background to monitoring the feedback mechanism for impact of tourism environmental. However, it's usually ignoring the effectiveness of monitoring and feedback mechanism [105,106]. EFT constructs the tourism ecological footprint according to various data of per capita consumption by bottom-up questionnaire and investigates statistics. The consequent was directly comparable based on productive land area; it is a global standards value [56]. It is suitable to research on a small-scale since the ecological burden is likely to be passed on by interregional trade in tourism destination [43,47,107]. LCA could identify the stage of development of the tourism destination and solve some problems [49]. It is difficult to quantify the environmental problems in sustainable development [108,109]. The theoretical framework for LAC include identifying the issues and concerns, and defining and describing the types of tourism opportunities, *etc.* in the planning area. It could solve the contradiction between development of tourism and conservation of resources, which is mostly influenced by the decisions of programmer makers and managers [42,110]. GIS are now recognized widely as a valuable tool for which applications for regional tourism planning have not mushroomed as in other fields. This is also reflected in the field of sustainable tourism. Nonetheless, sustainable tourism decision-making and carrying capacity estimation has a lot to benefit from using such technologies. GIS can be used for managing the various information needs, estimating indicators, and generally assisting decision making in the planning phase, as well as, in the monitoring and evaluation phases [111]. Therefore, researchers have used diverse methodologies with more quantitative analysis, as for each method there are certain advantages and disadvantages [112]. Current studies are using more comprehensive approaches. Previous literatures show that a tourist destination is a relatively complete artificial ecosystem; the ecological is a basis for sustainable development [113]. However, relevant research has failed to exhaustively analyze the structures, considering the functions and evolutionary mechanisms of the compound tourism destination ecosystem. This is indeed a shortcoming of those research studies. Entropy, as a measure of system dissipation or disorganization, has been used to analyze social systems in various contexts [114–118]. In relation to tourism [119–123], for example, Bailey noted that Social Entropy Theory (SET) was a very general macro sociological systems theory [114]. Kenneth and Bailey point out that the most recent applications of entropy are in social entropy theory and macro accounting theory [115]. Stepanic jr and Stefancic hold that the established level of analogy between certain characteristics of social systems and part of thermodynamic formalism in the simplified model encourage one to assume that even deeper analogies might be drawn to construct more complete and detailed models of social systems [116]. Wilson describes entropy in urban and regional modelling introducing a new framework for constructing spatial interaction and associated location models [117]. Cabral and Augusto summarized entropy multifaceted character with regard to its implications for urban sprawl, and propose a framework to apply the concept of entropy to urban sprawl for monitoring and management. Hao point out that the phenomenon of the increase of entropy also exists in the tourism destination's ecological system [121]. Zhao proposed the conception and mainly indicates that research can broaden insights on tourism systems' carrying capacities through entropy change analysis from the view of the tourism system's entropy principle under the tourism dissipative structure mechanism [122]. Qian noted that the tourism environment system was an open system. It is unceasingly exchanging material and energy with the outside. It is impossible to achieve the absolute balance through introduction of the negative entropy flow [123]. Relevant research has indicated that tourism destination ecosystems are a typical dissipative structure. Therefore, there is certain feasibility in analyzing its evolution and sustainable development potential from the perspective of entropy. Given the analysis above and based on the relevant former research, this study based on the structure, function and characteristic of tourism destination ecosystems, applied information entropy theory for Dunhuang city which combines analysis entropy with information entropy to establish a tourist destination quantitative ecosystem model for evaluating the potential of sustainable development of Dunhuang tourism. It could offset the disadvantage of indistinct strategies and lack of specificity in some degrees in the sustainable development of tourism destinations.

2. Study Area

Dunhuang City is located in the border area of Gansu, Qinghai and Xinjiang Provinces, as the western end of the Hexi Corridor in Gansu Province, China. It belongs to a typical arid oasis region with unique geographical, historical and cultural status. In history, Dunhuang was an important hub on the ancient Silk Road and the point of integration of Eastern with Western civilization. To some extent, it is known as "human Dunhuang" owing to the intersection and coordination of the world's four major cultural systems [124]. It is rich in tourism resources, with Mogao Grottoes called the "Pearl of Oriental Art", Mingsha Mountain known as "desert spectacle", Crescent Lake and other tourism resources. All these places promote the development of tourism resources. In 2012, the number of visitor arrivals was 0.312 million people-times and the total income was 2.687 billion Yuan [125]. However, this explosive growth has brought a series of severe problems for cultural heritage, such as, tourists' periodic overload. Because of those intensive human activities, the weak regional environment, and the global climate, there has been a shortage of water resources, as the core of the regional ecological problems, which continue to worsen [126]. Therefore, it is very necessary to study the potential of sustainable development of the Dunhuang tourism destination ecosystem.

3. Methodology

Entropy, firstly proposed by German physicist R Clausius [127,128], is the unique macroscopic quantity in thermodynamics and statistical physics. In 1948, Shannon introduced this concept into information theory and named it "information entropy" [129]. Generally speaking, information entropy theory is based on probability and statistics to reflect the degree of disorder and quantify the evolution direction of the system. When analyzing the complexity and uncertainty of problems, it can be used as a multi-dimensional method to quantify and determine the comprehensive benefits [130,131]. According to theory of dissipative structures by I. Prigogine [132–134], an open system, which is far from equilibrium during the process of exchanging matter and energy with outside environments, has the tendency of entropy growth; hence, this system, only by constantly introducing negative entropy from the outside flow in order to offset internal positive entropy, will finally have a new and ordered direction for further development. That means that the large entropy of the system corresponds to the low degree of order and *vice versa*.

3.1. Entropy Change and Dissipative Structure of the Tourism Destination Ecosystem

The tourism destination ecosystem is a special ecosystem of areas with rich tourism resources and occurrences, that is established based on the original nature or artificial ecosystem during tourism development [121]. As the spatial carrier between tourist activities and the ecological environment, tourism definition of ecosystems involves the continuous exchange of materials, energy and information with external environments. It also makes irreversible the non-equilibrium processes inside the system, which is always producing positive entropy, inflowing negative entropy with the characteristics of openness, which is far from equilibrium, nonlinearity and ordered fluctuation [122,123]. Tourism destination ecosystem is a typical dissipative self-organizing system that possesses dissipative characteristics and analyzes the evolutionary process of entropy changes.

The tourism destination ecosystem development and evolution is led mainly by the evolution of its socioeconomic ecosystem under normal conditions. Given this progression, this study analyzes the interactions between the tourism destination socioeconomic ecosystem and its natural ecosystem and other regions by analyzing the entropy change process of the tourism destination socioeconomic ecosystem. This involves analyzing the evolutionary process and developmental trend of the tourism destination ecosystem, as well as evaluating the sustainable development potential of the tourism destination ecosystem.

According to the dissipative structure theory [135], there are two parts of entropy changes in tourism destination ecosystems. The first one is the entropy flow caused by the tourism destination socioeconomic ecosystem's exchange of materials, energy and information with external environments *etc.*; it reflects the carrying capacity of the tourism destination nature ecosystem for its socioeconomic ecosystem. Another part is the entropy production caused by irreversible non-equilibrium processes inside the system, which reflects its regeneration potential and could indicate the vitality of the tourism destination ecosystem. The total entropy change of the system is the summation of entropy production and entropy flow, reflecting the overall level of development in the tourism destination ecosystem [136]. The environment is affected by the development of tourism and associated activities as part of the evolution in becoming a tourism destination ecosystem. This is a result of the increase in disorder of entropy production, which is caused by de-vegetation, water pollution, soil fertility, air quality degradation, biodiversity decline and the assimilation features of traditional culture, *etc.* inside of the tourism destination ecosystem. The total entropy changes of the system and disordered parameters will increase if the tourism destination ecosystem does not exchange moderate amounts of material, energy and information with external environments, so the entropy flow does not offset entropy production. This will result in some negative effects for the tourism destination ecosystem, such as an increase in disorder within the system, lack of power, regulatory failure and weaker functioning.

The analysis of entropy changes in tourism destination ecosystems describes the state of tourism system and changes during the exchange of recourses with an external system. The amount or size of entropy not only expresses the level of internal system resources' effective utilization, but also reflects the elastic changes of system affordability. The change in size of the system entropy refers to higher or lower effective utilization rates in the evolutionary process of the tourism destination ecosystem's exchange of materials and energy with external environments [122].

3.2. Indicator System Establishment

The indicator system is an effective tool for measuring and evaluating the tourism sustainable development level. There are lots of indicators that play a more important role for tourism sustainable development. Depending on the principles of scientific city, comprehensiveness, dynamics, hierarchy, maneuverability and perceptiveness [137,138], using the references from Indicators of Sustainable Development for Tourism Destinations: A Guidebook (WTO, 2004) [139], European Tourism Indicator System For Sustainable Destinations (EU, 2013) [140], ecological civilization city construction indicator system of Dunhuang city and related research results [23–41,141], the article establishes tourism destination ecosystem sustainable development analysis and indicators system evaluation according to three aspects. They are the structure, function and characteristic of tourism destination ecosystems; the entropy production and entropy flow in the process of system operation; and the ecological environment pollution and destruction during the tourism industry development within the system. The article selected two parts of the entropy production and entropy flow; four aspects that are the supportive entropy input index, the stressful entropy output index, the consumption metabolism index of entropy and the regenerate metabolism index of entropy; 29 representative index (Table 1).

Table 1. Index system hierarchy of sustainable development potential evaluation.

Criterion	Sub-Criterion	Indicators	Measurements
Entropy flow	Supportive entropy input index (A)	Number of travel agencies (A1)	unit
		Number of direct engaged persons in tourism industry (A2)	person
		Number of star-rated hotels (A3)	unit
		Number of beds in star-rated hotels (A4)	bed
		Passenger-kilometers by highways (A5)	10^4 passenger-km
		Passenger-kilometers by railways (A6)	10^4 passenger-km
		Passenger-kilometers by civil aviation (A7)	10^4 passenger-km
		Annual water supply (A8)	10^4 t
	Stressful entropy output index (B)	Number of visitor arrivals (B1)	10^4 person-times
		Transport expenditure as percentage of tourism expenditure (B2)	%
		sightseeing expenditure as percentage of tourism expenditure (B3)	%
		Hotels expenditure as percentage of tourism expenditure (B4)	%
		Catering expenditure as Percentage of tourism expenditure (B5)	%
		Water used by tourists (B6)	10^4 t
Entropy production	Consumption metabolism index of entropy (C)	Total wastewater discharged (C1)	10^4 t
		Industrial wastewater discharged (C2)	10^4 t
		Emission of disulfide (C3)	t
		solid wastes discharged (C4)	10^4 t
		waste discharge by tourists (C5)	t
		Carbon emission by tourism (C6)	t
	Regenerate metabolism index of entropy (D)	Number of training institutions for tourism (D1)	unit
		Direct engaged persons in tourism industry as percentage of employees (D2)	%
		Tourism GDP (D3)	10^4 Yuan
		Tourism GDP as percentage of GDP (D4)	%
		Investment in anti-pollution Projects as percentage of GDP (D5)	%
		Proportion of industrial solid waste treated and utilized (D6)	%
		Rate of harmless garbage disposal (D7)	%
		Green coverage rate in developed areas (D8)	%
		Gardens per capita (D9)	m^2

Supportive entropy input index: Mainly embodies attractiveness and bearing capacity of the tourism destination ecosystem. The tourism destination satisfied the tourists' demands by supporting resources, infrastructures and services, support foundation of tourism development, and promotes communication with the outside and for internal operations. Therefore, here we select the related index that can reflect tourism destination development status quo and potential of its development (basic infrastructure, available water resources, *etc.*).

Stressful entropy output index: Expresses tourism activities putting pressure on the tourism destination ecosystem. During the evolution process of tourism destination ecosystem and tourist industry development, tourists and their consumption (transport, hotels, sightseeing, catering) and tourist industry energy consumption have direct or indirect influence on the tourism local ecological environment, and puts some pressure on system development, as well as slowing down the positive evolution speed of the system.

Consumption metabolism index of entropy: During the evolution process of the tourism destination ecosystem, the discharge of wastes, pollutants produced and a series of ecological problems from tourism activities to some extent weaken the sustainable development potential of the system.

Regenerate metabolism index of entropy: Mainly shows human being's governance capacity of the tourism destination ecosystem. The waste discharge by tourists is above the system bearing capacity, so that some pollution cannot be purified by the system itself; therefore, the system must rely on artificial management policies and scientific technologies. That is why human beings invest into environmental pollution management as a recovery function of tourism destination ecosystem's sustainable development.

According to the established evolution indexes of the tourism destination ecosystem, there is the calculation formula for entropy flow, entropy production and total entropy changes (Table 2).

Table 2. Symbols and formulae of entropy flow, entropy production and total entropy change.

Objective	Symbols and Formula	Means
Supportive entropy input	$\Delta_e S_1$	Disorder of system
Stressful entropy output	$\Delta_e S_2$	Disorder of system
Consumption metabolism of entropy	$\Delta_i S_2$	Disorder of system
Regenerate metabolism of entropy	$\Delta_i S_1$	Disorder of system
Entropy flow	$\Delta_e S_2 - \Delta_e S_1$	Coordination of system
Entropy production	$\Delta_i S_2 - \Delta_i S_1$	Vitality of system
Total entropy change	$(\Delta_e S_2 - \Delta_e S_1) + (\Delta_i S_2 - \Delta_i S_1)$	Order and health of system

3.3. Depending on Entropy Information Evaluation Model Establishment

Based on information entropy's benefits, information entropy's evaluation model is widely used in many scientific fields [136,142]. The tourism research focuses on the following: measuring the weight of an indicator based on the entropy method [143]; analyzing the characteristics and development measures based on information entropy, theory of entropy and dissipative structure [121–123,126,144]. More qualitative analysis was be used and a few calculation methods for entropy of tourism systems. According to the information entropy of Shannon, if we used random variables X represents uncertainty in the system, the discrete random variable could be supposed as x and its value is $X = \{x_1, x_2, ..., x_n\}$ $(n \geq 2)$, the probability for each value is $P = \{p_1, p_2, ..., p_n\}$ $(0 \leq p_i \leq 1, i = 1, 2, ..., n)$. $\sum P_{i=1}1$. The information entropy can be described as follows [131,145,146]:

$$S = -\sum_{i=1}^{n} P_i \ln(P_i)$$

(1)

where S the is information entropy of an uncertain system, P_i is the probability of the random state variable X in the uncertain system.

3.3.1. Measurement for Entropy Flow and Entropy Production of Tourism Destination Ecosystem

According to the measurement models for information entropy theory, we compute a formula for the entropy flow and entropy production based on information entropy theory and models for each year, then analyzed the complexity, coordination order and health for tourism destination ecosystems. Measurement of n indictors in m years, ΔS represents the four types of entropy based on information entropy [130,131,145], $i.e.$, the input supportive type of entropy ($\Delta_e S_1$), the output pressure type of entropy ($\Delta_e S_2$), the consumption metabolic type of entropy ($\Delta_i S_2$) and the regeneration metabolic type of entropy ($\Delta_i S_1$).

$$\Delta S = -\frac{1}{\ln m} \sum_{i=1}^{n} \frac{q_{ij}}{q_j} \ln \frac{q_{ij}}{q_j}$$

(2)

where ΔS represents the four types of entropy, q_{ij} is the standardized value of calculated from the raw data, q_j is sum for standardized value of index in j year, m is sum for the number of appraisal events and n is the number of indicators, i is each index, $q_j = \sum_{i=1}^{n} q_{ij} (i = 1, 2, ..., n; j = 1, 2, ..., m)$.

If the number of index is n, and the number of appraisal events is m, then E_i denotes the indicator-based information of indicator i and can be derived thus:

$$E_i = -\frac{1}{\ln m} \sum_{j=1}^{m} \frac{q_{ij}}{q_i} \ln \frac{q_{ij}}{q_i}$$

(3)

where E_i is the information entropy of indicator, q_{ij} is the standardized value calculated from the raw data and q_i is sum for standardized value all appraisal events in i index, $q_i = \sum_{j=1}^{m} q_{ij} (i = 1, 2, ..., n; j = 1, 2, ..., m)$.

According to the entropy weighting method [142], the entropy weight of i indicator is defined as:

$$Q = (1 - E_i) \Big/ \left(n - \sum_{i=1}^{n} E_i \right)$$

(4)

where Q_i is the entropy weight of i, E_i is the indicator-based information entropy of indicator i, n is the number of indicator and $\sum_{i=1}^{n} Q_1 = 1$, $0 \le Q_i \le 1$, $n \ge 2$.

The entropy weight of an indicator is not the most important coefficient of the indicator in regard to decision-making issues. It is instead the relative degree of competition with other indicators when a set of evaluation objects is given and the evaluation indicators are determined, the entropy weighting value is closely related to the evaluation objects. From the information perspective, the entropy weight of an evaluation indicator represents how much useful information an indicator can provide [131,145]. When the entropy weighting of an indicator is larger than other indicators in the evaluation index system for the sustainable development of tourism destination ecosystems, the useful information provided by the indicator could have a greater impact on the system than the other indicators [131].

3.3.2. Sustainable Development Evaluation Model of the Tourism Destination Ecosystem

The index weight was calculated by information entropy, and then integrated to the value of normalization calculated a comprehensive score of values [130,131,146]:

$$G = \sum Q_i X_i \tag{5}$$

where: G is an appraisal score, Q_i is the weighting factor derived from information entropy (described below), X_i is the standardized value between 0 and 1 generated from raw data for each indicator. A larger value of G indicates a better state of the tourism destination ecosystem and a better potential of the tourism destination ecosystem for sustainable development.

4. Data Sources and Processing Method

4.1. Data Sources

Related data applied in this study were extracted from the 10th Five-Year Statistical Yearbook of Dunhuang City [147], 11th Five-Year Statistical Yearbook of Dunhuang City [148], Statistical Yearbook of Dunhuang City between 2011 and 2013 [149], Environmental Quality Bulletin of Dunhuang City between 2006 and 2012 [150]. Some data are obtained from the interview and questionnaire.

4.2. Data Processing Methods

This study adopted the standardize deviation to processing data and the score between [0, 1] when analyzing the evolution and development of the tourism destination socioeconomic ecosystem. The following aspects should be noted in processing data [131,146]: (a) As the entropy change model has used the four types of entropy for vector quantization, there is no need to distinguish between positive and negative indicators to standardize the data processing; (b) The assessment model for the sustainable development potential of the tourism destination ecosystem, which is based on information entropy, has not used vector quantization on different types of indicators, the data processing must distinguish between positive and negative indicators.

For the four indicators, the input supportive type of entropy ($\Delta_e S_1$) *and* regeneration metabolic type of entropy ($\Delta_i S_1$) are positive indicators, the bigger value means more coordination of the system. The output pressure type of entropy ($\Delta_e S_2$) and the consumption metabolic type of entropy ($\Delta_i S_2$) are negative indicators, the bigger value means less coordination of the system. The normalization methods for the positive and negative indicators are listed below:

Normalization method for positive indicators:

$$X'_{ij} = X_{ij} / Max(X_i) \tag{6}$$

Normalization method for negative indicators:

$$X'_{ij} = Min(X_i) / X_{ij} \tag{7}$$

where X'_{ij} is the normalized value of X_{ij}, X_{ij} is the raw data for indicator i in j year, X_i represents all of the original data for indicator i, and $Max(X_i)$ obtains the maximum of indicator i by function during the study period, and $Min(X_i)$ obtains the minimum of indicator i by function during the study period.

5. Results and Analysis

5.1. Entropy Change Analysis

The supportive entropy input showed a trend to remain stable during study period. The stressful entropy output was fluctuated within a slow upward trend (Table 3, Figure 1). For the stressful entropy output with the turning point in 2003 and 2008, the smallest value was in 2003, because the tourism industry was in a state of depression influenced by SARS. In addition, the global financial crisis and snow disaster of South China led to a smaller value in 2008. The burden on tourism destination ecosystems was decreased during those two years. However, the pressure of tourist destinations was increased with the recovery of the tourism industry. The burden of the Dunhuang tourism destination natural ecosystem was increased under the rapid development of tourism and increasing utilization of tourism resources, while the supportive entropy input experienced relatively slow growth. This shows that the pressure was increasing on the tourism destination socioeconomic ecosystem in some degree from 2008.

The consumption metabolism of entropy showed a slowed trend down between 2000 and 2012, the regenerate metabolism of entropy fluctuated with a sharply upward trend. The turning point of regenerate metabolism of entropy was in 2007, 2008 and 2011, which first increased and then decreased with the turning point in 2007 and slowed down sharply. The minimum value was in 2008 and then showed a slowed upward trend. The turning point was a sharp upward trend in 2011 (Table 3, Figure 2). This indicates that the ecological security was improved, the potential of metabolism was better and the vitality was improved gradually in Dunhuang tourism destination from 2008.

Table 3. Entropy production and total entropy change of the tourism destination socio-economic ecosystem in Dunhuang city on information entropy.

Year	Supportive Entropy Input	Stressful Entropy Output	Consumption Metabolism of Entropy	Regenerate Metabolism of Entropy	Entropy Flow	Entropy Production	Total Entropy Change
			Items				
2000	0.7846	0.6370	0.6243	0.8002	−0.1476	−0.1759	−0.3234
2001	0.7925	0.6418	0.6504	0.7838	−0.1507	−0.1334	−0.2840
2002	0.7958	0.6398	0.6429	0.7864	−0.1559	−0.1435	−0.2994
2003	0.7974	0.6356	0.6330	0.7865	−0.1618	−0.1535	−0.3153
2004	0.7793	0.6607	0.6400	0.7793	−0.1186	−0.1393	−0.2579
2005	0.7980	0.6711	0.6505	0.7893	−0.1269	−0.1389	−0.2658
2006	0.7981	0.6789	0.6357	0.7987	−0.1192	−0.1630	−0.2821
2007	0.7969	0.6850	0.6352	0.8332	−0.1119	−0.1980	−0.3099
2008	0.7980	0.6659	0.6360	0.7271	−0.1321	−0.0911	−0.2233
2009	0.7991	0.6747	0.6010	0.7357	−0.1245	−0.1346	−0.2591
2010	0.7993	0.6831	0.5837	0.7549	−0.1162	−0.1713	−0.2875
2011	0.8095	0.6925	0.5953	0.7581	−0.1171	−0.1628	−0.2798
2012	0.8101	0.6914	0.5933	0.8556	−0.1187	−0.2623	−0.3810

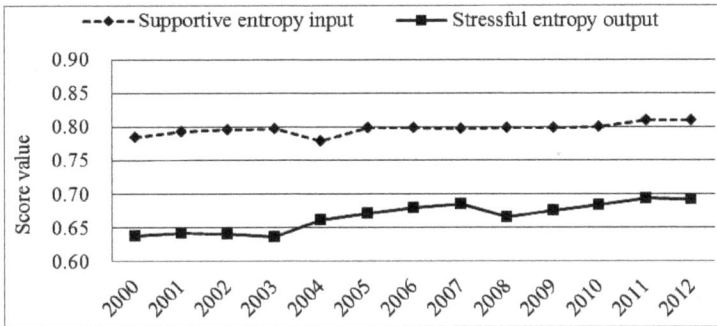

Figure 1. Score trends of the two entropy exchange types of the tourism destination.

The entropy flow showed fluctuation within a slow upward trend between 2000 and 2012, while the entropy production and total entropy change both sharply fluctuated. The turning points of entropy flow, entropy production and total entropy change were during the period of 2003 and 2008 because the tourism industry was mostly effected by the external environment, with the influences of SARS in 2003 and the global financial crisis and snow disaster of South China in 2008. The entropy flow fluctuated with a slow upward trend, which first decreased and then increased with the turning points in 2003 and 2008. The entropy production and total entropy change sharply fluctuated, first increasing and then decreasing with the turning point in 2003. Also, it first increased and then decreased with the turning point in 2008 (Table 3). This indicates that the Dunhuang tourism destination ecosystem was orderly and healthy during the study period.

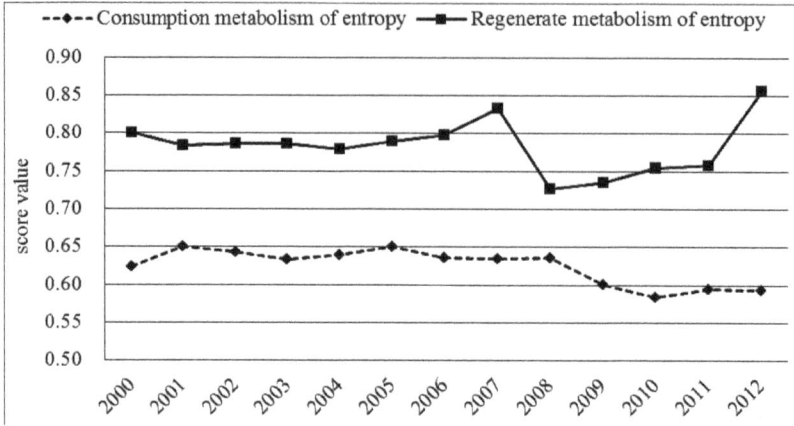

Figure 2. Score trends of the two entropy production types of tourism destination.

5.2. Analysis of Tourism Destination Ecosystem Sustainable Development Potential in Dunhuang

The values of supportive entropy input remained stable between 2000 and 2012. These values indicate the carrying capacity was relatively stable as a socio-economic ecosystem in Dunhuang. The values of stressful entropy output were decreased in 2003 and 2008, which showed that the pressure was increased on the nature ecosystem with the development of tourism (Figure 1). The values of consumption metabolism of entropy fluctuated with a slow downwards trend, and the values of regenerate metabolism of entropy fluctuated with a sharp upward trend. That indicated the metabolism of function was strengthened, which indicates some success in protecting the ecological environment and also its quality was improved (Figure 2). The values of sustainable development potential fluctuated with an upward trend (Figure 3). The lowest value was in 2008 and the highest was in 2012. This phenomenon may be related to external features of the tourism industry, which was in a status of trough influenced by the global financial crisis and snow disaster of Southern China in 2008. The stressful entropy output fluctuated with a downward trend; also the regenerate metabolism of entropy had the lowest value in 2008. The highest value is attributed to government attention and an improvement in ecological security, an increase in investment in anti-pollution projects as percentage of GDP and improved proportion of industrial solid waste treated and utilized.

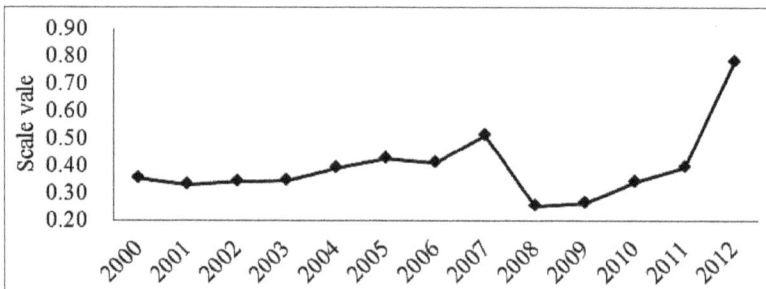

Figure 3. Score trends of tourism destination ecosystem sustainable development potential.

5.3. Analysis of Sustainable Development Measures Based on the Entropy Weights and Time Sequence Changes of the Indicators

The entropy weights of number of travel agencies, passenger-kilometers by highways and passenger-kilometers railways were largest among the supportive entropy input index in Dunhuang tourism destination ecosystem between 2000 and 2012 (Table 4). These indicate that significant increases of these three indexes had played an important role in strengthening the supportive entropy input system. However, the entropy weights of annual water supply were the smallest, and the entropy weights of passenger-km by civil aviation was smaller, showing some negative effects caused by the sharp decline of annual water supply and decrease of passenger-km by civil aviation in the Dunhuang tourism destination ecosystem. Those two indexes restricted the development of the supportive entropy input system. The shortage of water resources is a limiting factor for development of Dunhuang being located in the arid oasis area of northwest China. The transport passenger-kilometers focused on railways and highway, however, civil aviation supplemented the railway and highway with the development of the economy. The positive aspects of transport (railways, highway and civil aviation) and travel agencies should be improved for the supportive entropy input, as well as coordinating the relationship between water resources and to safeguard water supply by increasing effective use.

The entropy weights of number of visitor arrivals and water used by tourists were largest among the stressful entropy output index during the study period (Table 4). The value of those two indexes significantly increased the pressure of stressful entropy output index of Dunhuang tourism destination ecosystem. Those two indexes should be paid much attention, the number of visitor arrivals reasonably controlled and tourists guided on water use.

The entropy weights of industrial wastewater discharged and total wastewater discharged were largest among the consumption metabolism index of entropy between 2000 and 2012 (Table 4). Those indicate that the significant increases of wastewater discharged was strongly influenced by the consumption metabolism index of entropy and increasing the pressure on ecological environments of the tourism destination. The value of entropy weights for waste discharge by tourists was larger than that of carbon emissions, which indicates that the consumption metabolism of tourism destination is influenced more by increase of waste discharge by tourists than carbon emission by tourism. Given these circumstances, scientific controls should be put in place for the waste discharge of tourists.

The value of entropy weights for two indexes (proportion of industrial solid waste treated and utilized and investment in anti-pollution projects as percentage of GDP) were largest among regenerate metabolism index of entropy (Table 4). These show that those two indexes greatly impact the potential of regenerate metabolism system. Pollution should be controlled paying attention to the ecological security of the tourism destination, increasing investment in anti-pollution projects as a percentage of GDP and increasing proportion of industrial solid waste treated and utilized. The values of entropy weights were smallest in regenerate metabolism index of entropy, which include direct engaged persons in tourism industry as percentage of employees, rate of harmless garbage disposal and gardens per capita. These indicate improvement in the potential of regenerate metabolism system by increasing the direct engaged persons in the tourism industry as percentage of employees, improving the rate of harmless garbage disposal and increasing the gardens per capita.

Table 4. Information entropy and entropy weights of the sustainable development potential evaluation indicators for the tourism destination ecosystem in Dunhuang.

Indicator Type	Indicator	E_i	Q_i
Supportive entropy input index (A)	Number of travel agencies (A1)	0.9596	0.0295
	Number of direct engaged persons in tourism industry (A2)	0.9936	0.0047
	Number of star-rated hotels (A3)	0.9945	0.0040
	Number of beds in star-rated hotels (A4)	0.9941	0.0043
	Passenger-kilometers by highways (A5)	0.9669	0.0242
	Passenger-kilometers by railways (A6)	0.9674	0.0238
	Passenger-kilometers by civil aviation (A7)	0.9966	0.0025
	Annual water supply (A8)	0.9968	0.0023
Stressful entropy output index (B)	Number of visitor arrivals (B1)	0.9418	0.0426
	Transport expenditure as percentage of tourism expenditure (B2)	0.9919	0.0059
	sightseeing expenditure as percentage of tourism expenditure (B3)	0.9932	0.0050
	Hotels expenditure as percentage of tourism expenditure (B4)	0.9970	0.0022
	Catering expenditure as percentage of tourism expenditure (B5)	0.9906	0.0069
	Water used by tourists (B6)	0.9414	0.0429
Consumption metabolism index of entropy (C)	Total wastewater discharged (C1)	0.9153	0.0620
	Industrial wastewater discharged (C2)	0.8324	0.1226
	Emission of disulfide (C3)	0.9942	0.0043
	solid wastes discharged (C4)	0.9182	0.0598
	waste discharge by tourists (C5)	0.9414	0.0429
	Carbon emission by tourism (C6)	0.9627	0.0273
Regenerate metabolism index of entropy (D)	Number of training institutions for tourism (D1)	0.9683	0.0232
	Direct engaged persons in tourism industry as percentage of employees (D2)	0.9980	0.0014
	Tourism GDP (D3)	0.8609	0.1017
	Tourism GDP as percentage of GDP (D4)	0.9851	0.0109
	Investment in anti-pollution projects as percentage of GDP (D5)	0.8252	0.1279
	Proportion of industrial solid waste treated and utilized (D6)	0.7336	0.1948
	Rate of harmless garbage disposal (D7)	0.9972	0.0020
	Green coverage rate in developed areas (D8)	0.9799	0.0147
	Gardens per capita (D9)	0.9947	0.0039

6. Conclusions and Discussion

The analysis of the tourist destination ecosystem entropy change indicates an increase in the diversity and complexity of Dunhuang tourism destination's socio-economic ecosystem with the rapid development of the tourism industry; the demands placed on the natural ecosystem have increased. However, the pollution problems have been controlled, as shown by the overall upward trend for regenerate metabolism during the study period. The vitality of the tourism destination ecosystem was obviously strengthened from 2008. Based on the score of sustainable development potential for the tourism destination ecosystem between 2000 and 2012, the pressure on the natural ecosystem was increased, while the carrying capacity of its socio-economic system also strengthen. The regenerate metabolism system increased due to significant conservation achievements and development of the eco-environment in the Dunhuang tourism destination ecosystem. According to the entropy weight of this indicator and its impact on the sustainable development potential of Dunhuang tourism destination ecosystem, the countermeasures are as follows: Increase the potential of the supportive entropy input system by increasing the travel agencies and transportation; Reduce the pressure on the consumption metabolism system by decreasing the total wastewater discharged and industrial wastewater discharged; Enhance the potential of regenerate metabolism by focusing on the ecological security of the tourism destination, increase investment in anti-pollution projects as percentage of GDP and improve proportion of industrial solid waste treated and utilized.

The paper summarizes the former research results, and then further demonstrates by entropy change analysis, information entropy and negative entropy of dissipative structure system for evaluating the tourism destination ecosystem's sustainable development evolution feasibility. The numerical values show the orderly level of the tourism destination ecosystem demonstrating the system sustainable development potential. Combining the entropy weight of index and times series index will be more targeted for improving measures of Dunhuang tourism destination ecosystem's sustainable development. According to the data's availability, the article selects indexes focusing on supportive, stressful, consumed and regenerate indexes. The tourism destination ecosystem as a society-economics-environment artificial compound ecosystem, the tourists and local residents are significant participants and propellants of tourism sustainable development [27]. Their appreciation of tourism sustainable development plays an important role in system improvement; thus the indexes which are in line with their values should be chosen. On the other hand, using information entropy from the perspective of the development of tourism destination ecosystem evolution to analyze tourism destination sustainable development potential, it is beneficial to vertically analyze a single tourism destination. There are disadvantages in horizontally comparing tourism, and therefore this research must be improved.

Acknowledgments

This research is supported by China Scholarship Council (CSC), The Fundamental Research Funds for the Central Universities (lzujbky-2013-m02), Natural Science Foundation of China (41471462, 41461119).

Author Contributions

Huihui Feng, Xingpeng Chen, Peter Heck and Hong Miao designed the paper and all contributed to data collection and calculation.

Conflicts of Interest

The authors declare no conflict of interest.

References

1. General Assembly of the United Nations. The 19th Special Session for General Assembly of the United Nations resolutions. Available online: http://www.un.org/chinese/ga/spec/19/ar19_1.pdf (accessed on 19 September 1997).
2. Tang, C.; Zhong, L.; Cheng, S. A review on sustainable development for tourist destination. *Prog. Geogr.* **2013**, *32*, 984–992.
3. Butler, R. Sustainable tourism: A state-of-the-art review. *Tour. Geogr.* **1999**, *1*, 7–25.
4. Hardy, A.; Beeton, R.; Pearson, L. Sustainable tourism: An overview of the concept and its position in relation to conceptualizations of tourism. *J. Sustain. Tour.* **2002**, *10*, 475–496.
5. Lu, J.; Nepala, S. Sustainable tourism research: An analysis of papers published in the Journal of Sustainable Tourism. *J. Sustain. Tour.* **2009**, *17*, 5–16.
6. Buckley, R. Sustainable tourism: Research and reality. *Ann. Tour. Res.* **2012**, *39*, 528–546.
7. Liu, Z. Sustainable tourism development: A critique. *J. Sustain. Tour.* **2003**, *11*, 459–475.
8. Berno, T.; Bricker, K. Sustainable tourism development: The long road from theory to practice. *Int. J. Econ. Dev.* **2001**, *3*, 1–18.
9. Swarbrooke, J. *Sustainable Tourism Management*; Biddles Ltd., Guildford and King's Lyun: New York, NY, USA, 1999; pp. 4–5.
10. Bramwell, B.; Lane, B. Sustainable tourism: An evolving global approach. *J. Sustain. Tour.* **1993**, *1*, 1–5.
11. Smith, S. *Tourism Analysis: A Handbook*; Longman: London, UK, 1995; pp. 295–299.
12. Wight, P. Tools for sustainability analysis in planning and managing tourism and recreation in the destination. In *Sustainable Tourism: A Geographical Perspective*; Hall, C., Lew, A., Eds.; Addison Wesley Longman: New York, NY, USA, 1998; pp. 75–91.
13. Hunter, C. Sustainable tourism as an adaptive paradigm. *Ann. Tour. Res.* **1997**, *24*, 850–867.
14. Saarinen, J. Traditions of sustainability in tourism studies. *Ann. Tour. Res.* **2006**, *33*, 1121–1140.
15. Sharpley, R. Tourism and sustainable development: Exploring the theoretical divide. *J. Sustain. Tour.* **2000**, *8*, 1–19.
16. Gianna, M.; Laurie, M. There is no such thing as sustainable tourism: Re-conceptualizing tourism as a tool for sustainability. *Sustainability* **2014**, *6*, 2538–2561.
17. Aall, C. Sustainable tourism in practice: Promoting or perverting, the quest for a sustainable development. *Sustainability* **2014**, *6*, 2562–2583.
18. Saarinen, J. Critical sustainability: Setting the limits to growth and responsibility in tourism. *Sustainability* **2014**, *6*, 1–17.
19. Hunter, C. Aspects of the Sustainable Tourism Debate from a Natural Resources Perspective. In *Sustainable Tourism: A Global Perspective*; Harris, R., Griffin, T., Williams, P., Eds.; Butterworth-Heinemann: Oxford, UK, 2002; pp. 4–5.
20. Moyle, B.; McLennan, C.; Ruhanen, L.; Weiler, B. Tracking the concept of sustainability in Australian tourism policy and planning documents. *J. Sustain.* **2014**, *22*, 1037–1051.
21. Farrell, B.; Ward, L. Reconecptualizing tourism. *Ann. Tour. Res.* **2004**, *31*, 274–295.

22. Lansing, P.; Vries, P. Sustainable tourism: Ethical alternative or marketing ploy? *J. Bus. Ethics* **2007**, *72*, 77–85.

23. Delgado, A.; Saarinenc, J. Using indicators to assess sustainable tourism development: A review. *Tour. Geogr.* **2014**, *16*, 31–47.

24. Ko, T. Development of a tourism sustainability assessment procedure: A conceptual approach. *Tour. Manag. Issue* **2005**, *26*, 431–445.

25. Chris, H.; Sirakaya, E. Sustainability indicators for managing community tourism. *Tour. Manag.* **2006**, *27*, 1274–1289.

26. McElroy, J. Tourism penetration index in small Caribbean islands. *Ann. Tour. Res.* **1998**, *25*, 145–168.

27. McCool, S.; Moisey, R.; Nickerson, N.P. What should tourism sustain? The disconnect with industry perceptions of useful indicators. *J. Travel Res.* **2001**, *40*, 124–131.

28. Moore, S.; Polley, A. Defining indicators and standards for tourism impacts in protected areas: Cape Range National Park, Australia. *Environ. Manag.* **2007**, *39*, 291–300.

29. Castellani, V.; Sala, S. Sustainable performance index for tourism policy development. *Tour. Manag.* **2010**, *31*, 871–880.

30. Blancas, F.; Gonzalez, M.; Lozano-Oyola, M.; Pérez, F. The assessment of sustainable tourism: Application to Spanish coastal destinations. *Ecol. Indic.* **2010**, *10*, 484–492.

31. Twining, L.; Butler, R. Implementing STD on a small island: Development and use of sustainable tourism development indicators in Samoa. *J. Sustain. Tour.* **2002**, *10*, 363–387.

32. Roberts, S.; Tribe, J. Sustainability indicators for small tourism enterprises—An exploratory perspective. *J. Sustain. Tour.* **2008**, *16*, 575–594.

33. Schianetz, K.; Kavanagh, L. Sustainability indicators for tourism destinations: A complex adaptive systems approach using systemic indicator systems. *J. Sustain. Tour.* **2008**, *16*, 601–628.

34. Fernández, J.; Sánchez, R. Measuring tourism sustainability: Proposal for a composite index. *Tour. Econ.* **2009**, *15*, 277–296.

35. Blackstock, K.; White, V.; McCrum, G.; Scott, A.; Hunter, C. Measuring responsibility: An appraisal of a scottish national park's sustainable tourism indicators. *J. Sustain. Tour.* **2008**, *16*, 276–297.

36. Oyola, M.; Blancas, F.; González, M.; Caballero, R. Sustainable tourism indicators as planning tools in cultural destinations. *Ecol. Indic.* **2012**, *18*, 659–675.

37. Blancas, F.; Oyola, M.; González, M.; Guerrero, F.M.; Caballero, R. How to use sustainability indicators for tourism planning: The case of rural tourism in Andalusia (Spain). *Sci. Total Environ.* **2011**, *412–413*, 28–45.

38. Reddy, M. Sustainable tourism rapid indicators for less-developed islands: An economic perspective. *Int. J. Tour. Res.* **2008**, *10*, 557–576.

39. Miller, G. The development of indicators for sustainable tourism: Results of a Delphi survey of tourism researchers. *Tour. Manag.* **2001**, *22*, 351–362.

40. Wang, L. On the indicator system of sustainable development of tourism and the evaluating method. *Tour. Tribune* **2001**, *16*, 67–70.

41. Wan, Y. The method and indicates of evaluation on sustainable development for tourism. *Stat. Decis.* **2006**, *2*, 10–12.

42. Ahn, B.; Lee, B.; Shafer, C.S. Operationalizing sustainability in regional tourism planning: An application of the limits of acceptable change framework. *Tour. Manag.* **2002**, *23*, 1–15.

43. Hunter, C.; Shaw, J. The ecological footprint as a key indicator of sustainable tourism. *Tour. Manag.* **2007**, *28*, 46–57.

44. Gössling, S.; Hansson, C.; Hörstmeier, O.; Saggel, S. Ecological footprint analysis as a tool to assess tourism sustainability. *Ecol. Econ.* **2002**, *43*, 199–211.

45. Martín-Cejas, R.; Sánchez, P. Ecological footprint analysis of road transport related to tourism activity: The case for Lanzarote Island. *Tour. Manag.* **2010**, *31*, 98–103.

46. Mehdi, M.; Jerome, B. Ecotourism *versus* mass tourism: A comparison of environmental impacts based on ecological footprint analysis. *Sustainability* **2012**, *4*, 123–140.

47. Hunter, C. Sustainable tourism and the touristic ecological footprint. *Environ. Dev. Sustain.* **2002**, *4*, 7–20.

48. Yang, G.; Li, P. Touristic ecological footprint: A new yardstick to assess sustainability of tourism. *Acta Ecol. Dinica* **2005**, *25*, 1475–1480.

49. Castellani, V.; Sala, S. Ecological footprint and life cycle assessment in the sustainability assessment of tourism activities. *Ecol. Indic.* **2012**, *16*, 135–147.

50. Cui, F.; Liu, J.; **Li, Q**. Study of the theory and application of tourism bearing capacity index. *Tour. Tribune* **1998**, *22*, 41–44.

51. Xiao, J.; Yu, Q.; Liu, K.; Chen, D.; Chen, J.; Xiao, J. Evaluation of the ecological security of island tourist destination and island tourist sustainable development: A case study of Zhoushan Islands. *Acta Geogr. Sin.* **2011**, *66*, 842–852.

52. Salerno, F.; Viviano, G.; Manfredi, E.C.; Caroli, P.; Thakuri, S.; Tartari, G. Multiple Carrying Capacities from a management-oriented perspective to operationalize sustainable tourism in protected areas. *J. Environ.* **2013**, *128*, 116–125.

53. Zhong, L.; Deng, J.; Xiang, B. Tourism development and the tourism area life-cycle model: A case study of Zhangjiajie National Forest Park, China. *Tour. Manag.* **2008**, *29*, 841–856.

54. Yang, Y.; Luo, S.; Wang, X. Tourist destination life cycle early warning system research based on "Inflection Point" theory, China. *Popul. Resour. Environ.* **2009**, *19*, 110–113.

55. Rodríguez, J.; Parra-López, E.; Yanes-Estévez, V. The sustainability of island destinations: Tourism area life cycle and teleological perspectives: The case of Tenerife. *Tour. Manag.* **2008**, *29*, 53–65.

56. Yang, G. Targeting model of sustainable development in eco-tourism. *Hum. Geogr.* **2005**, *20*, 74–77.

57. Keitumetse, S. Cultural resources as sustainability enablers: Towards a community-based cultural heritage resources management (COBACHREM) model. *Sustainability* **2014**, *6*, 70–85.

58. Dina, R.; Geremia, G. A sustainable tourism paradigm: Opportunities and limits for forest landscape planning. *Sustainability* **2014**, *6*, 2379–2391.

59. Karatzogloul, B.; Spilanis, I. Sustainable tourism in Greek islands: The integration of activity-based environmental management with a destination environmental scorecard based on the adaptive resource management paradigm. *Bus. Strategy Environ.* **2010**, *19*, 26–38.

60. Lu, M. *Evaluation Pattern on the Sustainability of Tourist Destination*; Central China Normal University: Wuhan, China, 2001.

61. Xuan, G.; Lu, L.; Zhang, J. Residents' perception of tourism impacts in coast resorts: The case study of Haikou and Sanya Cities, Hainan Province. *Sciatica Geogr. Sonica* **2002**, *22*, 741–746.

62. Su, Q.; Lin, B. Classification of residents in the tourist attractions based on attitudes and behaviors: A case study in Xidi, Zhouzhuang and Jiuhua Mountain. *Geogr. Res.* **2004**, *23*, 104–114.

63. Zhao, Y.; Li, D.; Huang, M. On the residents' perceptions and attitudes towards tourism development in tourist destinations overseas: A review. *Tour. Tribune* **2005**, *20*, 85–92.

64. Huang, Y.; Huang, Z. A study on the structural equation model and its application to tourist perception for agri-tourism destinations: Taking southwest minority areas as an example. *Geogr. Res.* **2008**, *27*, 1455–1465.

65. Choi, H.; Sirakaya, E. Measuring residents' attitude toward sustainable tourism: Development of sustainable tourism attitude scale. *J. Travel Res.* **2005**, *43*, 380–394.

66. Wearing, S.; Wearing, M. Understanding local power and interactional processes in sustainable tourism: Exploring village-tour operator relations on the Kokoda Track, Papua New Guinea. *J. Sustain.* **2010**, *18*, 61–76.

67. Haukeland, J.; Grue, B.; Veisten, K. Turning national parks into tourist attractions: Nature orientation and quest for facilities, Scandinavian. *J. Hosp. Tour.* **2010**, *10*, 248–271.

68. Lucia, S.; Nicholas, L.; Thapa, B. Visitor perspectives on sustainable tourism development in the pitons management area world heritage site. *Environ. Dev. Sustain.* **2010**, *12*, 839–857.

69. Cottrell, S.; Vaske, J.; Roemer, J.M. Resident satisfaction with sustainable tourism: The case of Frankenwald Nature Park, Germany. *Tour. Manag. Perspect.* **2013**, *8*, 42–48.

70. Sörensson, A.; Friedrichs, Y. An importance–performance analysis of sustainable tourism: A comparison between international and national tourists. *J. Destin. Mark. Manag.* **2013**, *2*, 14–21.

71. Dorcheh, S.; Mohamed, B. Local perception of tourism development: A conceptual framework for the sustainable cultural tourism. *J. Manag. Sustain.* **2013**, *3*, 31–39.

72. Cottrell, S.; Vaske, J. Resident perceptions of sustainable tourism in Chongdugou, China. *Soc. Nat. Resour. Int. J.* **2007**, *20*, 511–525.

73. Miller, D.; Merrilees, B. Sustainable urban tourism: Understanding and developing visitor pro-environmental behaviors. *J. Sustain. Tour.* **2014**, *5*, 1–21.

74. Chen, C.; Chen, S.; Hong, T. The destination competitiveness of Kinsmen's tourism industry: Exploring the interrelationships between tourist perceptions, service performance, customer satisfaction and sustainable tourism. *J. Sustain. Tour.* **2011**, *19*, 247–264.

75. Bimonte, S.; Punzo, L. Tourism, residents' attitudes and perceived carrying capacity with an experimental study in five Tuscan destinations. *Int. J. Sustain.* **2011**, *14*, 242–261.

76. Waligo, V.; Clarke, J.; Hawkins, R. Implementing sustainable tourism: A multi-stakeholder involvement management framework. *Tour. Manag.* **2013**, *36*, 342–353.

77. Jamal, T.; Stronza, A. Collaboration theory and tourism practice in protected areas: Stakeholders, structuring and sustainability. *J. Sustain. Tour.* **2009**, *17*, 169–189.

78. Vellecco, I.; Alessandra, M. Sustainability and tourism development in three Italian destinations: Stakeholders' opinions and behaviors. *Serv. Ind. J.* **2010**, *30*, 2201–2223.

79. Spenceley, A. Requirements for sustainable nature-based tourism in transfrontier conservation areas: A southern African Delphi consultation. *Tour. Geogr.* **2008**, *10*, 285–311.

80. Holden, A. Exploring stakeholders' perceptions of sustainable tourism development in the Annapurna Conservation Area: Issues and challenge. *Tour. Hosp. Plan. Dev.* **2010**, *7*, 337–351.

81. Timur, S.; Getz, D. Sustainable tourism development: How do destination stakeholders perceive sustainable Urban Tourism? *Sustain. Dev.* **2009**, *17*, 220–232.

82. Dabphet, S.; Scott, N.; Ruhanen, L. Applying diffusion theory to destination stakeholder understanding of sustainable tourism development: A case from Thailand. *J. Sustain. Tour.* **2012**, *20*, 1107–1124.

83. Hardy, A.; Beeton, R. Sustainable tourism or maintainable tourism: Managing resources for more than average outcomes. *J. Sustain. Tour.* **2001**, *9*, 168–192.

84. Larson, P.; Poudyal, N. Developing sustainable tourism through adaptive resource management: A case study of Machu Picchu. *J. Sustain. Tour.* **2012**, *20*, 917–938.

85. Dodds, R. Sustainable tourism and policy implementation: Lessons from the case of Calviá, Spain. *Curr. Issues Tour.* **2007**, *10*, 296–322.

86. Dodds, R.; Butler, R. Barriers to implementing sustainable tourism policy in mass tourism destinations. *Int. Multidiscip. J. Tour.* **2010**, *5*, 35–53.

87. Farsari, Y.; Butler, R.; Prastacos, P. Sustainable tourism policy for Mediterranean destinations: Issues and interrelationships. *Int. J. Tour.* **2007**, *1*, 58–78.

88. Pigram, J. Sustainable tourism—Policy considerations. *J. Tour. Stud.* **1990**, *1*, 2–9.

89. Solstrand, M. Marine angling tourism in Norway and Iceland: Finding balance in management policy for sustainability. *Nat. Resour. Forum* **2013**, *37*, 113–126.

90. Xu, H.; Sofield, T. Sustainability in Chinese development tourism policies. *Curr. Issues Tour.* **2013**, *10*, 30–38.

91. Michael, C. *Local Government and Tourism Public Policy: A Case of the Hurunui District, New Zealand*; Lincoln University: Lincoln, UK, 2013.

92. Yasarataa, M.; Altinay, L.; Burns, P.; Okumus, F. Politics and sustainable tourism development—Can they co-exist? Voices from North Cyprus. *Tour. Manag.* **2010**, *31*, 345–356.

93. Whitford, M.; Ruhanen, L. Australian indigenous tourism policy: Practical and sustainable policies? *J. Sustain. Tour.* **2010**, *18*, 475–496.

94. Halla, C. Policy learning and policy failure in sustainable tourism governance: From first- and second-order to third-order change? *J. Sustain. Tour.* **2011**, *19*, 649–671.

95. Bahaire, T.; White, M. The application of geographical information systems (GIS) in sustainable tourism planning: A review. *J. Sustain. Tour.* **1999**, *7*, 159–174.

96. Boers, B.; Cottrell, S. Sustainable tourism infrastructure planning: A GIS-supported approach. *Tour. Geogr.* **2007**, *9*, 1–21.

97. Bunruamkaew, K.; Murayama, Y. Site suitability evaluation for ecotourism using GIS & AHP: A case study of Surat Thani province, Thailand. *Procedia Soc. Behav. Sci.* **2011**, *21*, 269–278.

98. White, V.; McCrum, G.; Blackstock, K.L.; Scott, A. Indicators and sustainable tourism: Literature review. *Tour. Manag.* **2006**, *27*, 1274–1289.

99. Ivan, K.; Mikulić, J. Research note: Measuring tourism sustainability: An empirical comparison of different weighting procedures used in modelling composite indicators. *Tour. Econ.* **2014**, *20*, 429–437.

100. Mikulića, J.; Kožićb, I.; Krešić, D. Weighting indicators of tourism sustainability: A critical note. *Ecol. Indic.* **2015**, *48*, 312–314.

101. Delgado, T.A.; Palomeque, F.L. Measuring sustainable tourism at the municipal level. *Ann. Tour. Res.* **2014**, *49*, 122–137.

102. Coccossis, H.; Mexa, A. *The Challenge of Tourism Carrying Capacity Assessment: Theory and Practice*; Ashgate, Basingstoke: Hampshire, UK, 2004; pp. 277–288.

103. Simón, F.; Narangajavana, Y.; Marqués, D.P. Carrying capacity in the tourism industry: A case study of Hengistbury Head. *Tour. Manag.* **2004**, *25*, 275–283.

104. Juradoa, E.; Tejadab, M.T.; García, F.A.; González, J.C.; Macías, R.C.; Peña, J.D.; Gutiérrez, F.F.; Fernández, G.G.; Gallego, M.L.; García, G.M.; *et al.* Carrying capacity assessment for tourist destinations. Methodology for the creation of synthetic indicators applied in a coastal area. *Tour. Manag.* **2012**, *33*, 1337–1346.

105. Green, H.; Hunter, C.; Moore, B. The environmental impact assessment of tourism development. In *Perspectives on Tourism Policy*; Johnson, P., Thomas, B., Eds.; Thomson Learning: Boston, MA, USA, 1992; pp. 29–47.

106. Green, H.; Hunter, C.; Moore, B. Assessing the environmental impact of tourism development: Use of the Delphi technique. *Tour. Manag.* **1990**, *11*, 111–120.

107. Peng, J.; Wu, J.; Jiang, Y.-Y.; Ye, M.-T. Shortcomings of applying ecological footprints to the ecological assessment of regional sustainable development. *Acta Ecol. Sin.* **2006**, *26*, 2716–2722.

108. Haywood, K. Can the tourist-area life cycle be made operational? *Tour. Manag.* **1986**, *7*, 154–167.

109. Guinee, J.; Heijungs, R.; Huppes, G.; Zamagni, A.; Masoni, P.; Buonamici, R.; Ekvall, T.; Rydberg, T. Life cycle assessment: Past, present, and future. *Environ. Sci. Technol.* **2011**, *45*, 90–96.

110. McCool, S. Planning for sustainable nature dependent tourism development: The Limits of Acceptable Change system. *Tour. Recreat. Res.* **1994**, *19*, 51–55.

111. Avdimiotis, S.; Mavrodontis, T.; Dermetzopoulos, A.S.; Riavoglou, K. GIS applications as a tool for tourism planning and education: A case study of Chalkidiki. *Tourism (Zagreb)* **2006**, *54*, 405–413.

112. Schianetz, K.; Kavanagh, L.; Lockington, D. Concepts and tools for comprehensive sustainability assessments for tourism destinations: A comparative review. *J. Sustain. Tour.* **2007**, *15*, 369–389.

113. Lacitignola, D.; Petrosillo, I.; Cataldi, M.; Zurlini, G. Modelling socio-ecological tourism-based systems for sustainability. *Ecol. Model.* **2007**, *206*, 191–204.

114. Bailey, K. Social entropy theory: An overview. *Syst. Pract.* **1990**, *3*, 365–382.

115. Swanson, G.A.; Kenneth, D.B. Social Entropy Theory, Macro Accounting, and Entropy Related Measures, 2006. Available online: http://www.isssbrasil.usp.br/isssbrasil/pdfs/2006-247.pdf (accessed on 30 July 2014).

116. Stepanic, J.; Stefancic, H.; Zebec, M.S.; Perackovic, K. Approach to a quantitative description of social systems based on thermodynamic formalism. *Entropy* **2000**, *2*, 98–105.

117. Wilson, A. Entropy in urban and regional modelling: Retrospect and prospect. *Geogr. Anal.* **2010**, *42*, 364–394.

118. Cabral, P.; Augusto, G.; Tewolde, M.; Araya, Y. Entropy in urban systems. *Entropy* **2013**, *15*, 5223–5236.

119. Leiper, N. Industrial entropy in tourism systems. *Ann. Tour. Res.* **1993**, *20*, 221–226.

120. Sabkbar, H.; Ahmad, A.; *Karimian, T.* Spatial tourism planning in Isfahan by entropy model. *Agrochimica* **2014**, *58*, 67–82.

121. Hao, C. A Tentative analysis of entropy and dissipative structure in tourist destination ecological system. *J. Taiyuan Norm. Univ. (Soc. Sci. Ed.)* **2010**, *9*, 66–68.

122. Zhao, L.; Zhang, W. Applied research of dissipative structure theory on tolerance threshold of tourism system-on the basis of tourism system entropy principle. *Tour. Forum* **2010**, *3*, 10–16.

123. Qian, Y. Application of the dissipative structure theory in the traveling environment management. *J. Anhui Agric. Sci.* **2006**, *34*, 5006–5007.

124. Ministry of Agriculture of the People's Republic of China. This Is the Time to Be Excavated of Dunhuang Culture: The Report for Culture Industry of Dunhuang, Gansu Province. Available online: http://www.agri.gov.cn/dfv20/GS/ncwh/dtzx/201306/t20130603_3480954.htm (accessed on 3 June 2013).

125. Statistics Report for Economic and Social Development of Dunhuang City in 2012. Available online: http://zwgk.dunhuang.gov.cn/ReadNews.asp?NewsID=2367 (accessed on 28 July 2014).

126. It is Urgent Time to Combating Desertification in Dunhuang by Expert's Opinions. Available online: http://news.xinhuanet.com/local/2010-09/11/c_12542051.htm (accessed on 14 September 2010).

127. Xing, X. Evolution equation for physical entropy and information entropy. *Sci. China (Ser. A)* **2001**, *31*, 78–84.

128. Li, X. The entropy in Socio-economic system. *J. Syst. Dialectics* **1996**, *10*, 84–87.

129. Shannon, C. A mathematical theory of communication (1–2). *Bell Syst. Tech. J.* **1948**, *27*, 623–656.

130. Zhang, Y.; Yang, Z.; Li, W. Measurement and evaluation of interactive relationships in urban complex ecosystem. *Acta Ecol. Sin.* **2005**, *25*, 1734–1740.

131. Lin, Z.; Xia, B. Analysis of sustainable development ability of the urban ecosystem in Guangzhou City in the perspective of entropy. *Acta Geogr. Sin.* **2013**, *68*, 45–57.

132. Leife, R. Understanding organizational transformation using a dissipative structure model. *Hum. Relat.* **1989**, *42*, 899–916.

133. Kondoh, Y. Function for dissipative dynamic operators and the attractor of the dissipative structure. *Phys. Rev. E* **1993**, *48*, 2975–2979.

134. Dooley, K. A complex adaptive systems model of organization change nonlinear dynamics. *Psychol. Life Sci.* **1997**, *1*, 69–97.

135. Zhan, K.; Sheng, X. *Prigogine and Theory of Dissipative Structure*; Shanxi Science & Technology Press: Xi'an, China, 1998.

136. Lu, L.; Bao, J. The course and mechanism of evolution about Qiandao Lake based on the theory of dissipative structure. *Acta Geogr. Sin.* **2010**, *65*, 755–768.

137. Valentin, A.; Spangenberg, J. A guide to community sustainability indicators. *Environ. Impact Assess. Rev.* **2000**, *20*, 381–392.

138. Dai, Y. Sustainable development indexes systems in tourist cities. *Yunnan Geogr. Environ. Res.* **2006**, *18*, 35–39.

139. Indicators of Sustainable Development for Tourism Destinations: A Guidebook. Available online: http://www.e-unwto.org/content/x53g07/fulltext?p=a6a03d7c67d24258bf5c4ae9e96b6530&pi=0#section=890050&page=2&locus=40 (accessed on 1 October 2004).

140. European Tourism Indicator System Toolkit for Sustainable Destinations. Available online: http://ec.europa.eu/enterprise/sectors/tourism/sustainable-tourism/indicators/documents_indicators/eu_toolkit_indicators_en.pdf (accessed on 10 February 2013).

141. Niu, Y. The study on index system of sustainable tourism. *Chn. Pop. Resour. Environ.* **2002**, *12*, 42–45.

142. Zhang, J.; Singh, V. *Theory and Applications of Information Entropy*; China Water Power Press: Beijing, China, 2012; pp. 79–80.

143. Yang, X. Evaluation research of sustainable development of tourism based on the information entropy and AHP. *Sci. Technol. Eng.* **2008**, *22*, 6176–7004.

144. Jiang, C.; Zhang, Y. Study on the mechanism of evolution about island resorts system based on the theory of dissipative structure: A case study of Meizhou Island. *Chin. Agric. Sci. Bull.* **2013**, *29*, 213–220.

145. Di, Q.; Han, X. Sustainable development ability of China's marine ecosystem in the perspective of entropy. *Sci. Geogr. Sin.* **2014**, *34*, 6664–6671.

146. Chen, Y.; Liu, J. An index of equilibrium of urban land-use structure and information dimension of urban form. *Geogr. Res.* **2001**, *20*, 146–152.

147. Dunhuang City Bureau of Statistics. *10th Five-Year Statistical Yearbook of Dunhuang City*; Wang, W., Ed.; Juelun Culture Communication Co., Ltd.: Xiamen, China, 2006; pp. 252–305.

148. Dunhuang City Bureau of Statistics. *11th Five-Year Statistical Yearbook of Dunhuang City*; Ren, B., Ed.; Sanmenxia Culture Communication Co., Ltd.: Sanmenxia, China, 2011; pp. 258–309.

149. Dunhuang City Bureau of Statistics. *2011 Statistical Yearbook of Dunhuang City*; Dunhuang City Bureau of Statistics: Dunhuang, China, 2012.
150. Dunhuang Municipal Environmental Protection Bureau. *Environmental Quality Bulletin of Dunhuang City (2006–2012)*; Dunhuang Municipal Environmental Protection Bureau: Dunhuang, China, 2013.

Sustainability Assessment of Solid Waste Management in China: A Decoupling and Decomposition Analysis

Xingpeng Chen, Jiaxing Pang, Zilong Zhang and Hengji Li

Abstract: As the largest solid waste (SW) generator in the world, China is facing serious pollution issues induced by increasing quantities of SW. The sustainability assessment of SW management is very important for designing relevant policy for further improving the overall efficiency of solid waste management (SWM). By focusing on industrial solid waste (ISW) and municipal solid waste (MSW), the paper investigated the sustainability performance of SWM by applying decoupling analysis, and further identified the main drivers of SW change in China by adopting Logarithmic Mean Divisia Index (LMDI) model. The results indicate that China has made a great achievement in SWM which was specifically expressed as the increase of ISW utilized amount and harmless disposal ratio of MSW, decrease of industrial solid waste discharged (ISWD), and absolute decoupling of ISWD from economic growth as well. However, China has a long way to go to achieve the goal of sustainable management of SW. The weak decoupling, even expansive negative decoupling of ISW generation and MSW disposal suggests that China needs timely technology innovation and rational institutional arrangement to reduce SW intensity from the source and promote classification and recycling. The factors of investment efficiency and technology are the main determinants of the decrease in SW, inversely, economic growth has increased SW discharge. The effects of investment intensity showed a volatile trend over time but eventually decreased SW discharged. Moreover, the factors of population and industrial structure slightly increased SW.

Reprinted from *Sustainability*. Cite as: Chen, X.; Pang, J.; Zhang, Z.; Li, H. Sustainability Assessment of Solid Waste Management in China: A Decoupling and Decomposition Analysis. *Sustainability* **2014**, *6*, 9268-9281.

1. Introduction

The production of solid waste (SW) is an inevitable consequence of population boom, economic growth, rapid urbanization and the rise of human living standards, especially for developing countries due to incomplete institutional arrangements for solid waste management (SWM) [1,2]. SWM is one of the most challenging issues faced by developing countries [3]. Due to incomplete institutional arrangement and improper handling of SWM, developing countries are suffering from serious pollution problems caused by growing quantities of SW, such as contamination of water, soil and atmosphere, negative impacts on human health, and its contribution to climate change [2,3]. The sustainable management of SW will become necessary at all phases of impact from planning to design, to operation, and to decommissioning in the 21st century [4].

As the second largest economy in the world, China has become the largest SW generator in the world since 2004 owing to the unprecedented rate of urbanization, industrialization and steadily improving living standards [5,6]. According to the projection of the World Bank (2005), the total amount of SW in China will be over 480 million tons in 2030 [7]. In order to resolve the environmental issues induced by SW disposal, China has devoted considerable efforts to managing SW and issued a series of regulations and policies for SWM, such as the Law of PR China on the Prevention of Environmental pollution, which is the main legislation specifically pertaining to SWM and pollution control, the Law of the People's Republic of China on The Prevention and Control of Environmental Pollution by Solid Waste issued in 1996 and amended in 2004, the Law on Circular Economy Promotion issued in 2009, and has established a legal framework on SW reduction, reuse and recycling (3Rs) *et al.* The investment in SW treatment equipment and infrastructure increased 7.94 times, and over 3.95 times more SW was treated or disposed of safely from 1991–2010. However, compared with developed countries, such as Germany, Japan, Sweden, The Netherlands and Sweden, China still has a long way to go in the sustainable management of SW (especially for municipal solid waste, MSW) with respect to reduction, recycling, reuse and safely treatment technology and strategy [5,8,9].

There are lots of studies which have focused on China's SWM at the city level [8,10–14], provincial level [6,15] and even at the country level [5,7,9,16–20], including the trend of SW generation, the composition of SW, the barriers of SWM [7,9,18,21], and the impacts of SW on the environment [15,22] and contribution to climate change [13]. However, to our knowledge, few studies have focused on the sustainability assessment of China's SWM which is very important for designing relevant policy for further improving the overall efficiency of SWM. The paper integrated the approach of decoupling and decomposition analysis to assess the sustainability status of SWM by analyzing the decoupling of SW from economic growth, and identified the main driving factors of changes in SW during 1991–2010, to serve as the basis for future policy scenarios.

2. Methodologies

Sustainability assessment has been developed conceptually and through practical applications, and has increasingly become associated with the family of impact assessment tools [23,24]. Ness *et al.* conceptually reviewed several tools of sustainability assessment based on three main categories such as indicators/indices, product-related assessment, and integrated assessment tools [24]. With regard to the approaches for sustainability assessment of SMM, many scholars applied environmental performance indicators [25], zero waste index [26,27], life cycle assessment [28,29], material flow analysis [30] and emergy (or exergy) analysis [15] to assess the sustainability of SWM, and other scholars have carried out the sustainability assessment by means of computer based multiple sustainability assessment models and sustainability assessment by success and efficiency factors [30]. Except for the approaches mentioned above, from the perspective of macroeconomics, decoupling of economic activity from environmental or waste impacts has been proposed as the inter-linked objectives for enhancing cost-effective and operational environmental policies in the context of sustainability [31,32]. Due to the problems of availability of a sufficient quantity of reliable data, especially for detailed data on SW composition, it is very difficult to conduct process-based sustainability analysis (e.g., LCA, MFA) of China at the macro level. So, the paper proposes to apply the decoupling analysis (including decoupling SW generation and discharge from economic growth) to evaluate the sustainability performance of SWM in China during

1991–2010. The paper also used decomposition analysis to investigate the underlying determinant effects that influence the change of total SW generation and discharge.

2.1. Decoupling Analysis

The definition of decoupling environmental pressures (E) from economic growth (take GDP for example) is shown as in Figure 1 [33]. Decoupling status could be estimated by the GDP elasticity values of environmental pressure which shows in Equation (1):

$$\text{GDP elasticity of E} = \%\Delta E / \%\Delta GDP \qquad (1)$$

When using economic output per capita as the X-axis and environmental impact as the Y-axis, eight logical possibilities can be distinguished. In order to not over-interpret slight changes as significant, a ±20% variation of the elasticity values around 1.0 (0.8–1.2) is still regarded as coupling [33,34]. Compared to this diagram, actually, our research results mostly fell in the right part, which is indicated as zone 1 to zone 4 in Figure 1.

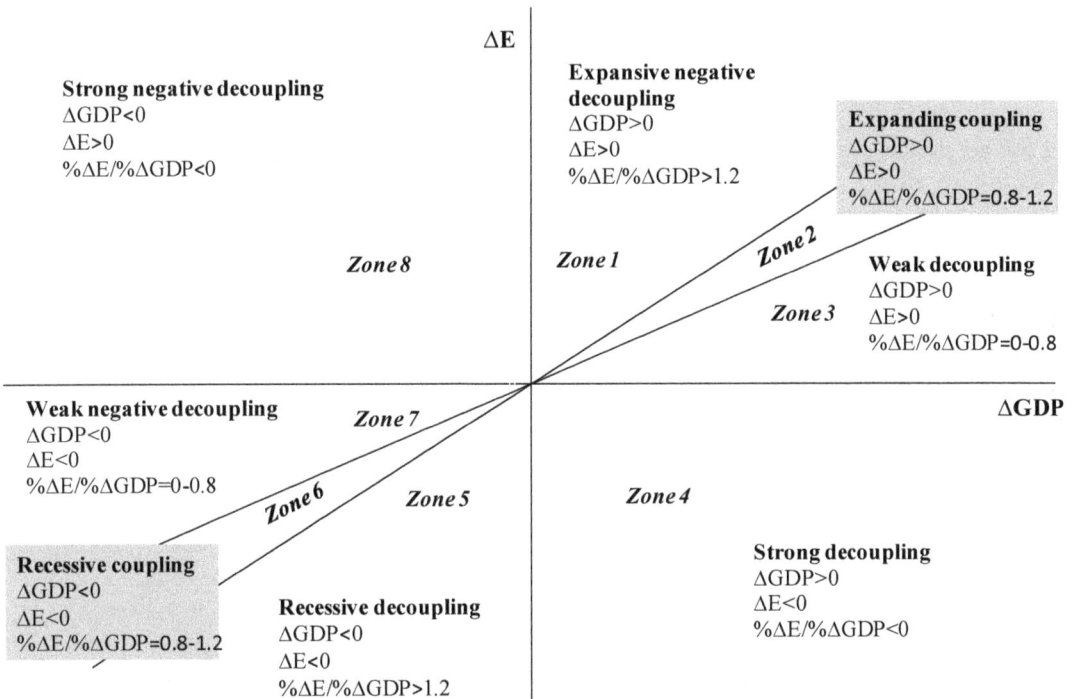

Figure 1. Degree of coupling and decoupling zones ([33]).

The paper conducted the decoupling analysis of industrial solid waste (includes both the non-hazard and hazard industrial solid waste) generation (ISWG), industrial solid waste discharge (ISWD), and municipal solid waste generation (MSWG) from economic growth by calculating the GDP elasticity of ISWG, ISWD, MSWG according to Equation (1). As for the MSWG, since urban population and demographic trend are the key drivers for urban garbage generation and disposal, the role of population in MSWG decoupling allows the capturing of the dynamic effect of consumption and population growth, provides the relative direction of the change and the scale of the change, and also allows comparing real variations in MSWG and population [35,36]. So, the paper carried out the decoupling analysis of MSWG from population increase by calculating the population (Pop) elasticity of MSWG according to Equation (2):

$$\text{Pop elasticity of MSWG} = \%\Delta\text{MSWG}/\%\text{Pop} \qquad (2)$$

2.2. Index Decomposition Analysis

Decomposition methodology which is a technique has been widely applied in environmental analysis, provides a linkage between an aggregate and the original raw data whereby information of interest is captured in a concise and usable form [37,38]. The paper applied logarithmic mean divisia index (LMDI), which is one of index decomposition analysis methods and has been widely applied to analyze resource consumption and waste emission, to identify major driving factors influencing changes in ISWD and the harmless disposal of municipal solid waste (MSWHD) of China. The ISWD in China was decomposed into six influencing factors: population, per capita GDP, industrial structure, investment intensity in treatment of ISW, investment efficiency of ISW recycle, and ISWD intensity. The MSWHD in China was divided into three influencing factors: urban population, per capita investment in sanitation (includes collection, transportation and disposal), investment efficiency of MSW treatment and disposal (Table 1). Based on the LMDI decomposition model [38–40], the paper decomposed the ISWD into six influencing factors, and MSWHD into three influencing factors. The meaning of each factor is shown in Table 1.

Table 1. The fundamental equations of LMDI model.

Equations	The Meaning of Factors in Equations
$ISW = P \times \dfrac{G}{P} \times \dfrac{G_2}{G} \times \dfrac{ISW_{(IN)}}{G_2} \times \dfrac{ISW_{(RECYCLE)}}{ISW_{(IN)}} \times \dfrac{ISW}{ISW_{(RECYCLE)}}$ $= P \times a \times s \times e \times b \times d \quad (3)$	P: population scale; G: gross domestic production (GDP); G_2: value added of the secondary industry; $ISW_{(IN)}$: investment in treatment of industrial solid waste; $ISW_{(RECYCLE)}$: the amount of ISW recycled; ISW: the amount of ISW discharged; a: per capita GDP, s: the proportion of the second industry to GDP refers to industrial structure; e: investment intensity of ISW management; b: investment efficiency of ISW recycling; d: ISW discharged intensity refers to technology.
$MSWHD = UP \times \dfrac{A}{UP} \times \dfrac{MSWHD}{A}$ $= UP \times m \times n \quad (4)$	UP: urban population refers to urban scale; A: investment in sanitation; MSWHD: the amount of harmless disposal of MSW; m: per capita investment in sanitation; n: investment efficiency in MSW treatment and disposal.

The paper focused on the period of 1991–2010. All the socio-economic data used in this paper was collected from China Statistical Yearbook (1992–2011) [41]. The GDP data measured and GDP of secondary industry as real GDP value in purchase power parity (at constant price in 1978) to eliminate the impact of price factors on the data. The data related to ISW, MSW and investment in SWM comes from China Environment Yearbook (1992–2011) [42]. Due to the difficulty in estimating the amount of municipal garbage generation in practical statistics in China, the municipal garbage generation is often replaced by municipal garbage disposal, which refers to the amount of municipal garbage collected and transported to disposal factories or sites and includes the garbage produced from households, commercial activities, markets, street cleaning, public sites *et al.* in the urban area [42].

3. Results

3.1. Overview of SW Generation in China

With fast economic growth driven by rapid industrialization, the total amount of ISWG has increased (with an average annual increase rate of 11.5%) from 587.59 million tons in 1990 to 2409.44 million tons in 2010 (Figure 2). Due to the constant increase of ISW disposed and utilized (the amount of increase was 4.9 and 4.58 times, respectively, during 1991–2010), the amount of ISWD in the last couple years showed a decreasing trend which can be divided into two phases: during 1991–1997, it showed decreasing trend, after 1997, the amount of discharge leaped forward to 70.48 million tons in 1998, then the continuous decreasing trend appeared again during 1998–2010. The reason for the sudden increase in ISWD in 1998 is that the amount of ISWG in the mining and quarrying department rapidly increased from 311.9 million tons in 1997 to 408.23 million tons in 1998. In virtue of the implementation of clean production and circular economy policies, the ratio of ISW utilized increased from 37.92%–67.14%. The benefits of ISW utilization in China is not only shown by the decreased ISWD, but also the output values from utilization, which were 31.05 billion in 2000 and 177.85 billion in 2010. In general, China has made great progress in ISW management which is represented by the significant decrease of discharge and increase of utilization ratio. However, the increase rate of ISW utilized was smaller than ISWG (Figure 2) which means that there is an increasing amount of non-utilized ISW. Besides, the utilization ratio is still at a very low level, especially compared with developed countries (e.g., USA, Germany, Japan, *et al.*).

With rapid urbanization and living standard improvements, the MSWG has been constantly increasing [43]. As the main part of MSW, municipal garbage and its management dramatically influences urban sustainable development. Due to the data unavailability of municipal garbage generation, the paper only analyzes the disposed and harmless disposed municipal garbage. The amount of municipal garbage disposal increased from 68.94 million tons in 1990 to 155.77 million tons in 2005, then decreased from 155.77 million tons in 2005 to 148.41 million tons in 2006, then slowly increased from 148.41 million tons in 2005 to 158.05 million tons in 2010, and its average annual growth rate is 6.16% during the whole period (Figure 3). The proportion of harmless disposed MSW ranged from 50%–60%, but after 2006, increased rapidly during 2006–2010. The gap between MSWG and MSWHD remained stable in the period of 1990–2001, then enlarged from 2001–2003 and eventually narrowed due to the fast increase in MSWHD. In 2010, 34.93 million tons MSW still was not treated harmlessly. The per capita MSWG remained stable which indicates that the effect of population on MSWG is linear.

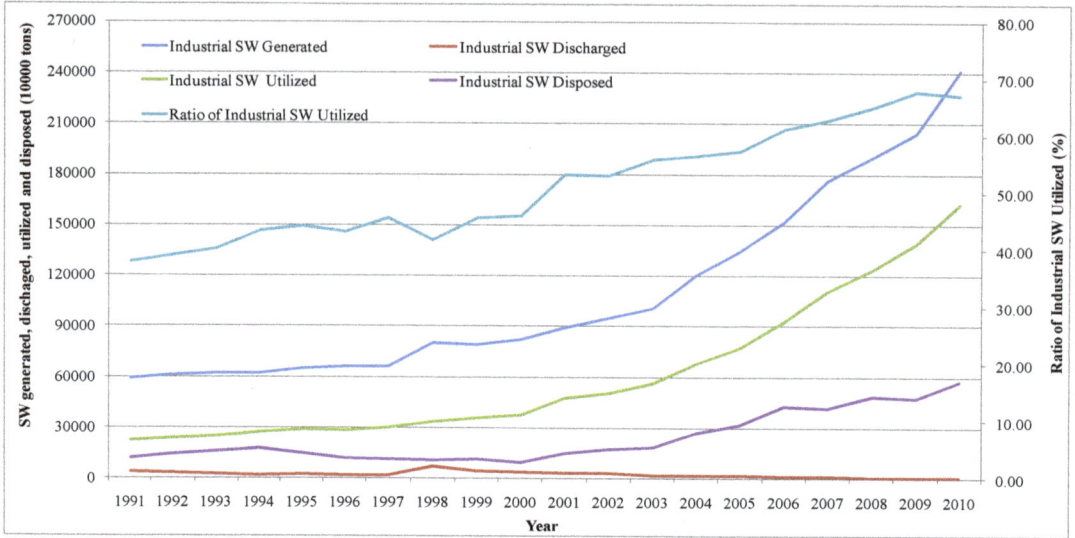

Figure 2. The amount of ISW generated, discharged and utilized in China during 1991–2010.

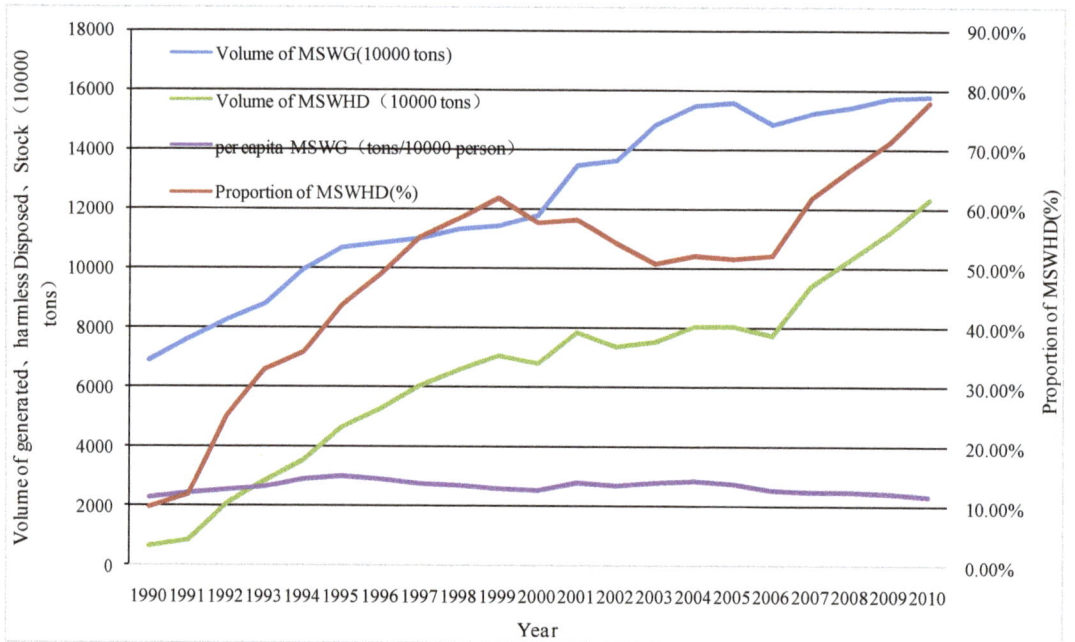

Figure 3. The amount of MSW disposal in China during 1990–2010.

3.2. Decoupling Analysis

In order to investigate the temporal variations in the decoupling of SW discharge from economic growth and population expansion specifically, the paper calculated the decoupling indicators in each time period by Equations (1) and (2). The results of the decoupling analysis of ISWG, ISWD, and MSWG from GDP, and MSWG from population in the 19 periods (1991–1992, 1992–1993, ..., 2009–2010) is shown in Figures 4 and 5.

The increasing amount of ISWG could lead to pressure for ISW utilization and recycling at a certain technical level, and thus, potentially contribute to an increase in the discharge amount of ISW. Besides, if the amount of ISW cannot be controlled with economic growth, this means that the socio-economic cost for reducing the environmental pressure induced by ISW could be very large. Therefore, the paper conducts a decoupling analysis for both ISWG and ISWD (Figure 4). Referring to the ISWG, strong decoupling of ISWG from economic growth happened only in the periods of 1993–1994, 1996–1997 and 1998–1999, and weak decoupling happened in each period of 1990–1993, 1994–1996, 1999–2000, 2001–2003, and 2008–2009. Non-decoupling of ISWG happened in the periods of 1997–1998, 2000–2001, 2003–2008 and 2009–2010, and non-decoupling in the period of 2000–2001, and 2004–2008 shows expanding decoupling status, while the periods of 1997–1998, 2003–2004 and 2009–2010 even showed expansive negative decoupling status. The results suggest the need for policies to reduce the ISW from source controlling.

As for the ISWD, the GDP elasticity was negative, which reflects that the strong decoupling happened at the national level during the period 1991–2010 except the period of 1994–1995 which showed weak decoupling status, and 1997–1998 which showed very strong expansive negative decoupling status due to the discharge of industrial solid waste in 1998 being 4.55 times of that in 1997 due to the fast increase in ISWD of the mining and quarrying sector (from 17.85–59.62 million tons).

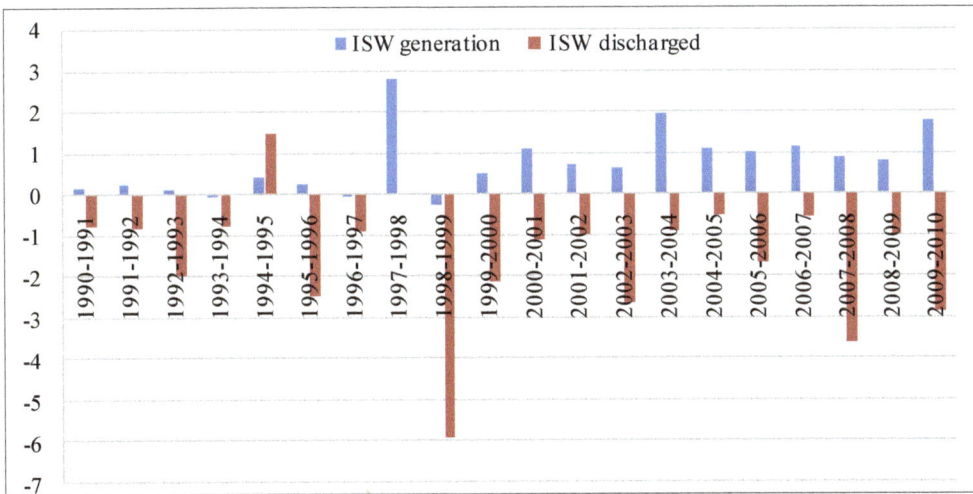

Figure 4. The GDP elasticity of ISWG and ISWD (%ΔE/%ΔGDP) in China, notes: the GDP elasticity of ISWD in the period of 1997–1998 is not displayed in the figure due to its value (45.32) being too large when compared with other periods.

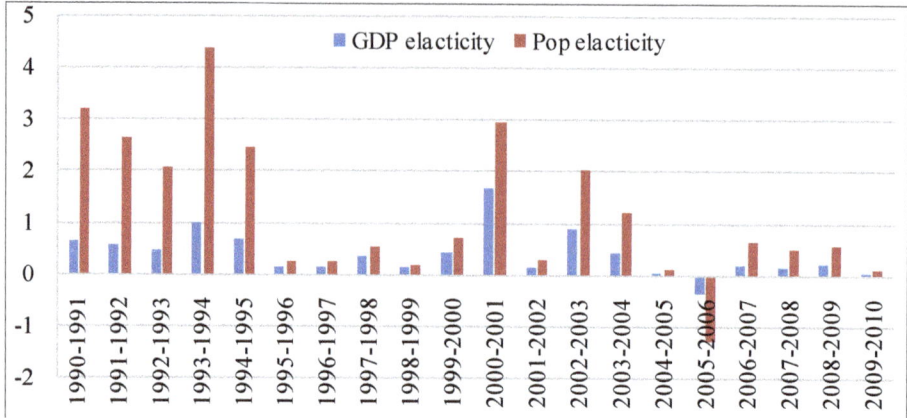

Figure 5. The GDP and population (Pop) elasticity of MSWG in China.

With regard to the decoupling of MSWG from economic growth, as depicted in Figure 5, the expansive negative decoupling was observed in the periods of 1993–1994 and 2000–2001 due to the MSWG growing faster than the economy by more than 1.2 times. Furthermore, the absolute decoupling happened only in the period of 2005–2006. Most of the investigated period showed a weak decoupling status.

As for the decoupling of MSWG from population growth, the situation was even worse. The population elasticity of MSWG is bigger than the GDP's except for in the period of 2005–2006. The expansive negative decoupling is observed in first five periods and the periods of 2000–2001 and 2002–2004.

3.3. Decomposition Analysis of SW Change

In order to understand the main drivers of the changes in SW and thus design appropriate measures to decrease SW, the paper conducted the decomposition analysis according to equations in Table 1. The decomposition result of ISWD is shown in Figure 6 with reference to the year 1991 and MSWHD is shown in Figure 7 with reference to the year 2000.

Referring to the ISWD (Figure 6), the decomposition analysis indicates that the factors of investment efficiency and technology were the main determinants to decrease the ISWD. However, the factor of economic growth was the most important factor that led to the increase in ISWD. The factor of investment in ISW treatment fluctuated to a large extent, while finally decreasing ISWD. The factors of population growth and industrial structure also contributed to increasing ISWD, but the contribution was less than economic growth.

With regards to MGH, the factors of urban population and per capita investment in sanitation were the main drivers of the MSWHDG increasing (Figure 7). The positive effect of urban population on MSWHD indicates that the capability of MSW harmless treatment and disposal was promoted with the expansion of urban scale. The change in contribution of per capital investment over the periods indicates that the investment in MSW management had lag effects on MSWHD (the contribution of per capital investment was negative in the first period and then switched to positive in subsequent periods). However, the factor of investment efficiency showed negative effects on MSWHD due to the decrease in efficiency over the whole period (the investment efficiency decreased 2.67 times). All the results indicate that the bigger cities have higher capacity to dispose MSW harmlessly. Additionally, the investment in MSW management can promote the harmless disposal amount of MSW. However, the investment efficiency was still at a low level and even decreased which suggests the need for policies to promote the utilization level of current MSW treatment and disposal facilities, and optimize MSW management strategies (e.g., reduce the MSWG though clarification and recycling, recycle the organic waste to produce biological energy, develop the urban mining to reuse the metal waste, *etc.*).

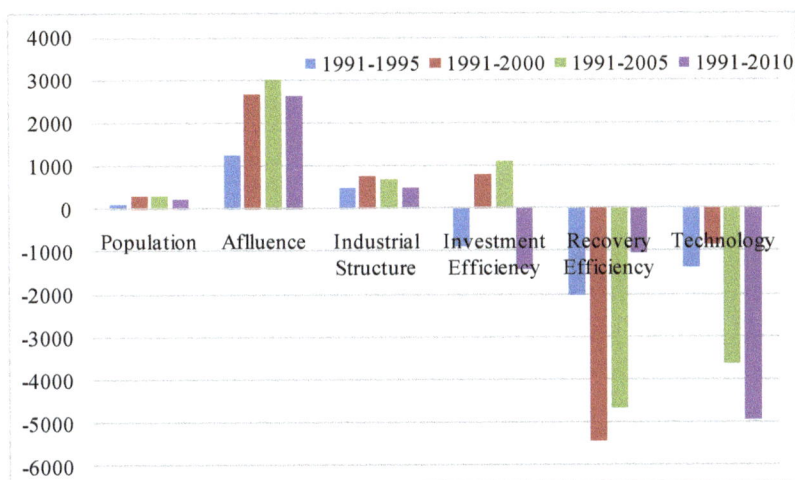

Figure 6. LMDI decomposition of change of total ISWD in China during 1991–2010.

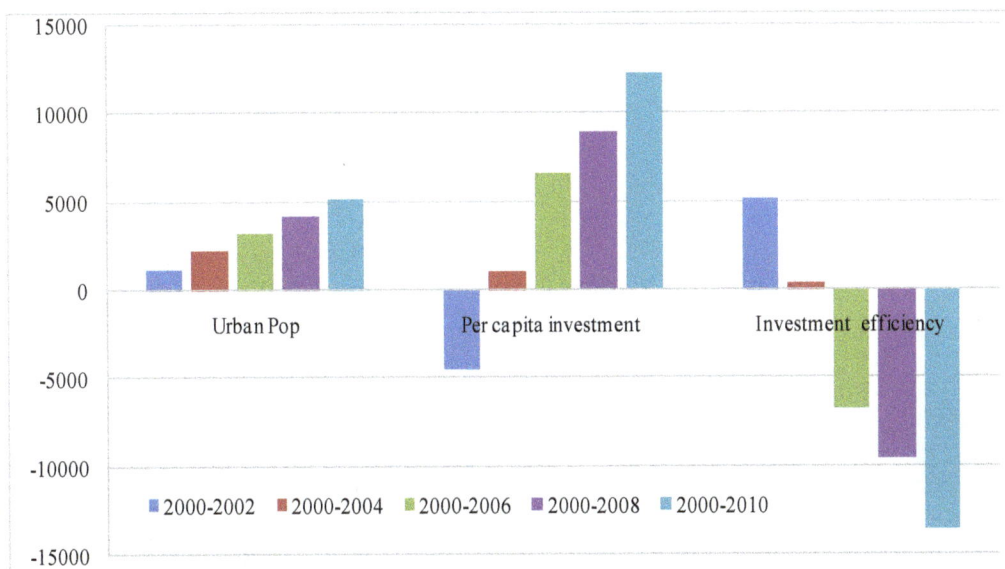

Figure 7. LMDI decomposition of change of MSWHD in China during 2000–2010.

4. Summaries and Conclusions

The paper investigated the sustainability of SWM by analyzing the decoupling of SW generation from economic growth and population increasing during 1990–2010 in China, and further identified the driving factors of ISWG and MSWHD by adopting the LMDI method to serve as the basis for future policy scenarios.

In general, with fast economic growth fueled by urbanization and industrialization, the amount of SW generation has been continuously increasing which will lead to great pressure on SWM. The ISWD decreased due to the increasing amount of utilization of ISW. However, the ratio of utilization is still low, especially compared with developed countries. As for the MSW, both the amount of generation and harmless disposal increased. The amount of MSWG and MSWHD also increased fast, and the gap between MSWG and MSWHD still remains large. The gap between SW generated and utilized suggests that a series of policies should be adopted to improve the capacity in MSWHD.

The paper provided evidence of the absolute decoupling of ISWD from economic growth except for the period of 1997–1998, and weak decoupling, even non-decoupling (including expanding and expansive negative decoupling) of ISWG in most investigated periods. The conflicting results of ISWG and ISWD indicate that China needs timely technology innovation and rational institutional arrangement to reduce the intensity of solid waste emission from the source. Regarding the MSW, the absolute decoupling of disposal from economic growth only can be found in the period of 2005–2006. Most of the investigated period showed the weak decoupling status. With regard to the decoupling of MSWG from urban population growth, the situation is even worse. The expansive negative decoupling was observed in seven of 20 periods investigated, and weak decoupling occurred in 11 periods.

The decomposition analysis helps to understand how the specific driving factors (*i.e.*, population scale, economic growth, industrial structure, urban scale, technology and investment in SWM) affected the changes in ISWD and MSWHD. Results confirm that the economic growth, industrial structure and population scale are the main drivers of increases in ISWD. Meanwhile, investment in ISW management and technology contribute to decreases in the discharge amount, but still cannot offset the positive contribution of the driving factors mentioned above. The negative effects of investment and technology suggest the need for policies to increase investment and promote technology for raising the utilization ratio further. As for the MSW, the urban scale and investment in harmless disposal helps increase the amount of harmless disposal, which means that larger cities will possess a greater capability to dispose MSW harmlessly, due to infrastructure and financial resources for MSW treatment is better in large cities than small cities. Inversely, the investment efficiency has a negative effect on MSWHD which suggests need for policies to promote the utilization level of current MSW treatment and disposal facilities, and optimize MSW management strategies (e.g., reduce the MSWG through classification and recycling, recycle the organic waste to produce biological energy, develop urban mining to reuse metal waste, *etc.*).

The approaches applied in the paper are suggested as a tool to evaluate the sustainability of SWM and identify the drivers of changes in SW to design future policy scenarios for sustainable management of SW in the near future. However, it is worth noting that the quantity of data (e.g., we use the amount of MSW disposal to replace generation, and investment in sanitation to represent investment in MSW management) may lead to estimation bias of the results to a certain extent. Nevertheless, we believe our findings provide solid and meaningful results with useful implications for policy makers.

Acknowledgments

The authors would like to acknowledge the financial support from the Natural Science Foundation of China (41471462, 41301652, 41101126 and 41261112), the Specialized Research Fund for the Doctoral Program of Higher Education (20120211120026), and the Fundamental Research Funds for the Central Universities (lzujbky-2013-132).

Author Contributions

Jiaxing Pang contributed to data collection, data processing and draft paper; Xingpeng Chen and Zilong Zhang conceived and designed the study, conducted data analysis as well as paper revision; Hengji Li contributed to data analysis and paper revised.

Conflicts of Interest

The authors declare no conflict of interest.

References

1. Vergara, S.E.; Tchobanoglous, G. Municipal solid waste and the environment: A global perspective. *Annu. Rev. Environ. Resour.* **2012**, *37*, 277–310.
2. Guerrero, L.A.; Maas, G.; Hogland, W. Solid waste management challenges for cities in developing countries. *Waste Manag.* **2013**, *33*, 220–232.
3. Al-Khatib, I.A.; Monou, M.; Abu Zahra, A.S.F.; Shaheen, H.Q.; Kassinos, D. Solid waste characterization, quantification and management practices in developing countries. A case study: Nablus district-palestine. *J. Environ. Manag.* **2010**, *91*, 1131–1138.
4. Pires, A.; Martinho, G.; Chang, N.B. Solid waste management in european countries: A review of systems analysis techniques. *J. Environ. Manag.* **2011**, *92*, 1033–1050.
5. Chen, X.; Geng, Y.; Fujita, T. An overview of municipal solid waste management in China. *Waste Manag.* **2010**, *30*, 716–724.
6. Xue, B.; Geng, Y.; Ren, W.X.; Zhang, Z.L.; Zhang, W.W.; Lu, C.Y.; Chen, X.P. An overview of municipal solid waste management in inner mongolia autonomous region, China. *J. Mater. Cycles Waste Manag.* **2011**, *13*, 283–292.
7. Hoornweg, D.; Lam, P.; Chaudhry, M. Waste management in China: Issues and recommendations. *Urban Development Working Papers*; World Bank: Washington, DC, USA, 2005.
8. Yuan, H.; Wang, L.; Su, F.; Hu, G. Urban solid waste management in chongqing: Challenges and opportunities. *Waste Manag.* **2006**, *26*, 1052–1062.

9. Zhang, D.Q.; Tan, S.K.; Gersberg, R.M. Municipal solid waste management in china: Status, problems and challenges. *J. Environ. Manag.* **2010**, *91*, 1623–1633.

10. Li, Z.-S.; Yang, L.; Qu, X.-Y.; Sui, Y.-M. Municipal solid waste management in Beijing city. *Waste Manag.* **2009**, *29*, 2596–2599.

11. Chung, S.S.; Poon, C.S. A comparison of waste management in Guangzhou and HongKong. *Resour. Conserv. Recycl.* **1998**, *22*, 203–216.

12. Ju, M.W.; Li, A.M.; Liu, Z.; Ma, Z.Z.; Xu, X.X. Municipal solid waste management in Dalian municipality, China. *Waste Manag.* **2011**, *31*, 809–810.

13. Zhao, W.; Huppes, G.; van der Voet, E. Eco-efficiency for greenhouse gas emissions mitigation of municipal solid waste management: A case study of Tianjin, China. *Waste Manag.* **2011**, *31*, 1407–1415.

14. Chung, S.S.; Poon, C.S. Characterisation of municipal solid waste and its recyclable contents of guangzhou. *Waste Manag. Res.* **2001**, *19*, 473–485.

15. Liu, G.Y.; Yang, Z.F.; Chen, B.; Zhang, Y.; Su, M.R.; Zhang, L.X. Emergy evaluation of the urban solid waste handling in Liaoning province, China. *Energies* **2013**, *6*, 5486–5506.

16. Yang, L.; Chen, Z.L.; Liu, T.; Wan, R.; Wang, J.; Xie, W.G. Research output analysis of municipal solid waste: A case study of China. *Scientometrics* **2013**, *96*, 641–650.

17. Wei, J.B.; Herbell, J.D.; Zhang, S. Solid waste disposal in China—Situation, problems and suggestions. *Waste Manag. Res.* **1997**, *15*, 573–583.

18. Chung, S.S.; Lo, C.W.H. Local waste management constraints and waste administrators in China. *Waste Manag.* **2008**, *28*, 272–281.

19. Wang, H.T.; Nie, Y.F. Municipal solid waste characteristics and management in China. *J. Air Waste Manage. Assoc.* **2001**, *51*, 250–263.

20. Jiang, Y.; Kang, M.Y.; Liu, Z.; Zhou, Y.F. Urban garbage disposal and management in China. *J. Environ. Sci.* **2003**, *15*, 531–540.

21. Troschinetz, A.M.; Mihelcic, J.R. Sustainable recycling of municipal solid waste in developing countries. *Waste Manag.* **2009**, *29*, 915–923.

22. Herva, M.; Neto, B.; Roca, E. Environmental assessment of the integrated municipal solid waste management system in Porto (Portugal). *J. Clean. Prod.* **2014**, *70*, 183–193.

23. Mori, K.; Christodoulou, A. Review of sustainability indices and indicators: Towards a new city sustainability index (CSI). *Environ. Impact Assess. Rev.* **2012**, *32*, 94–106.

24. Ness, B.; Urbel-Piirsalu, E.; Anderberg, S.; Olsson, L. Categorising tools for sustainability assessment. *Ecol. Econ.* **2007**, *60*, 498–508.

25. Huang, Y.T.; Pan, T.C.; Kao, J.J. Performance assessment for municipal solid waste collection in Taiwan. *J. Environ. Manag.* **2011**, *92*, 1277–1283.

26. Zaman, A.U. Measuring waste management performance using the 'zero waste index': The case of Adelaide, Australia. *J. Clean. Prod.* **2014**, *66*, 407–419.

27. Zaman, A.U. Identification of key assessment indicators of the zero waste management systems. *Ecol. Indicat.* **2014**, *36*, 682–693.

28. Del Borghi, A.; Gallo, M.; del Borghi, M. A survey of life cycle approaches in waste management. *Int. J. Life Cycle Assess.* **2009**, *14*, 597–610.

29. Menikpura, S.N.M.; Gheewala, S.H.; Bonnet, S. Framework for life cycle sustainability assessment of municipal solid waste management systems with an application to a case study in Thailand. *Waste Manag. Res.* **2012**, *30*, 708–719.

30. Zurbrugg, C.; Caniato, M.; Vaccari, M. How assessment methods can support solid waste management in developing countries—A critical review. *Sustainability* **2014**, *6*, 545–570.

31. Zhang, L.M.; Xue, B.; Geng, Y.; Ren, W.X.; Lu, C.P. Emergy-based city's sustainability and decoupling assessment: Indicators, features and findings. *Sustainability* **2014**, *6*, 952–966.

32. Wagner, J. Incentivizing sustainable waste management. *Ecol. Econ.* **2011**, *70*, 585–594.

33. Tapio, P. Towards a theory of decoupling: Degrees of decoupling in the EU and the case of road traffic in Finland between 1970 and 2001. *Transport Policy* **2005**, *12*, 137–151.

34. Vehmas, J. Europe in Global Battle of Sustainability: Rebound Strikes Back? In *Advanced Sustainability Analysis*; Turun kauppakorkeakoulu: Turku, Finland, 2003.

35. Recalde, M.Y.; Guzowski, C.; Zilio, M.I. Are modern economies following a sustainable energy consumption path? *Energy Sustain. Dev.* **2014**, *19*, 151–161.

36. Ziolkowska, J.R.; Ziolkowski, B. Product generational dematerialization indicator: A case of crude oil in the global economy. *Energy* **2011**, *36*, 5925–5934.

37. Zha, D.L.; Zhou, D.Q.; Ding, N. The contribution degree of sub-sectors to structure effect and intensity effects on industry energy intensity in China from 1993 to 2003. *Renew. Sust. Energ. Rev.* **2009**, *13*, 895–902.

38. Ang, B.W.; Zhang, F.Q. A survey of index decomposition analysis in energy and environmental studies. *Energy* **2000**, *25*, 1149–1176.

39. Zha, D.L.; Zhou, D.Q.; Zhou, P. Driving forces of residential CO_2 emissions in urban and rural China: An index decomposition analysis. *Energy Policy* **2010**, *38*, 3377–3383.

40. Xu, X.Y.; Ang, B.W. Index decomposition analysis applied to CO_2 emission studies. *Ecol. Econ.* **2013**, *93*, 313–329.

41. National Bureau of Statistics of the People's Republic of China. *China Statistical Yearbooks 1991–2011*; China Statistical Press: Beijing, China, 1991–2011.

42. Ministry of Environmental Protection of the People's Republic of China. *China Environment Yearbook 1991–2011*; China Environment Yearbook Press: Beijing, China, 1991–2011.

43. Cheng, H.; Hu, Y. Mercury in municipal solid waste in China and its control: A review. *Environ. Sci. Technol.* **2011**, *46*, 593–605.

Impact Analysis of Air Pollutant Emission Policies on Thermal Coal Supply Chain Enterprises in China

Xiaopeng Guo, Xiaodan Guo and Jiahai Yuan

Abstract: Spurred by the increasingly serious air pollution problem, the Chinese government has launched a series of policies to put forward specific measures of power structure adjustment and the control objectives of air pollution and coal consumption. Other policies pointed out that the coal resources regional blockades will be broken by improving transportation networks and constructing new logistics nodes. Thermal power takes the largest part of China's total installed power generation capacity, so these policies will undoubtedly impact thermal coal supply chain member enterprises. Based on the actual situation in China, this paper figures out how the member enterprises adjust their business decisions to satisfy the requirements of air pollution prevention and control policies by establishing system dynamic models of policy impact transfer. These dynamic analyses can help coal enterprises and thermal power enterprises do strategic environmental assessments and find directions of sustainable development. Furthermore, the policy simulated results of this paper provide the Chinese government with suggestions for policy-making to make sure that the energy conservation and emission reduction policies and sustainable energy policies can work more efficiently.

Reprinted from *Sustainability*. Cite as: Guo, X.; Guo, X.; Yuan, J. Impact Analysis of Air Pollutant Emission Policies on Thermal Coal Supply Chain Enterprises in China. *Sustainability* **2015**, *7*, 75-95.

1. Introduction

According to the Air Quality Report issued by the Chinese Environmental Protection Ministry, in 2013 China's annual haze days reached the highest level on record (19.5 days) owing to unsustainable development and an unreasonable energy structure. In addition, the national annual concentration of PM2.5 has reached 72 micrograms per cubic meter. Only three cities reached the standard, which accounts for 4.1 percent of all 74 cities. The Intergovernmental Panel on Climate Change (IPCC) found that coal combustion is the main source of air pollutants. In addition, the research done by the Chinese Academy of Sciences (CAS) showed that: the sources of Beijing's particulate matter (PM10) emission are secondary inorganic aerosols (26%), industrial pollution (25%), coal consumption (18%), soil dust (15%), biomass burning (12%), and automobile exhaust gas and waste incineration (4%). The secondary inorganic aerosols, industrial pollution, and coal consumption are all caused by the burning of fossil fuels. Therefore, China's worsening air quality is probably caused by extensive coal consumption and a high energy consumption development pattern.

Coal consumption is considered to be the second most important anthropogenic contributor to global air pollution [1,2]. From the perspective of coal consumption structure, the electric power industry is the biggest contributor to the coal resources consumption. The power generation industry produces more than 40 percent of total air pollution emissions in China [3]. China's past electricity development policies have led the power industry to make excessive investment in thermal power installed capacity and low operational efficiency [4]. If a long-term transition towards a low carbon economy is not carried out, China's CO_2 emissions could rise by 160%–250% from 2010 to 2050 [5]. By the year 2020, China's thermal power generation may reach over 7 trillion kilowatt-hours, and air-pollution intensity will be nearly twice the 2005 level [6].

To accelerate the control of air pollution, the Chinese government has introduced many policies managing the air quality and adjusting the energy structure since 2009. These air pollution emission policies will surely change the operating environment of coal supply chain enterprises. So it is urgent to analyze the policy impact and optimize the development patterns for coal and power industries under new policy restraints [7]. From the perspective of the power industry, coal-fired power will translate into clean and efficient energy power after the implementation of sustainable development policies [5,8]. China's hydropower, wind power, and nuclear power industry will meet a tremendous need in the following decades under the encouragement of energy structure adjustment and an emission reduction policy [9,10]. Nuclear and hydropower may play a dominant role in contributing to China's air pollution reduction in the long term [11]. Also, the promotion of renewable energy utilization will surely have a great effect on China's coal and power industry [12]. Further, the air pollution emission reduction policies can also accelerate the technological advancement and clean coal utilization of thermal coal supply chain enterprises [13,14].

The thermal coal supply chain system is the whole system of coal enterprises, power enterprises, and coal transportation enterprises, which guide the process from coal production to coal consumption. In addition, air pollutant emission reduction policies' impact on the thermal coal supply chain is a complex, dynamic evolution process concerning many fields such as policies related to energy conservation and emission reduction, economic development, power production and consumption, resource exploitation and utilization, and energy price. So it is obvious that the power structure adjustment has nonlinear characteristics. The system dynamics (SD) method not only models the market's real behavior but also properly explains the relationship between the main variables of the system [15]. Considering the advantages of integrity and dynamics that system dynamics has in analyzing complex dynamic problems, this paper set up a complete system dynamics model by analyzing: the air pollution emitted during coal combustion, coal washing technology, installed capacity, unit transform, and new energy power generation, under the constraint of the new atmospheric pollutant emission policy, to seek a development pattern for the thermal coal supply chain.

The SD approach has been applied in investigating the sustainable management of electric power systems. Some scholars have set up an SD-based model to investigate the distributed energy resource expansion planning [15–17] and energy efficiency improvement [18], considering both energy states and production constraints. Other scholars use SD methodology to simulate the behavior of the renewable energy sectors such as nuclear [19] and photovoltaic energy [20]. SD models are also widely built to explore the effects of energy consumption and CO_2 emission reduction policies [21–24]. Previous models have structured the investment, dispatch, pricing heuristics, and electricity generation resource factors with common emission reduction policies such as feed-in-tariffs, investment subsidies, and carbon taxes. In order to deeply investigate the new development pattern of the power and energy industry, some scholars have qualitative explored the link between transportation systems and air pollution reduction policies using the SD approach [25,26]. However, the dynamic behavior of the thermal power system and the complicated feedback of a coal system under pollution reduction policies were not taken into account in previous studies.

As policy refinement increases, China's air pollutant emission reduction policies select the specific goals of pollution emission, coal cleaning proportion, and desulfurated capacity proportion. Under the new policy situation, the former system dynamic model is not suitable to measure the effect on China's thermal coal supply chain member enterprises.

The objectives of this paper are as follows:

(1) To analyze China's recent air pollution reduction policies systematically.
(2) To apply some descriptive and effective methodologies to simulate the fundamental and dynamic development process of thermal coal supply chain member enterprises driven by emission reduction policies.
(3) Taking China as a case study, to develop a system dynamics model based on Vensim PLE for Windows Version 5.10a software (Ventana Systems, Inc., Cambridge, MA, USA), which offers a realistic platform for predicting the development trends of China's thermal coal supply chain enterprises from 2012 to 2022.

The simulated results are significant in predicting energy proportion, power proportion, and development routes of thermal coal supply chain member enterprises. In addition, the simulations of China's air pollution emission reductions have meaning for policy-making institutions.

2. China's Air Pollution Emission Reduction Policies

China is the world's biggest energy consumer [27,28]. According to the information distributing platform of National Statistics, China's coal consumption accounts for about 70% of the total energy consumption in 2013. The thermal coal consumption of China's electric industry is 1967 million tons, accounting for 64.29% of the total coal consumption. China's growing atmospheric pollution problem is caused by the long-term accumulation of multiple factors such as unsustainable development patterns and unreasonable energy electric power industry structures. In order to better reduce air pollution, the Chinese government has introduced many laws involving concrete measures since 2009.

In 2009, after realizing the importance of sustainable development the Chinese government first proposed to reduce carbon dioxide emissions per unit of GDP by 40%–45% in 2020, as of the 2005 level. This restrictive goal has also been included in the medium and long term planning of China's national economic and social development.

In 2010, the fifth plenary session of the 17th CPC Central Committee emphasized that China should make the construction of a resource-saving and environment-friendly society an important focus despite the acceleration of economic growth and the readjustment of the economic structure. Prime Minister Wen Jiabao promised that China would energetically promote energy conservation and raise the efficiency of energy consumption, especially in industry and transportation. Wen also said that China would add 80 million tons of standard coal energy saving ability annually, and that all additional and reconstructive coal-fired units must build and run flue gas desulfurization facilities synchronously.

Since 2012, hazy weather in the Beijing-Tianjin-Hebei and Yangtze River Delta areas is becoming more and more frequent, so the importance of air pollutant prevention is becoming more and more noticeable. In October 2012, the Chinese National Development and Reform Commission, the Ministry of Environmental Protection, and the Ministry of Finance jointly issued The 12th Five-Year Plan for the prevention and control of atmospheric pollution in key areas. The 12th Five-Year Plan pointed out that we should focus on the optimization of industrial layout and energy structure and the enhancement of clean energy under the current serious situation of air pollution. China should adhere to a diversified energy development strategy, strive to improve the proportion of clean low-carbon fossil energy and non-fossil energy, promote the efficient utilization of clean coal, implement the alternatives of traditional energy, and speed up the optimization of energy production and consumption structure. By the end of the 12th period, the proportion of non-fossil energy consumption should increase to 11.4% and the proportion of coal consumption should decrease to 65%. The proportion of non-fossil energy generation installed capacity should reach 30%.

In June 2013, the premier, Li Keqiang, introduced ten measures for the control of air pollution in the state council executive meeting. The main content of these measures is the reduction of air pollutant emission, the control of high energy-consuming enterprises, the adjustment of energy structures, and the new energy conservation and emissions reduction mechanisms of incentive and constraint. We should speed up the adjustment of energy structure by implementing the interregional transmission project, controlling coal consumption reasonably, and promoting the use of clean coal. By the end of 2014, China should complete the elimination of backward installed capacity ahead of time. The Chinese government should follow the guiding and incentive rule of taxes and subsidies, in order to push for sustainable development.

In September 2013, The Action Plan for the Control of Air Pollution, issued by the state council, clearly pointed out that in 2017 the inspirable particle concentrations of China's prefecture-level cities should be reduced by 10% from 2012 levels and the inspirable particle concentrations of Beijing-Tianjin-Hebei, the Yangtze River Delta area, and the Pearl River Delta area should be reduced by about 25%, 20%, and 15%, and the PM10 concentrations of Beijing especially should be reduced to under 60 micrograms per cubic meter. The Action Plan for the Control of Air Pollution also mentioned the adjustment of energy structure, the optimization of industrial layout, the improvement of environmental economic policies, and some other measures. Measures optimizing the industrial layout covered the capacity limitation of energy-intensive and highly polluting industries, the acceleration of backward capacity elimination, and the compression of excessive capacity. China should take a sustainable development pattern by controlling coal consumption, increasing the proportion of washed coal, accelerating the use of clean energy, and raising the energy usage effectiveness. An energy conservation and emissions reduction mechanism of incentive and constraint should be promoted actively in order to improve sustainable environmental economic policies.

In addition, some local governments introduced policies to cooperate with the implementation of The Action Plan for the Control of Air Pollution, such as an action plan implementing rules for the control of air pollution in Beijing-Tianjin-Hebei and surrounding areas. These rules made more detailed targets for the provinces and cities of the Beijing-Tianjin-Hebei region. The plan pointed out that by the end of 2017 the proportion of Beijing's coal consumption will drop below 10% and high-quality energy such as electricity and natural gas will account for more than 90%. Beijing will cut 13 million tons of raw coal by using many comprehensive measures such as eliminating backward production capacity, clear violations capacity, strengthening energy conservation and emissions reduction, implementing clean energy replacement, safe development of nuclear power, and strengthening the new energy-efficient utilization.

Overall, these air-pollution reduction policies focused on optimizing the industrial structure and energy structure, and giving impetus to industrial transformation and upgrading. Along with the restriction of coal consumption and the development of clean energy generation, the business environment and development modes of coal enterprises and power generation enterprises must change significantly. It is a big challenge for thermal coal supply chain member enterprises, including coal enterprises, power generation enterprises, and thermal coal transportation enterprises, to make the right decisions in response to the policy influence [29]. Completing the analysis of policy implications will be of great significance in researching the policy mechanism and predicting the developing direction for thermal coal supply chain node enterprises under the policy background. The policy implications are complex and dynamic, so it is difficult for a general model to simulate the changing process of each factor in the thermal coal supply chain under the influence of air pollutant emission reduction policies. Considering the advantages of the SD model on integrity and dynamics during complex analysis, this paper plans to establish a system dynamics model of thermal coal supply chain member enterprises' development processes under the impact of air pollutant emission reduction policies to figure out the policy mechanism and assist decision-making in node enterprises.

3. A System Dynamics Model of Policy Impact on Thermal Coal Supply Chain Member Enterprises

3.1. System Dynamics Methodology

SD is a systems modeling and dynamic simulation methodology for analysis of dynamic complexity in socioeconomic and biophysical systems [30]. Based on the principles of system thinking and feedback control theory, system dynamics helps in understanding the time-varying behavior of complex systems [31]. Our SD model is divided into two parts: qualitative analysis and quantitative analysis. The causality diagram is mainly used for the qualitative analysis of the model and the system dynamics flow diagram is used to realize the quantitative analysis. We used a causality diagram to qualitatively analyze the transfer process of impact of air pollution reduction policies on the thermal coal supply chain. The causality diagram achieves a qualitative analysis of complex correlations and influences within various factors of the thermal coal supply-chain system by drawing system factors and the positive or negative causal chains connecting factors. Furthermore, we used a flow diagram to analyze the quantitative relationship between the three variables of the coal subsystem, the power subsystem, and the transportation subsystem. The dynamic flow diagram builds up the quantitative analysis model to simulate and analyze the system's behavior by drawing the visual state variable, the rate variable, and the auxiliary variable, and setting function relationships and initial values among variables through the Vensim_PLE software. The SD model simulated the changing trend of various internal variables under the influence of relevant policies. In addition, we predicted the power structure adjustment and the developing route of coal enterprises by summarizing the simulated results.

The thermal coal supply chain is a system with obvious boundaries, and policy impact on it will be both dynamic and complex. Although other types of dynamic, quantitative modeling can do the impact analysis, an SD model has an advantage in solving dynamic problems because it is the only method that can better reflect forward and reverse policy impact processes along the thermal coal supply chain, and reveal the extent of the influence air pollution reduction policy has on factors in the supply chain system [32]. So this study uses system dynamics to analyze air pollutant emission policies' impact on the development of thermal coal supply chain member enterprises in China, simulating the dynamic transfer process of policy influence. We believe that this SD model can properly analyze the impact of air pollution reduction policies on the development of thermal coal supply chain member enterprises, and simulate the dynamic transfer process of relevant policy influence.

3.2. Model Structure Analysis

In order to analyze the transfer process of policy impact within the supply chain system, this paper firstly sorted out possible policy impact types by combining specific action plans and reduction targets mentioned in the above air pollutant emission reduction policies. The air pollutant emission reduction policies are aimed at strengthening the management of air pollution and reducing the concentration of PM10. Coal, as the primary source of atmospheric pollutants, is a key regulatory object of these policies. Policies such as The Action Plan for the Control of Air Pollution will impact the scale, costs, and benefits of coal enterprises, power generation enterprises, and transport enterprises through a variety of means. At present, air pollutant emissions policies such as The Action Plan for the Control of Air Pollution and the action plan implementing rules for the control of air pollution in the Beijing-Tianjin-Hebei area will impact the following aspects of thermal coal supply chain node enterprises:

3.2.1. Industrial Structure Adjustment

The Action Plan for the Control of Air Pollution (or The Action Plan, for short) required the full or partial elimination of small coal-fired boilers in central heating, to be replaced by the use of electricity or natural gas, and forbade new coal boilers in key areas such as Beijing and Tianjin. New projects in Beijing-Tianjin-Hebei, the Yangtze River Delta, and the Pearl River Delta are banned from having supporting self-provided coal-fired power stations. In addition to cogenerations, new thermal-power generation projects are banned. We must speed up the development of hydropower, wind power, solar power, and nuclear power on the basis of safety and efficiency. Nuclear power installed capacity will reach 50 million kilowatts by 2017. Such policies will reduce the capacity of new thermal power generating units and increase the coal enterprise's strategic and operational risk. The Action Plan highlighted how to govern small coal-fired boilers comprehensively, and encouraged the accelerated construction of "coal to gas" and "coal to electricity" projects. Small coal-fired boilers under 10 tons will be eliminated by 2017.

3.2.2. Energy Structure Adjustment

The Action Plan clearly pointed out that the proportion of China's coal consumption will decrease to less than 65% by 2017. China should focus on the orderly development of hydropower, the efficient utilization of geothermal energy, wind energy, solar energy, biomass energy, and the safe development of nuclear power. It must increase the proportion of new power generation and the consumption of renewable energy power by developing projects of wind power and nuclear power generation. Coal should gradually be replaced by increasing both the proportion of outside transmission in the gas supply and the non-fossil energy intensity. The Chinese government recommends the use of electricity and other clean energy instead of coal. Such policies are intended to reduce the proportion of coal consumption in the total energy consumption and limit the relative demand for coal resources.

In addition, The Action Plan stopped imports of inferior foreign coal and restricted sales of bulk coal with high ash and sulfur. These policies would affect the demand-supply condition of coal and the market price of thermal coal, and would bring demand-supply risks to coal enterprises and fuel price risks to power generation companies. The Action Plan also pointed out that by 2017 the proportion of washed coal should rise to 70%, in order to reduce air pollutant emissions in the process of coal combustion. This policy will increase the washing equipment investment and running cost of coal enterprises, thus raising its investment risk.

3.2.3. Technical Renovation

The Action Plan pointed out that coal-fired power generation enterprises should reduce harmful gas emissions by speeding up the desulfuration renovation of thermal power units and formulating strict desulfuration capacity targets. In 2012, the proportion of desulfuration capacity is 56%. This proportion will increase to 70% by 2017. Key areas such as Beijing and Tianjin have to complete the pollution control work of coal-fired power plants before the end of 2015. Meanwhile power generation enterprises should strengthen the development of desulfurization and denitration technology and intensify the exchange of air-pollution control management experience. These policies force coal enterprises to put a lot of money into the desulfurization reform and the development of related emissions reduction technology. These measures will occupy much of the working capital and greatly increase the operational risk, the cash flow risk, and the cost risk for power generation companies.

3.2.4. Market Mechanism Adjustment

The market mechanism adjustment policy includes two aspects of incentives and punishment. In terms of incentives, The Action Plan pointed out that the government will adjust the sales price and perfect the denitration electricity price policy considering the cost of denitration and the local characteristics. Existing thermal power units that adopt new dedusting facilities' renovation technology should be given price supports. In terms of punishment, The Action Plan pointed out that the government should increase the intensity of atmospheric pollutant emission levies and improve atmospheric pollution standards. This kind of market adjustment policy will increase the price risk and cost risk of thermal coal supply chain enterprises.

3.2.5. Vehicle Environmental Management

The Action Plan made it clear that in the future China will strengthen the environmental protection management of vehicles and forbid sales of environmental substandard vehicles. This policy had a limit on coal transportation vehicles. It may affect coal supplies, and even cut down the thermal coal supply chain.

The above policies will impact node enterprises of the thermal coal supply chain, and further impact upstream and downstream enterprises driven by principal–agent relationships, benefit distributions, and decision implementations of member enterprises. In the thermal coal supply chain, coal enterprises, power generation enterprises, and coal transportation enterprises are adjoined to one another and jointly realize the processes of thermal coal production, transportation, and consumption. So when one of these three parties is affected by air pollutant emission reduction policies, the other parties will be affected as well.

For thermal power generation enterprises, the policy of shutting down small thermal power plants and the prohibition on new coal-fired power projects will shrink the scale of the thermal power industry, thus reducing coal demand and influencing the income of the coal industry. The reduction in the coal supply will limit power generation enterprises' coal selection. The deviation between the actual burning coal and the design coal can reduce power plant boiler efficiency and improve fuel costs of power generation enterprises. The increasing air pollutant emission pressure will encourage power generation enterprises to conduct the desulfurization and denitration renovation of units. These investments will reduce power generation enterprises' profits. In addition, the rapid rise of clean energy power generation inevitably leads to a decrease in the thermal power installed capacity proportion. This shrinkage will surely decrease the proportion of thermal power generation and the profitability of the thermal power industry. If the thermal power industry pays a fixed proportion of its income towards investment in new construction, with no obvious changes in the unit installed costs of thermal power, the new thermal power installed capacity will be further reduced.

For coal enterprises, there are two kinds of air pollution reduction methods: one is coal-production reduction, and the second is coal washing [33]. The increase of coal washing equipment investment will enhance the coal enterprise investment risk, but it can also lower atmospheric pollutant emissions, such as sulfur dioxide and nitrogen oxides, in the process of coal combustion, and cut down the cost risk to thermal power generation enterprises by reducing atmospheric pollutant discharge. Meanwhile, the increase in washed coal will reduce the power supply coal consumption rate, thus reducing the fuel costs of the thermal power industry. Coal washing technology can remove about 50%–80% of the coal ash content. Every 1% reduction in the ash content in thermal coal brings 2–5 grams' reduction in the standard power supply coal consumption rate. With respect to coal supply and demand, the reduction of thermal power generation will cut the coal demand, and thus affect the income and construction investment of the coal industry. The reduction of the coal production growth rate and the import limitation on inferior foreign coal will reduce the growth rate of the coal supply. The coal supply/demand situation tends to loosen, under air pollutant emission reduction policies. The falling coal price will further affect the benefits of coal enterprises.

For coal transportation enterprises, coal-conveying vehicles as important mobile pollution sources will probably be restricted by environmental protection indexes. Vehicles under environmental protection standard could not travel on the road. The policy will block coal transportations and influence the fuel supply of thermal power plants.

After analyzing policies, this study uses a causality loop to do the qualitative analysis of air pollution policies' impact on the development of thermal coal supply chain member enterprises in China. The causality loop can be seen in Figure 1.

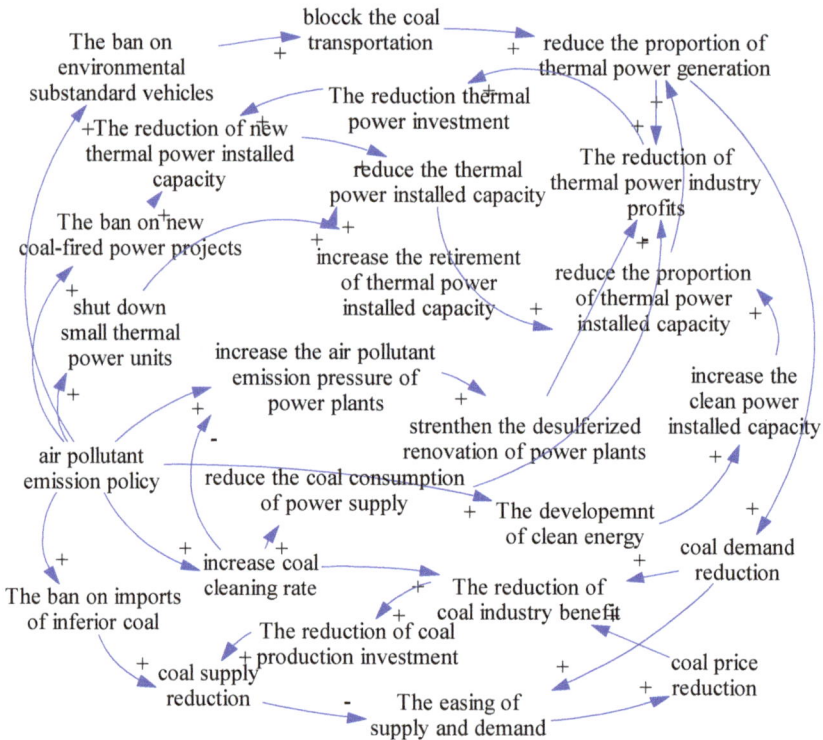

Figure 1. Causality loop of air pollutant emission policies' impact.

A causality loop is a directed graph applied to analyze the interaction relationships between internal variables of a dynamic system [34]. It consists of several single causality chains. The causality chain representing positive effects is called the positive chain, with "+" next to the arrows. The causality chain representing negative effects is called the negative chain, with "−"next to the arrows. A causality loop is composed of multiple causal chains.

On the basis of the model structure analysis and causality loop, this study divided the system dynamics model of air pollution reduction policies' impact on the development of thermal coal supply chain member enterprises into three modules according to the composition of the thermal coal supply chain. The main factors of the coal module are: amount of washed coal, increasing rate of washed coal, the coal washing cost, proportion of washed coal, proportion goal for washed coal, coal industry profits, coal construction investment, domestic coal production, the coal production rate, net coal import, coal supply, coal demand, coal supply/demand ratio, coal supply/demand ration factor, coal prices, coal consumption proportion of the power industry, and coal consumption of the power supply. The main factors of the power generation module are: thermal-power installed capacity, increase or decrease in the thermal power installed capacity, clean-energy power installed capacity, changing rate of clean-energy power installed capacity, total generated energy, thermal-power generation, proportion of thermal power generation, thermal power profits, thermal power construction investment, grid purchase of thermal power, coal consumption of the power industry, SO_2 emission of thermal power coal consumption, SO_2 emission goal of thermal power coal consumption, desulfuration capacity, desulfuration capacity changing rate, and desulfuration cost. The main factor in the coal transportation module is coal-conveying vehicles' environmental success rate.

The confirmation of model factors and the qualitative analysis using causality loop lays a foundation for further quantitative analysis.

3.3. Model Design

After determining the causality loop and the main factors involved in it, this study begins the quantitative analysis of air pollution reduction policies' impact on the node enterprises of the thermal coal supply chain in China by drawing a dynamic flow diagram. A dynamic flow diagram is used to depict the logical relationship between system factors with symbols [35]. This study uses a dynamic flow diagram to clear the feedback form and control the law of the system. Firstly, this paper classified the main factors determined by the causality loop according to their characteristics. Variables representing cumulative results are set as state variables. Variables representing the changing speed rate of state variables are set as rate variables. The rest of the relevant variables are set as auxiliary variables. This study uses Vensim_PLE software to establish the flow graph of air pollutant emission policies' impact on the development of thermal coal supply chain member enterprises. The flow graph is shown in Figure 2.

The above system dynamics model of air pollutant emission policies' impact on the development of thermal coal supply chain member enterprises contains five state variables, six rate variables, and 28 auxiliary variables including time. The upper part of Figure 2 is the coal module of this system dynamics model, and the lower part is the power generation module and the transport module. In Figure 2, the arrow direction indicates the transfer process of air pollutant emission policies' impact between thermal coal supply chain member enterprises. The impact of policies is transmitted through the proportion goal of washed coal, the sulfur dioxide emission of coal-fired power plants, the shutting down capacity of small coal-fired boilers, and the desulfurization grid purchase price influencing the thermal power installed capacity and the coal supply/ demand situation. The Action Plan and other policies have clear plans relative to some key factors such as the goal of air pollutant emission and the installed capacity. So these four factors are expressed as time functions in the quantitative analysis of the system dynamics model. This study used the already built dynamics model of air pollutant emission policies' impact on the development of thermal coal supply chain member enterprises to simulate the transfer process of air pollutant emission policy impact within the thermal coal supply chain during the period 2013–2022 and analyzed changing trends of various factors under the influence of policies during these decades.

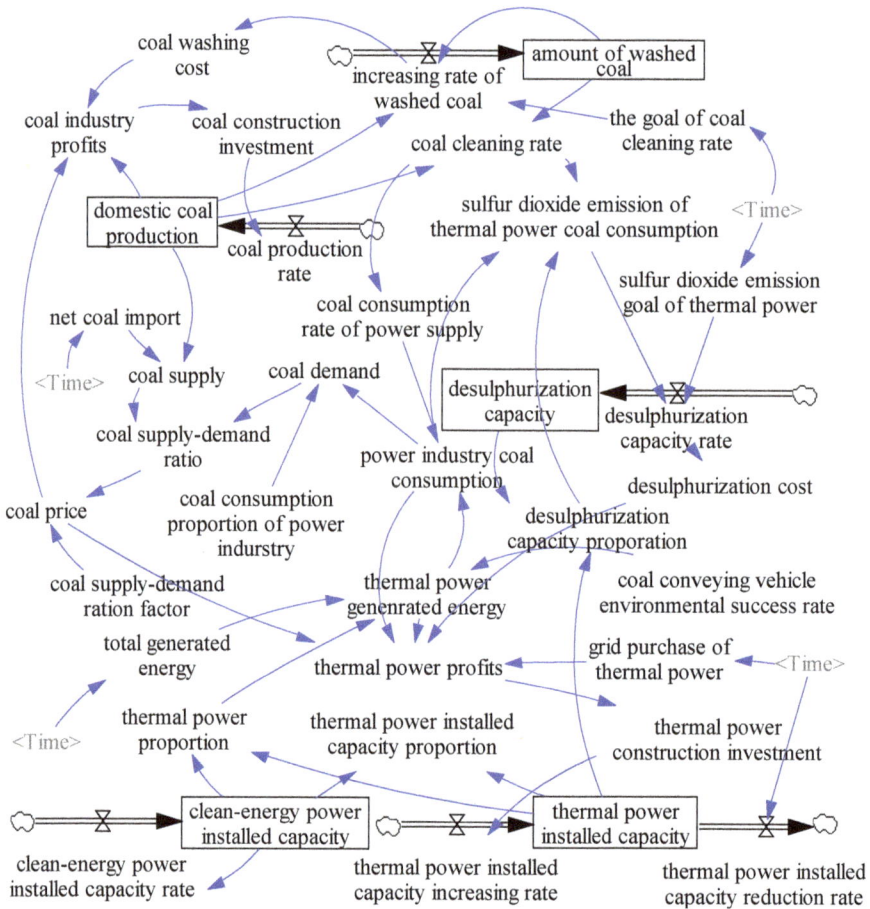

Figure 2. Flow graph of air pollutant emission policies' impact.

In order to perform quantitative analysis on the transmission of atmospheric pollutant emission policies' impact in the thermal coal supply chain and simulate the changing trends of various internal factors under this influence, this study needed to further determine the function of variables on both ends of the arrow in the flow graph. Before determining the functional relationships, the initial value and unit of each factor have to be settled according to statistical data and actual situations. The variable category, variable name, initial value, and unit of variables in this system dynamics model are all shown in Table 1.

Table 1. Variable settings of system dynamics model.

Variable Category	Variable Name	Initial Value	Unit
state variable	amount of washed coal	205,000	10^4 Ton
auxiliary variable	coal washing cost	-	10^8 Yuan
rate variable	increasing rate of washed coal	-	10^4 Ton
auxiliary variable	proportion of washed coal	-	-
auxiliary variable	proportion goal of washed coal	-	-
auxiliary variable	coal industry profits	-	-
auxiliary variable	coal construction investment	-	10^8 Yuan
state variable	domestic coal productions	357,357	10^4 Ton
rate variable	coal production change rate	-	10^4 Ton
auxiliary variable	net coal import volume	-	10^4 Ton
auxiliary variable	coal supply	-	10^4 Ton
auxiliary variable	coal demand	-	10^4 Ton
auxiliary variable	coal supply/demand ratio	-	-
auxiliary variable	coal supply/demand ration factor	-	-
auxiliary variable	coal price	-	Yuan/Ton
auxiliary variable	coal consumption proportion of power industry	-	-
auxiliary variable	coal consumption of power supply	-	g/KWH
state variable	thermal power installed capacity	819,000	MW
rate variable	thermal power installed capacity increasing rate	-	MW
rate variable	thermal power installed capacity reduction rate	-	MW
state variable	clean-energy power installed capacity	323,940	MW
rate variable	clean-energy power installed capacity change rate	-	MW
auxiliary variable	total generated energy	-	10^8 KWh
auxiliary variable	thermal power generated energy	-	10^8 KWh
auxiliary variable	proportion of thermal power generation	-	-
auxiliary variable	proportion of thermal power installed capacity	-	-
auxiliary variable	thermal power generation profits	-	10^8 Yuan
auxiliary variable	thermal power construction investment	-	10^8 Yuan
auxiliary variable	grid purchase of thermal power	-	Yuan/KWH
auxiliary variable	power industry coal consumption	-	10^4 Ton
auxiliary variable	sulfur dioxide emissions of thermal power coal consumption	-	10^4 Ton
auxiliary variable	sulfur dioxide emission goal of thermal power coal consumption	-	10^4 Ton
state variable	desulfurization capacity	753,480	MW
rate variable	desulfurization capacity changing rate	-	MW
auxiliary variable	proportion of desulfurization capacity	-	-
auxiliary variable	desulfurization cost	-	10^8 Yuan
auxiliary variable	coal-conveying vehicle environmental success rate	-	-

In Table 1, the initial value of the amount of washed coal, the dominant coal productions, the thermal power installed capacity, the clean-energy power installed capacity, and the desulfurization capacity are from China's current situation as of 2012. The value of the proportion of washed coal, the sulfur dioxide emission goal of thermal power coal consumption, and the thermal power installed capacity reduction rate are controlled by time according to the relevant requirement of China's air pollutant emission reduction policies. For example, the proportion of washed coal in 2012 is 56%, and The Action Plan pointed out that this proportion will reach 70% by 2017. This model assumes that the proportion of washed coal rises at a constant speed during the period 2012–2017, and by maintaining this rate continues to rise. In the simulation period of 2013–2022, the proportions of washed coal range respectively from 59%, 62%, 65%... 71%... 83% to 86%. For thermal power sulfur dioxide emissions targets, The Action Plan pointed out that the PM10 concentration will drop by more than 10% in 2017 compared to 2012. According to the relevant data, in 2012, China's total carbon dioxide emission was 502 tons, and thermal power's contribution to sulfur dioxide emission accounted for about 50%. This study also assumes that sulfur dioxide emission declines at a constant speed during these five years, and by maintaining this speed continues to decline. So, during the simulation period of 2013–2022, the sulfur dioxide emission goal of thermal power (10,000 tons) ranges from 250, 245, 240... 205 to 200. The Action Plan pointed out that China plans to weed out 50 MW of small thermal power units in 2012–2017. So the rate of thermal power unit elimination for these five years is 10 MW per year. The eliminating speed is expected to accelerate from 2018. So the eliminating capacity of thermal power unit during the period 2018–2022 is 12 MW per year. The total generating capacity and the coal imports during the simulation period are estimated from the regression analysis according to the historical data of 2000–2012. The prediction equations are:

$$G_t = 3218.8 \times (Year - 1999) + 7613.9 \tag{1}$$

$$I_t = 2.8 + 0.4 \times (Year - 2012) \tag{2}$$

where G_t is the total generating capacity in year t and I_t is the coal imports in year t.

The grid purchase of thermal power is estimated according to the desulfurization thermal power electricity price in 2012. So the grid purchase of thermal power is 0.45 yuan per kilowatt-hour in the first five year of simulation, and it decreases to 0.435 yuan per kilowatt-hour after 2017.

The system dynamics model of air pollutant emission reduction policies' impact on the development of thermal coal supply chain member enterprises contains more than 30 functions among relevant variables. Due to the length limitation of this article, we only enumerate those functions with obvious characteristics and great significance to this study. Important functions of impact transfer are introduced as follows:

$$IR = CP \times WG - WC \tag{3}$$

where IR is the increasing rate of washed coal, CP is the domestic coal productions, WG is the proportion goal of washed coal, and WC is the amount of washed coal. Equation (3) shows that the increasing rate of washed coal is the difference between the proportion goal of washed coal and the actual amount of China's washed coal.

$$TS = (1.6 \times TC \times 1.5\%) \times (1 - PW \times 0.3) \times (1 - DS \times 0.9) \qquad (4)$$

where TS is the sulfur dioxide emissions of thermal power coal consumption, TC is the thermal power industry coal consumption, PW is the proportion of washed coal, and DS is the desulfurization capacity proportion.

$$SE = 1.6 \times CC \times CS \times (1 - DS \times DE) \qquad (5)$$

where SE is the SO_2 emission during the coal combustion of thermal-power industry, CC is the power industry coal consumption, CS is the coal sulfur content (the sulfur content of China's coal is about 1.5%), DS is the desulfuration capacity proportion, and DE is the desulfuration efficiency (China's desulfuration efficiency is about 90%).

$$TP = GP \times TG - CP \times TC / 10000 / 0.7 - DC \qquad (6)$$

where TP is the thermal power profits, GP is the grid purchase of thermal power, TG is the thermal power generated energy, CP is the coal price, TC is the thermal power industry coal consumption, and DC is the desulfuration cost. Fuel costs account for 70% of the total costs of thermal power plants.

4. Results and Discussion

This study analyzes the air pollution reduction policies' impact on the node enterprises of the thermal coal supply chain by comparing trends of key model factors with and without policy influence. The Vensim_PLE software is used to set up the above system dynamics model and simulate tendencies of the variables during the period 2013–2022. The trend graph of Vensim_PLE software can reflect changing trends more intuitively. This article selects several important factors such as SO_2 emissions of thermal power coal consumption, desulfurization capacity proportion, and the changing rate of thermal power installed capacity to measure the impacts of two typical kinds of air pollution reduction policies, and analyzes the changing developing modes brought by these two policies from multiple perspectives.

4.1. The Impact of Sulfur Dioxide Emission Target Policy

The Action Plan pointed out that the PM10 concentration will drop by more than 10% in 2017 compared to 2012. In the preceding part of this study, the SO_2 emission goals of thermal power during the simulation period 2013–2022 have been settled according to the relevant data. The new desulfurization installed capacity of the thermal power industry is fixed to 20,000 MW per year without the influence of this policy. When affected by this emission target policy, the new desulfurization installed capacity of the thermal power industry equals 20,000 (MW per year) multiplied by the ratio of the actual SO_2 emission and the emission goal of the thermal power industry. This paper analyzes the impact of emissions targets policy by comparing the changes of various factors with and without the constraints of this policy. The screenshots of simulation results using Vensim_PLE software are shown in Figure 3. Figure 3 displays the comparison results of SO_2 emissions of thermal power coal consumption (curve 1), the desulfurization capacity proportion (curve 2), and the increasing rate of desulfuration capacity (curve 3). Part (a) shows the simulated results under the impact of SO_2 emissions target policy and the annual SO_2 emission target (curve 4) during the decade simulation period. Part (b) shows the simulated results without policy impact.

Without policy constraints on the thermal power industry, SO_2 emission during coal combustion increases to the highest point of 5.7854 million tons in 2022. The desulfurization capacity proportion remains at around 89% with no obvious changes. The new desulfurization capacity increases from 17,566 in 2012 to 32,227 MW in 2022. Under the constraint of pollution emission targets, the SO_2 emission reduced greatly in 2013–2017 and then started to level off. It decreased to the lowest point of 2.8284 million tons in 2019. The desulfurization capacity proportion was greatly increased in 2013–2017. This proportion will reach 99% since 2017. There is no obvious difference between the two simulation results of thermal power installed capacity increasing rate with and without the constraint of pollution emission policies.

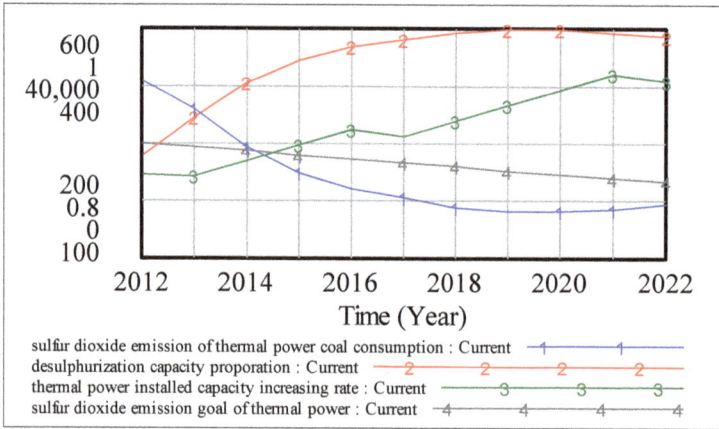

(a)

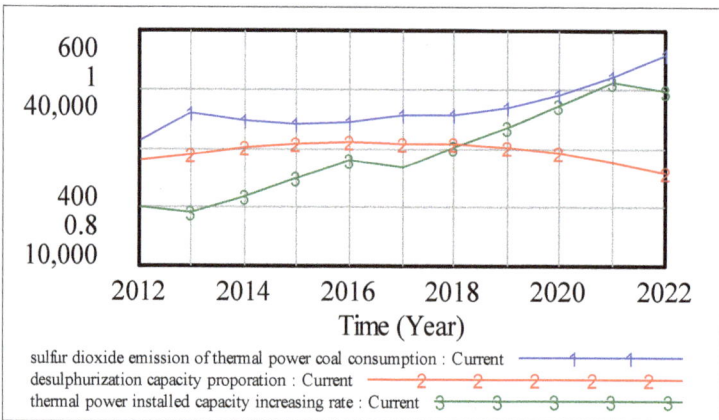

(b)

Figure 3. (a) Simulated results under the influence of SO_2 emissions target policy; (b) Simulated results without the influence of SO_2 emissions target policy.

We can tell from the compared results that SO_2 emission target policies will play an important role in reducing air pollutant emissions. SO_2 emission during coal combustion of the thermal power industry will be reduced by about 3 million tons per year. This emission target policy also can lead to a significant increase in the desulfurization capacity proportion. It is estimated that China will nearly complete the desulfurized renovation of the thermal power industry in 2017 under the influence of SO_2 emission target policy. As the result of space limitations, this paper only contrasts the changes in key factors. The SO_2 emission target policy can also decrease the proportion of thermal power installed capacity by reducing coal industry profits and thermal power construction investment. In 2012, China's thermal power installed capacity proportion is 71.6%. According to the simulated result, this proportion will decline steadily during the period 2013–2022. It will decrease to the lowest point of 62% in 2022. These structural changes are driven by the need for air pollution mitigation, not only in China but also in Europe and many developed countries [36]. After adjusting the parameters and variables based on the actual situation, this SD model can be used to simulate the policy impacts in all these countries. With the increasing of energy conservation and emission reduction policy's intensity, countries like China will implement sustainable development strategies and accelerate the transformation of the power industry.

4.2. The Impact of Coal Washing Proportion Target Policy

The Action Plan clearly set up the target that washed coal should account for more than 70% of total coal production in 2017. In the preceding part of this study, the coal washing proportion goals of thermal power during the simulation period of 2013–2022 have been settled according to the relevant data. This paper assumes that the coal washing proportion target will increase at a constant speed from 56% in 2012 to 71% in 2017 and keep increasing at the same speed during the rest of the simulated period under the impaction of the coal washing proportion target policy. Without the policy's influence, the coal washing proportion target will be fixed at 56% during the whole simulated period. The coal washing proportion goal influences the actual amount of washed coal by determining the increasing rate of washed coal. The increasing rate of washed coal equals the domestic coal production multiplied by the coal washing proportion goal. The amount of washed coal is the integral of washed coal's increasing rate. This paper analyzes the impact of proportion targets policy by comparing the changes of various factors with and without the constraints of this policy. The screenshots of simulation results using Vensim_PLE software are shown in Figure 4. Figure 4 displays the comparison results of actual coal washing proportion (curve 1), SO_2 emission during the coal combustion of the thermal power industry (curve 2) and the increasing rate of thermal power installed capacity (curve 3). Part (a) shows the simulated results under the impact of coal washing target policy and the annual goal of coal washing proportion (curve 4) during the simulation period. Part (b) shows the simulated results without the policy's impact.

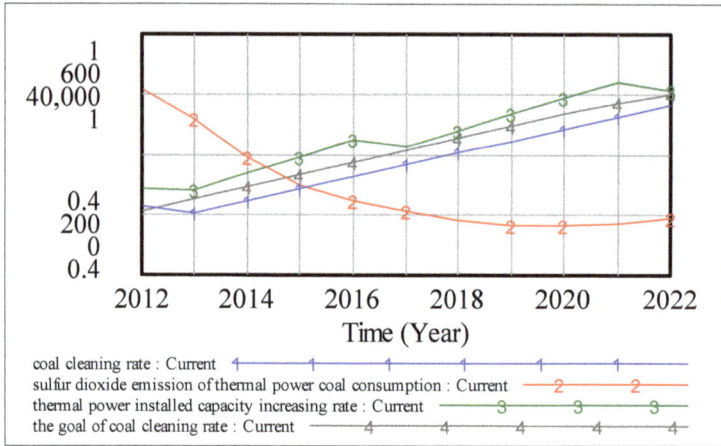

(a)

Figure 4. *Cont.*

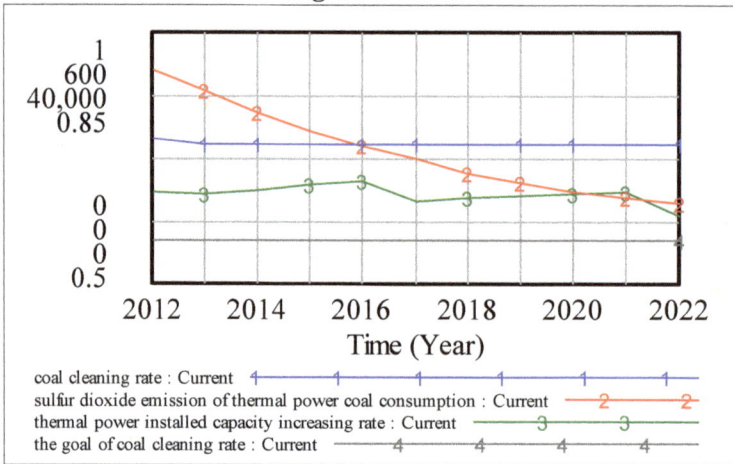

(b)

Figure 4. (**a**) Simulated results under the impact of washing coal proportion target policy; (**b**) Simulated results without the impact of washing coal proportion target policy.

Without the constraints of coal washing proportion target policy, the actual coal washing proportion is almost running at 56% during the simulated period. SO_2 emission from the coal combustion of the thermal power industry has a slow decline of about 2 million tons. In addition, the new thermal power installed capacity is fluctuating around 14,500 MW for the simulated period. Under the constraint of the coal washing proportion target, the actual proportion of washed coal increases at a constant speed from 56% in 2012 to 82% in 2022. SO_2 emission from the coal combustion of the thermal power industry has a sharp decline of more than 2 million tons and eventually stabilizes at about 2.9 million tons during the period 2013–2022. The new thermal power installed capacity will increase from 14,479 MW in 2012 to 30,720 MW in 2022.

We can tell from the simulated results shown in Figure 4 that the proportion target of washed coal established by air pollution reduction policies plays an important part in reducing SO_2 emissions. SO_2 emissions from the coal combustion of the thermal power industry will be reduced by about 2 million tons per year, which is less than the emission reduction driven by the emission target policy mentioned above. The formulation of a coal washing proportion goal makes the actual amount of washed coal increase by nearly 30%. In addition, the increase in washed coal can reduce the coal consumption rate in the power supply and improve the profits of the thermal power industry, and eventually lead to a slight rise in new thermal power installed capacity. At the same time, the sharp rise in clean energy power capacity will ensure that the power structure can shift in the sustainable development direction. The coal washing proportion goal policy can reduce the thermal power enterprise's fuel costs and improve its profits so that the thermal power industry can put more money into the desulfurization and denitration reform of coal-fired units. In general, a coal washing proportion goal policy can promote energy conservation and the emissions reduction technologies of the coal and thermal power industries by economic means. More applications of energy conservation and emission reduction technology can reduce the energy intensity of the industry and reduce air pollutant emissions fundamentally.

5. Conclusions

This paper simulated the air pollution reduction policies' impact on the thermal coal supply chain members in China by establishing SD models. These policies will greatly impact the development patterns of coal enterprises, power enterprises, and coal transportation enterprises. Moreover, the influence will transmit to the upstream and downstream enterprises along the thermal coal supply chain driven by business transactions and decision implementation. Besides China, many other countries (such as Japan and EU-27) have also made air pollution control policies, just like The Action Plan, to set emission targets and restrict coal consumption [37]. According to the simulated results, air pollution reduction policies can significantly improve air quality by promoting power structure adjustment and improving energy efficiency. As the main way of developing and utilizing non-fossil energy, the power industry will take the clean development route to coordinate sustainable development strategy. Under the pressure of air pollution reduction, the thermal power industry will implement the transformation of energy-saving and emission reduction ahead of time. These policies also provide coal washing technology and new energy power generation with good development opportunities. Renewable energy and nuclear electricity generation will have to a develop quickly to accelerate the structure adjustment of the electric power industry and realize the transformation of energy from fossil fuels towards clean energy. Therefore our simulation analysis of different policy interventions has meaning for countries that have not yet established their own air pollution control policies (such as India). In the future, in the promotion of long-distance transmission of electricity and coal, the distribution of coal in the power industry will be further optimized and the development of thermal coal supply chain members will be affected constantly. We can continue to analyze their development path under the new policy environment by adjusting the parameters and variables of the existing SD model.

Acknowledgments

Project supported by the Fundamental Research Funds for the Central Universities of China (No. MS201439).

Author Contributions

Xiaopeng Guo designed the study and revised the manuscript. Xiaodan Guo participated in designing the study, interpreted the data, wrote the manuscript and revised it until its final version. Jiahai Yuan provided good advices for conclusions and revised the manuscript.

Conflicts of Interest

The authors declare no conflict of interest.

References

1. Xue, B.; Geng, Y.; Katrin, M.; Lu, C.; Ren, W. Understanding the Causality between Carbon Dioxide Emission, Fossil Energy Consumption and Economic Growth in Developed Countries: An Empirical Study. *Sustainability* **2014**, *6*, 1037–1045.
2. Frances, C.M. Climate Change and Air Pollution: Exploring the Synergies and Potential for Mitigation in Industrializing Countries. *Sustainability* **2009**, *1*, 43–54.
3. Zhao, X.; Ma, Q.; Yang, R. Factors influencing CO_2 emissions in China's power industry: Co-integration analysis. *Energy Policy* **2013**, *57*, 89–98.
4. Zhao, X.; Thomas, P.L.; Cui, S. Lurching towards markets for power: China's electricity policy 1985–2007. *Appl. Energy* **2012**, *94*, 148–155.
5. Peggy, M.; Kenneth, B.K. Modelling tools to evaluate China's future energy system—A review of the Chinese perspective. *Energy* **2014**, *69*, 132–143.
6. Liu, L.; Zong, H.; Zhao, E.; Chen, C.; Wang, J. Can China realize its carbon emission reduction goal in 2020: From the perspective of thermal power development. *Appl. Energy* **2014**, *124*, 199–212.
7. Tan, Z.; Zhang, H.; Shi, Q.; Xu, J. Joint optimization model of generation side and user side based on energy-saving policy. *Electr. Power Energy Syst.* **2014**, *57*, 135–140.
8. Cai, L.; Guo, J.; Zhu, L. China's Future Power Structure Analysis Based on LEAP. *Energy Sources Part A Recovery Util. Environ. Eff.* **2013**, *35*, 2113–2122.
9. Zheng, M.; Zhang, K.; Dong, J. Overall review of China's wind power industry: Status quo, existing problems and perspective for future development. *Renew. Sustain. Energy Rev.* **2013**, *24*, 379–386.
10. Hirota, K. Comparative Studies on Vehicle Related Policies for Air Pollution Reduction in Ten Asian Countries. *Sustainability* **2010**, *2*, 145–162.
11. Cai, W.; Wang, C.; Zhang, Y.; Chen, J. Scenario analysis on CO_2 emissions reduction potential in China's electricity sector. *Energy Policy* **2007**, *35*, 6356–6445.

12. Mathews, A.J.; Tan, H. The transformation of the electric power sector in China. *Energy Policy* **2013**, *52*, 170–180.

13. Wen, Z.; Li, H. Analysis of potential energy conservation and CO_2 emissionsreductionin China's non-ferrous metals industry from a technology perspective. *Int. J. Greenh. Gas Control* **2014**, *28*, 45–56.

14. Wang, K.; Wang, C.; Lu, X.; Chen, J. Scenario analysis on CO_2 emissions reduction potential in China's iron and steel industry. *Energy Policy* **2007**, *35*, 2320–2335.

15. SheikhiFini, A.; Parsa Moghaddam, M.; Sheikh-el-Eslami, M.K. A dynamic model for distributed energy resource expansion planning considering multi-resource support schemes. *Electr. Power Energy Syst.* **2014**, *60*, 357–366.

16. Zhu, H.; Huang, G. Dynamic stochastic fractional programming for sustainable management of electric power systems. *Electr. Power Energy Syst.* **2013**, *53*, 553–563.

17. Salman, A.; bin Razman, M.T. Using system dynamics to evaluate renewable electricity development in Malaysia. *Renew. Electr. Dev.***2013**, *43*, 24–39.

18. Li, L.; Sun, Z. Dynamic Energy Control for Energy Efficiency Improvement of Sustainable Manufacturing Systems Using Markov Decision Process. *Cybern. Syst.* **2013**, *43*, 1195–1205.

19. Garcia, E.; Mohanty, A.; Lin, W.; Cherry, S. Dynamic analysis of hybrid energy systems under flexible operation and variable renewable generation-Part II: Dynamic cost analysis. *Energy* **2013**, *52*, 17–26.

20. Santiago, M.; Luis, J.M.; Felipe, B. A system dynamics approach for the photovoltaic energy market in Spain. *Energy Policy* **2013**, *60*, 142–154.

21. Ali, K.; Mustafa, H. Exploring the options for carbon dioxide mitigation in Turkish electricpower industry: System dynamics approach. *Energy Policy* **2013**, *60*, 675–686.

22. Feng, Y.; Chen, S.; Zhang, L. System dynamics modeling for urban energy consumption and CO_2 emissions: A case study of Beijing, China. *Ecol. Modell.* **2013**, *252*, 44–52.

23. Li, F.; Dong, S.; Li, Z.; Li, Y.; Wan, Y. The improvement of CO_2 emission reduction policies based on system dynamics method in traditional industrial region with large CO_2 emission. *Energy Policy* **2012**, *51*, 683–695.

24. Nastaran, A.; Abbas, S. A system dynamics model for analyzing energy consumption and CO_2 emission in Iranian cement industry under various production and export scenarios. *Energy Policy* **2013**, *58*, 75–89.

25. Özer, B.; Görgün, E.; Incecik, S. The scenario analysis on CO_2 emission mitigation potential in the Turkish electricity sector: 2006–2030. *Energy* **2013**, *49*, 395–403.

26. Frederick, A.A.; David, O.Y.; Alex, A.P. A Systems Dynamics Approach to Explore Traffic Congestion and Air Pollution Link in the City of Accra, Ghana. *Sustainability* **2010**, *2*, 252–265.

27. Lin, B.Q.; Moubarak, M. Renewable energy consumption—Economic growth nexus for China. *Renew. Sustain. Energy Rev.* **2014**, *40*, 111–117.

28. Bloch, H.; Rafiq, S.; Salim, R. Economic growth with coal, oil and renewable energy consumption in China: Prospects for fuel substitution. *Econ. Modell.* **2015**, *44*, 104–115.

29. Shen, J.F.; Xue, S.; Zeng, M.; Wang, Y.; Wang, Y.J.; Liu, X.L.; Wang, Z.J. Low-carbon development strategies for the top five power generation groups during China's 12th Five-Year Plan period. *Renew. Sustain. Energy Rev.* **2014**, *34*, 350–360.

30. Weller, F.; Cecchini, L.A.; Shannon, L.; Sherley, R.B.; Robert, J.M.; Altwegg, R.; Scott, L.; Stewart, T.; Jarre, A. A system dynamics approach to modelling multiple drivers of the African penguin population on Robben Island, South Africa. *Ecol. Modell.* **2014**, *277*, 38–56.

31. Jose, B.C.; Tan, R.R.; Culaba, A.B.; Ballacillo, J.A. A dynamic input–output model for nascent bioenergy supply chains. *Appl. Energy* **2009**, *86* (Suppl. 1), S86–S94.

32. Haghshenas, H.; Vaziri, M.; Gholamialam, A. Evaluation of sustainable policy in urban transportation using system dynamics and world cities data: A case study in Isfahan. *Cities* **2014**, in press.

33. Mao, X.Q.; Zeng, A.; Hu, T.; Xing, Y.K.; Zhou, J.; Liu, Z.Y. Co-control of local air pollutants and CO_2 from the Chinese coal-fired power industry. *J. Clean. Prod.* **2014**, *67*, 220–227.

34. Wang, S.; Xu, L.; Yang, F.L.; Wang, H. Assessment of water ecological carrying capacity under the two policies in Tieling City on the basis of the integrated system dynamics model. *Sci. Total Environ.* **2014**, *472*, 1070–1081.

35. Rehan, R.; Knight, M.A.; Unger, A.J.A.; Haas, C.T. Financially sustainable management strategies for urban wastewater collection infrastructure–development of a system dynamics model. *Tunnell. Undergr. Space Technol.* **2014**, *39*, 116–129.

36. Bollen, J.; Brink, C. Air pollution policy in Europe: Quantifying the interaction with greenhouse gases and climate change policies. *Energy Econ.* **2014**, *46*, 202–215.

37. Kanada, M.; Fujita, T.; Fujii, M.; Ohnishi, S. The long-term impacts of air pollution control policy: Historical links between municipal actions and industrial energy efficiency in Kawasaki City, Japan. *J. Clean. Prod.* **2013**, *58*, 92–101.

GIS-Based Synthetic Measurement of Sustainable Development in Loess Plateau Ecologically Fragile Area—Case of Qingyang, China

Chenyu Lu, Chunjuan Wang, Weili Zhu, Hengji Li, Yongjin Li and Chengpeng Lu

Abstract: Synthetic measurement of regional sustainable development has been one of the key issues in the research field of sustainability. In this paper, Qingyang City located in the Loess Plateau ecologically fragile area of Northwest China is used for a case study, and the present study aims to investigate the degree of sustainable development by conducting temporal- and spatial-scale based analysis, with the assessment index system, assessment model and GIS approach well integrated. The results show that the development pattern of Qingyang generally fits the mode of unsustainable development, even in the presence of certain levels of spatial differences. The sustainable development state in ecologically fragile area of China's Loess Plateau is non-optimistic, which is an uncoordinated status among subsystems of regional sustainable development. Although the level and tendency of regional sustainable development keeps increasing, such enhancement is abnormal. With the rapid deterioration of environmental and natural resources, their inhibitory effect on the economy and society would expand, eventually leading to the slow development rate or the recession of the entire system. The only solution is to change the traditional mode of economic development, to follow the guide of ecological economic conception so that the goal of achieving regional sustainable development strategies could be met ultimately. Meanwhile, the characteristics of different regions should be taken into account in order to achieve optimal spatial structure.

Reprinted from *Sustainability*. Cite as: Lu, C.; Wang, C.; Zhu, W.; Li, H.; Li, Y.; Lu, C. GIS-Based Synthetic Measurement of Sustainable Development in Loess Plateau Ecologically Fragile Area— Case of Qingyang, China. *Sustainability* **2015**, *7*, 1576-1594.

1. Introduction

Ever since the Industrial Revolution took off in the 18th century, humanity's infinite material and spiritual needs have been satisfied through continuous expansion of traditional economic scale, characterized by "mass production, mass consumption and mass abandonment" [1]. Under such circumstances, a lot of problems have appeared to exert negative feedback on the normal functioning of social and economic activities [2]. Therefore, how to achieve sustainable development has become a common life aspiration of all human beings [3]. In China, environmental problems have become increasingly prominent, and the worsening conditions of the environment haven't been well controlled [4]. Therefore, sustainable development has become the basic guiding philosophy and national development strategy in China for coordinated development in social, economic and environmental aspects [5].

Studies of sustainable development in other countries can be divided into two phases: phase I (1962–1992) represents the phase when the concept of sustainable development was proposed and its associated theories explored. United Nations Conference on Environment and Development was held in Rio de Janeiro in 1992, after which the implementation phase of sustainable development strategy was officially initiated. Since the concept of sustainable development has been put forward, how to define this concept and how to link this abstract concept with cognitive and practical methods become two major challenges [6]. Meanwhile, the word sustainability with quantitative value, together with its related theoretical and practical studies, have become key components of study on sustainable development [7]. Therefore, the assessment of sustainable development has become a hot and central research topic in the field of sustainable development [8], and a series of theoretical and practical studies related to sustainability assessment have been done. In general, two types of assessment methods have been developed. The first type uses a single or composite index, whereas the second type applies multi-index system. Also, both types are still in the exploratory stage. Overall, the specific relationship between indexes and sustainable development goals, the determination of index weight and threshold, and the development of an integrated assessment method remains the core and challenge of any assessment system [9]. In particular, how to establish a reasonable sustainability assessment system with an integrated assessment model, and how to study the contributions of human activities to the realization of the goal of sustainable development represent key and hot topics of current sustainable development research [10].

In China, studies on sustainable development have started to track the trajectory of research on global sustainable development since the mid-1980s [11,12], which could be further divided into two parts, namely, theoretical studies and practical applications. Theoretical studies mostly focus on aspects such as the establishment of sustainability assessment system [13–15], the analysis of coordination development in PRED (population, resources, environment and development) system [16–18], and the assessment and determination of sustainability [19–21]. By contrast, practical studies mostly focus on aspects such as regional coordination development in PRED system [22–24], and the evolution of regional sustainability over time [25–27]. Although the work of some scholars involves the spatial distribution aspect of sustainability [28,29], most work is simply static description with the lack of time content. Currently, research on China's sustainable development is developing rapidly with a consistent deepening trend. As to regional practice, accompanying with the construction advance of national main functional zones, particularly with increasing environmental constraints that guide China's future development, many counties and districts have put an emphasis on promoting the process of sustainable development, and quantitative characterization of sustainable development level and its development process is urgently needed [30]. In order to promote the process of sustainable development towards a new height and to better draw lessons from experience, research on sustainable development field has displayed a gradual shift, namely, from a qualitative approach towards a quantitative approach, with an emphasis on the combination of such shift when factors of regional geography and environment are explicitly considered [31]. Therefore, such process reflects a shift from traditional conceptual models towards specific regional models [32], and a shift from static pattern study towards dynamic process study [33].

Overall, relevant studies have achieved a lot, although some problems and deficiencies still exist. First, studies that focus on the integration of spatial and temporal aspects of sustainable development are still lacking. Also, among existing research results, studies that have taken spatial and temporal dimensions into account simultaneously in order to conduct an integrated study on the evolution of sustainable development are extremely rare. By contrast, most current studies are testing the evolution patterns of sustainable development at temporal scale, although research on sustainable development should ultimately focus on specific geographic regions, or, in other words, research on the evolution of sustainable development at spatial scale is equally important. Second, many problems have been found regarding the assessment system of sustainable development, including mismatches between theory and practice, huge number of index systems, overemphasis on economic assessment, overlook of resource and environment assessment, and the failure of index selection process to truly implement the concept of sustainable development. To a certain extent, the present paper attempts to solve some of these problems by providing some complementary work.

In the present paper, Qingyang City is used as the case study region. Qingyang is located in Loess Plateau zone of eastern Gansu Province, and represents a typical gully region of China's Loess Plateau. The administrative division in Qingyang City includes eight sub-regions, seven counties and one district (Figure 1). It belongs to semiarid region due to the lack of surface water. Thanks to natural conditions and human economic activities, the ecological environment is fragile in this region [34]. In terms of economic structure, agriculture has played a dominant role in Qingyang. In recent years, as Qingyang becomes an important component of Ordos-based national energy industrial base and the energy and chemical industry base of eastern Gansu Province, the contribution of industry to national economy has exceeded that of agriculture. Meanwhile, traditional agriculture-based industrial structure has been broken, leading to the transition between agriculture and industry, as well as the rapid development of local economic structure [35]. However, with the deepening of the transition between the agriculture and industry, contradictions among aspects such as regional population, economy, society, resources and environment become increasingly acute, and man-land relationship has deteriorated rapidly in Qingyang. Also, accompanied with the emergence of a series of ecological and environmental problems, many issues begin to arise from both natural and anthropogenic disturbance. It is not surprising that both the complexity of the integrated system and regional sustainable development have met severe challenges [36]. Worse still, constraints imposed by resources and environment on economic and social development become more and more severe, and contradictions between local resource development and environmental pressure become more and more acute. Equally, the capacity for sustainable development is worrisome. In particular, the lack of water resources and quality in Qingyang has been worsened by a groundwater pollution problem due to oil extraction, which has caused increasingly grim situation of water resource, and seriously restricted economic, social and sustainable development. In fact, a series of core issues related to sustainable development, such as the transition of industry structure, the deterioration of ecological environment, and the transformation of social structure, has been commonly found in the ecologically fragile area of China's Loess Plateau [37]. Therefore, we must stand on the height of sustainable development, and wisely handle the coordinated and balanced development in terms of regional population, economy, society, resources and environment.

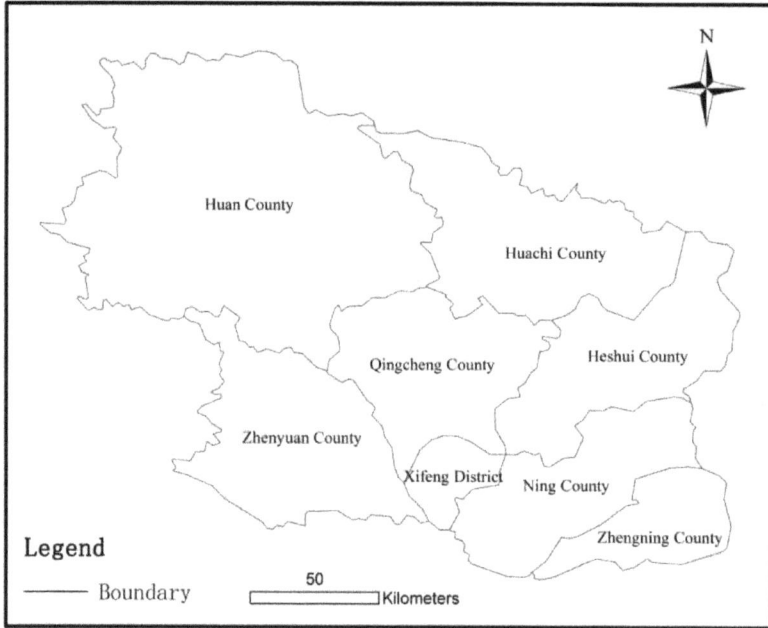

Figure 1. The administrative map in Qingyang City.

In this paper, the assessment index system and the integrated assessment models with temporal and spatial scale explicitly considered are established, with existing research results and regional characteristics as the basis. Also, the application of GIS technology allows a successful integration of both spatial and temporal dimensions so that a study on synthetic measurement of sustainable development evolution in ecologically fragile area of China's Loess Plateau could be facilitated, the process of two-dimensional analysis and integration could be carried out, and deficiencies in the existing research could be compensated. Overall, the present paper has strong theoretical and practical significance, and could contribute to the understanding, revealing and solving of a series of problems that exist during the process of sustainable development, just as what Qingyang is experiencing now. In this sense, this paper could provide theoretical support and decision-making basis for achieving sustainable development strategies.

2. Methodology

2.1. The Establishment of the Index System

The assessment index system of sustainable development is the basis for an integrated assessment of regional sustainable development. As it concerns the division of regional sustainable development, it is generally considered an improvement based on PRED system. According to the conception of PRED, the system is divided into two subsystems (*i.e.*, the economy and the society) in the present study so that the characteristic of regional sustainable development could be reflected more clearly. In a word, based on geographical features as well as actual situations, draw on existing achievements, and followed the integrated, systematic, representative, concise, accessible and practical principles [38,39], the present paper chooses corresponding indexes from five aspects, namely, the population, economy, society, environment and resources, and then establishes the four-category index system that reflects the sustainability level of ecological fragile area in China's Loess Plateau.

According to the criterion layers of scale, structure and health level of the population, 5 indexes are selected for the population system. In addition to these criterion layers, the quality of the population is always counted as the criterion layer, which could reflect the educational status of local people. Since the educational status of local people is reflected in a social system, it is not selected for the population system. Since all the selected indexes are reasonable, the criterion layers could be embodied by index layers, and the characteristics of the population system could be reflected accurately by these criterion layers and index layers.

According to the criterion layers of scale, structure, efficiency and openness of economy, 9 indexes are selected for the economic system. Since 4 criterion layers and 9 index layers are commonly selected for the economic system in the literature, they are both representative and reasonable. The criterion layers can be embodied by index layers relatively. Overall, the characteristics of the economic system could be reflected accurately by these criterion layers and index layers.

According to the criterion layers of quality of people's lives, social development level, and social activity, 9 indexes are selected for the social system. Since 3 criterion layers are commonly selected for the social system in the literature, they are both representative and reasonable. Also, the 9 selected index layers are reasonable. Due to the lack of sufficient data, several indexes are not selected as the criterion layers of social development level (e.g., the percentage of social insurance coverage). Overall, the characteristics of society system can be reflected clearly by these criterion layers and index layers.

According to the criterion layers of natural resources and social resources, 7 indexes are selected for the resource system. Two criterion layers are selected based on the basis of resource classification, and they are both representative and reasonable. However, the index layers are not ideal for the resource system. Due to the lack of sufficient data, several significant indexes are not selected (e.g., the indexes of water resource and energy consumption). Overall, the characteristics of the resource system could be reflected accurately by these criterion layers and index layers.

According to the criterion layers of environmental pollution and protection, 8 indexes are selected for the environmental system. Since 2 criterion layers are commonly used in the literature, they are representative and reasonable. Also, the selected 8 index layers are reasonable. Due to the lack of sufficient data, several indexes are not selected as the criterion layers of environmental protection (e.g., the harnessing percentage of soil and water loss). Overall, the characteristics of the environmental system could be reflected accurately by these criterion layers and index layers.

Meanwhile, index attributes are categorized. If the larger an index value is, regional sustainable development is more likely, such index is then named a positive index, notated as "+". By contrast, if the larger an index value is, regional sustainable development is more unlikely, such index is then named a negative index, notated as "–" (Table 1).

Table 1. The assessment index system of sustainable development.

System Layer	Criterion Layer	Index Layer	Attributes (+ or −)
Population system	Population scale	Population size (A1)	−
		Population density (A2)	−
		Natural population growth rate (A3)	−
	Population structure	The percentage of non-agricultural populations (A4)	+
	Health level	The average life expectancy (A5)	+
Economy system	Economic scale	GDP per capita (B1)	+
		Economic density (B2)	+
		Per capita investment in fixed assets (B3)	+
		Per capita financial income (B4)	+
		GDP growth rate (B5)	+
	Economic structure	The share of primary industries in GDP (B6)	−
		The share of third industries in GDP (B7)	+
	Economic efficiency	Input-output ratio (B8)	+
	Economic openness	Total imports and exports (B9)	+
Society system	The quality of people's lives	Per capita net income of farmers (C1)	+
		Per capita living expenditure of farmers (C2)	+
		Per capita disposable income of urban residents (C3)	+
		Per capita consumption expenditure of urban residents (C4)	+
	Social development level	The number of people engaged in scientific research every 10,000 people (C5)	+
		The number of students enrolled every 10,000 people (C6)	+
		Per capita dwelling area (C7)	+
		Per capita telecommunications services (C8)	+
	Social activity	Total retail sales of social consumer goods (C9)	+
Resource system	Natural resources	Per capita arable land area (D1)	+
		Per capita garden land area (D2)	+
		Per capita forest land area (D3)	+
		Per capita grassland area (D4)	+
		Per capita food production (D5)	+
	Social resources	Available health technical personnel every 10,000 people (D6)	+
		Available hospital beds every 10,000 people (D7)	+
Environment system	Environmental pollution	Total amount of wastewater discharge (E1)	−
		Total amount of industrial solid waste emissions (E2)	−
		Total amount of industrial waste gas emissions (E3)	−
		The fertilizing intensity of chemical fertilizer (E4)	−
	Environmental protection	Industrial wastewater discharge-related compliance rate (E5)	+
		Integrated utilization rate of industrial solid waste (E6)	+
		Smoke and dust emissions related compliance rate (E7)	+
		Capacities of waste gas treatment facilities (E8)	+

In order to ensure that the data could reflect actual situations of the studied area as much as possible, and to eliminate possible effects of different index dimensions on assessment results, extreme values are standardized so that assessment indexes are dimensionless [40]. The calculation formula is:

Positive index: the larger the index value is, the better it is.

$$Z_{ij} = \frac{X_{ij} - X_{min}}{X_{max} - X_{min}} \tag{1}$$

Negative index: the smaller the index value is, the better it is.

$$Z_{ij} = \frac{X_{max} - X_{ij}}{X_{max} - X_{min}} \tag{2}$$

In this formula, Z_{ij} is the standardized value of X_{ij}; X_{ij} is the actual value of the administrative unit i and indexical stage j; X_{max} and X_{min} represent the maximum and minimum value, respectively, of the same index at different administrative unit and stage.

2.2. Integrated Assessment Model

There are lots of models for the assessment of regional sustainable development, and they could be generally divided into 2 types. The first type puts an emphasis on sustainability measurement, which is commonly used to evaluate the level or the capacity of regional sustainable development. The second type puts an emphasis on coordination measurement, which is used to evaluate the coordination development status among different indexes or different systems. Generally speaking, both types tend to put an emphasis on certain aspects of regional sustainable development, and thus are unilateral or non-comprehensive. In this study, an integrated assessment model based on these 2 types is used for a comprehensive evaluation of regional sustainable development. One obvious advantage of this model is that it can be used for a comprehensive assessment of regional sustainable development based on the aspects of sustainability and coordination measurement, and the results of comprehensive measurement could be quantified. Theoretically and empirically, this model is more reasonable than other alternatives. Therefore, this integrated assessment model with 4 components is selected for the present study.

2.2.1. Index of Sustainable Development

Principal component analysis is applied to eliminate possible redundant information among various indexes, to extract important factors, and to calculate the contribution of each index. Then, by using a weighted linear model, the composite score, namely, the index of sustainable development level in each subsystem is calculated [41].

$$F_m = \sum_{j=1}^{n} W_j X_j \tag{3}$$

In this formula, F_m represents the index of sustainable development level in each subsystem, m represents the number of subsystems; n represents the number of principal components in each subsystem; W_j represents the contribution level of each principal component in each subsystem; X_j represents the score of all principal components in each subsystem.

The integrated index value of sustainable development level could reflect the integrated level of regional sustainable development. Based on obtained index values of sustainable development in various subsystems, weighted linear models are used for measurement purpose [41]:

$$SDI = \sum_{j=1}^{m} \alpha_j F_m \tag{4}$$

In this formula, SDI represents the integrated index of regional sustainable development; F_m represents sustainable development level of each subsystem α_j represents the weighted value of each subsystem. In the present study, the coordination degree needs to be measured. Since the conception of coordination degree is based on the equal importance of each subsystem, here it is assumed that all five subsystems are equally important, and each subsystem is assigned the value 0.2.

2.2.2. Sustainability Degree

Sustainability degree is the criterion used to measure regional sustainable development and the sustainability of each subsystem. Usually, the developmental rate of each subsystem is used as the calculation criterion [42]:

$$SI = \frac{1}{T} \sum_{i=0}^{T-1} S(t-i) \tag{5}$$

In this formula, SI represents the sustainability of regional sustainable development; $S(t-i)$ represents the developmental rate of the system during the time span of t-i~; $i = 0,1,\cdots,T-1$; T represents the reference period.

2.2.3. Coordination Degree

Since the developmental process of a system is dynamic, the coordination degree is used to quantify the coordinated development degree of interactive factors within or among systems. Therefore, the coordinated development degree among regional systems could be quantified and determined through the calculation of coordination degree. In the present paper, principal component analysis, simulate regression model and the membership function of fuzzy mathematics are used to assess the coordination degree among subsystems. Based on the development index of each subsystem, regression curve fitting process could provide the predictive value of sustainable development index of each subsystem, namely, the coordination value, and the coordination degree between subsystems could be calculated then using both the coordination value and the membership function of fuzzy mathematics [43].

$$u(i/j) = \exp\left\{-\frac{(F_i - F')^2}{S^2}\right\} \tag{6}$$

In this formula, $u(i/j)$ is the state coordination degree of system i relative to system j; F_i is the development index of system i (the actual value); F' is the predictive value of best development degree of system j relative to system i (the coordinated value); S^2 is the variance of system i.

The coordination degree between two systems, $U(i, j)$, could be calculated via coordination degree, with the specific formula shown as following [43]:

$$U(i,j) = \frac{\min\{u(i/j), u(j/i)\}}{\max\{u(i/j), u(j/i)\}} \qquad (7)$$

In this formula, $U(i, j)$ represents the coordination degree between system i and system j; $u(i/j)$ represents the state coordination degree of system i relative to system j; $u(j/i)$ represents the state coordination degree of system j relative to system i. $U(i,j) \in [0,1]$, and the more it is close to 1, it is more coordinated between system i and system j. Otherwise, it is more non-coordinated.

The coordination degree among more than three systems could be calculated as shown below [43]:

$$U_k(1,2,...,k) = \frac{\sum_{i=1}^{k} U(i/\bar{i}_{k-1}) \times U_{k-1}(\bar{i}_{k-1})}{\sum_{i=1}^{k} U_{k-1}(\bar{i}_{k-1})}, k = 3, 4, ..., m \qquad (8)$$

In this formula, m is the number of subsystem; $\bar{i}_{k-1}$ is the set of any $k-1$ subsystem except for system i; $U(i/\bar{i}_{k-1})$ is the state coordination degree imposed by subsystem i on any $k-1$ subsystem; $U_{k-1}(\bar{i}_{k-1})$ is the coordination degree of any $k-1$ subsystem except for system i.

According to the obtained formula used for calculating the coordination degree, when the actual value is more close to the coordinated value, the larger the coordination degree is, and the entire system is with higher coordination degree. Otherwise, the system is with lower coordination degree. Also, with reference to relevant literature [40], the coordination degree of a system could be categorized as four levels, namely, extremely uncoordinated, uncoordinated, barely coordinated and coordinated (Table 2).

Table 2. Metrics for the coordination degree.

0~0.50	0.50~0.85	0.85~0.95	0.95~1
Extremely uncoordinated	Uncoordinated	Barely coordinated	Coordinated

2.2.4. Tendency Degree

Tendency degree refers to the maintaining or sustainable ability of a regional system, and is commonly used to measure and determine whether a regional system is developed towards a healthy direction or not. Therefore, it is an integrated indicator and an inherent union of system development, continuity and coordination [42]:

$$S = \alpha_1 \square SDI + \alpha_2 \square SI + \alpha_3 \square U \tag{9}$$

In this formula, S is the tendency degree of sustainable development; SDI, SI and U represent the integrated index of sustainable development degree, sustainability degree and coordination degree. Since the present paper assumes that all three are equally important, α_1, α_2 and α_3 is with same weighted value, and the weighted average method is used to calculate the weighted average with $\alpha_1 = \alpha_2 = \alpha_3 = 1/3$.

2.3. A Model Based Approach for Analysis of Spatial Structure

The regional centroid refers to the equilibrium point of an element's torque on its space plane. In practice, with the assumption that a specific region consists of n space units, the centroid coordinate of space unit i is (x_i, y_i), M_i is the element value of space unit i, and x, y represent the spatial coordinate of such element within the entire region. The calculation method is as shown following:

$$x = (\sum_{i=1}^{n} M_i x_i) / (\sum_{i=1}^{n} M_i) \tag{10}$$

$$y = (\sum_{i=1}^{n} M_i y_i) / (\sum_{i=1}^{n} M_i) \tag{11}$$

In this study, M_i represents element attributes of county i, x_i and y_i represents the spatial coordinate of the administrative centroid of county; and i, x and y represents the centroid coordinate with relevant attributes. When the centroid attributes of an element are moving through time, the moving direction will point at high-density area, and the deviation distance will indicate the degree of un-equilibrium [44,45]. Also, it can be seen from centroid formula that if the geological locations of every site remain unchanged (*i.e.*, such consistence is assumed unchanged throughout the research period), then factors that directly affect the centroid would be the speed and degree of the centroid attributes. Since the development and change of such attributes would affect regional change of centroid, the latter could clearly reflect the developmental trajectories and spatial differences of such attributes.

3. Results and Discussion

The chosen assessment system and integrated assessment model is used to thoroughly access and analyze the sustainable development status of each county in Qingyang, which, combines with the evolution of spatial structure analysis, reveals the spatial and temporal evolution characteristics and patterns of its regional sustainable development. This present study selects five years, namely, 2000, 2003, 2006, 2009 and 2012 for further analysis. The origin of data used in this study is some local statistical information, including statistical yearbook, statistic bulletin and government report provided by local government.

3.1. Integrated Assessment and Analysis

As to the aspect of integrated indexes that reflect regional sustainable development, each county in Qingyang has shown a gradual rising trend in recent decade since 2000 (Table 3). Specifically, the rate of increase is relatively fast during the early stage with a high magnitude. However, the climbing space is constrained during the late stage, and the rate of increase becomes slow. This is because the speed of current economic and social development is relatively fast. Although both resources and environmental systems have displayed some degradation trend, the exponential growth rate of sustainable development indexes regarding economic and social systems temporarily exceeds the decreasing rate of sustainable development indexes regarding resources and environmental systems, thus causes such observed phenomena. However, the increase of index values is just temporary with crises hidden behind, which indicates that the increase of integrated degree of regional sustainable development in Qingyang is an abnormal pattern because such development is at the expense of environmental resources.

Table 3. The integrated index of sustainable development level.

	Xifeng District	Qingcheng County	Huan County	Huachi County	Heshui County	Zhenyuan County	Ning County	Zhengning County
2000	−0.38	−0.39	−0.36	−0.31	−0.34	−0.45	−0.43	−0.42
2003	0.02	0.04	0.02	0.13	0.07	−0.03	−0.04	0.01
2006	0.61	0.65	0.51	0.53	0.56	0.58	0.48	0.50
2009	1.18	1.12	1.01	1.05	1.06	1.09	0.98	0.97
2012	1.46	1.41	1.24	1.32	1.35	1.27	1.33	1.38

In terms of the sustainability degree of regional sustainable development, each county in Qingyang has displayed a first increasing and then decreasing trend (Table 4). Before 2009, the overall sustainability degree of every county in Qingyang displayed a rising pattern with an increase of amplitude. After 2009, however, the sustainability degree of every county in Qingyang showed certain level of decrease with a decrease of amplitude. This indicates that during the early stage every county in Qingyang continued to show a certain level of non-deceleration development that is both stable and sustainable. However, during the late stage, the status of sustainable development has been shifted towards the status of unsustainable development due to increasing pressure imposed by rapid economic and social development. Overall, the development in Qingyang is unsustainable.

Table 4. The sustainability degree.

	Xifeng District	Qingcheng County	Huan County	Huachi County	Heshui County	Zhenyuan County	Ning County	Zhengning County
2003	0.40	0.43	0.38	0.44	0.41	0.42	0.39	0.43
2006	0.50	0.52	0.44	0.42	0.45	0.52	0.46	0.46
2009	0.52	0.50	0.46	0.45	0.47	0.51	0.47	0.46
2012	0.46	0.45	0.40	0.41	0.42	0.43	0.44	0.45

As to the coordination aspect of regional sustainable development, although slight fluctuations have been observed in recent decade since 2000, each county in Qingyang has shown varying levels of a declining trend (Table 5). Before 2009, the overall coordination degree of the entire area showed a significant decreasing trend. Specifically, County Huachi and Heshui was changed from the status of coordinated to barely coordinated, whereas the rest was changed from the status of barely coordinated to uncoordinated. Obviously, rapid economic growth has brought enormous pressure on natural resources and environment, and the developmental direction of economic society was opposite to that of natural resources and environment. As a result, coordination degree has been weakened in general, and the coordinated development between some subsystems was gradually deteriorated in particular. Since 2009, the decreasing trend of coordination degree among counties has been constrained to a certain degree, and some counties even have started to show an increasing trend. However, such situation still belongs to the uncoordinated status. This is because in recent decade, Qingyang City government has begun to implement the circular economy as a development strategy. To some extent, such practice has successfully curbed the degradation trend of resources and environmental systems, and thus contributed to a much more coordinated development between subsystems, although it is far less than enough to completely reverse the worsening trend. Overall, the uncoordinated development trend found among Qingyang City's population, economic, social, resources and environmental systems has imposed severe challenges to its regional sustainable development.

Table 5. The coordination degree.

	Xifeng District	Qingcheng County	Huan County	Huachi County	Heshui County	Zhenyuan County	Ning County	Zhengning County
2000	0.93	0.89	0.87	0.95	0.96	0.90	0.86	0.92
2003	0.91	0.88	0.88	0.92	0.95	0.90	0.88	0.91
2006	0.84	0.84	0.85	0.91	0.93	0.87	0.85	0.89
2009	0.76	0.78	0.79	0.85	0.86	0.80	0.82	0.81
2012	0.78	0.79	0.81	0.84	0.90	0.84	0.81	0.85

As far as the tendency degree of regional sustainable development is concerned, each county in Qingyang has shown a gradual upward trend in recent decade since 2000 (Table 6). Specifically, the rate of increase is relatively fast during the early stage with a high magnitude. However, the rate of increase becomes slow during the late stage. This is because economic and social development is relatively fast, and thus plays an important supportive role in promoting regional sustainable development. Although both resources and environmental systems have displayed some degradation trend, the rapid development of the economy and society has masked the reality, resulting in a gradual increase in the integrated indexes that reflects the status of regional sustainable development. However, the increase of such index values is just temporary, and the increasing trend could not be maintained persistently over time. Actually, the rate of increase becomes slow during the early stage, which has fully demonstrated this point. On the surface, Qingyang City is developing towards sustainable development status, although the truth is that a lot of crises are hidden behind, or, in other words, the current development is by no means sustainable.

Table 6. The tendency degree.

	Xifeng District	Qingcheng County	Huan County	Huachi County	Heshui County	Zhenyuan County	Ning County	Zhengning County
2003	0.44	0.45	0.43	0.50	0.48	0.43	0.41	0.45
2006	0.65	0.67	0.60	0.62	0.65	0.66	0.60	0.62
2009	0.82	0.80	0.75	0.78	0.80	0.80	0.76	0.75
2012	0.90	0.88	0.82	0.86	0.89	0.85	0.86	0.89

3.2. Spatial Structure Analysis

On the basis of geographic information system and ArcGIS software, the present paper applies centroid displacement method to analyze the evolution of spatial structure regarding regional sustainable development in Qingyang, with spatial gaps of regional sustainable development visualized and quantified. The detailed procedures include: (1) the establishment of spatial databases. On the basis of ArcGIS software, Qingyang administrative map is digitalized (including administrative boundaries and the administrative center), and projection transformation is converted in order to determine the spatial plane coordinate of administrative center for each district. Meanwhile, the attributes of all elements in each district through years are imported; (2) By using the centroid calculation formula and on the basis of relevant element attributes, the spatial centroid of each element through years in Qingyang is calculated. Based on obtained centroid results, the trajectory of the moving centroid for each element is drawn in order to help analyze temporal and spatial evolution patterns.

In Qingyang region, the centroid distribution width of sustainable development integrated index is relatively large with a high magnitude of fluctuations over years (Figure 2). Between 2000 and 2003, the centroid was shifted towards the northeast direction with a high magnitude, and the displacement distance was around 50 km. This is due to the rapid growth rate of regional sustainable development in Huachi and Heshui counties located in the northeastern part of Qingyang. Also, both counties are known for their relatively high level of sustainable development. Between 2003 and 2006, the centroid was shifted towards the southwest direction with a high magnitude. During this period, the growth rate of regional sustainable development in Xifeng, Qingcheng and Zhengyuan is relatively high, all of which are located in the southwestern part of Qingyang with relatively high level of sustainable development. Between 2006 and 2009, the centroid displacement distance was relatively short (e.g., several tens of meters with a high level of overlap). During this period, the spatial distribution pattern of every county and district regarding regional sustainable development did not change significantly. Between 2009 and 2012, the centroid was shifted slightly towards the southeastern direction. This is due to the rapid growth rate of sustainable development in Ning and Zhengning located in the southeastern area of Qingyang.

The centroid of sustainability degree regarding regional sustainable development in Qingyang is with a certain level of fluctuations but limited fluctuation range (Figure 3). All the centroids of sustainability degree are distributed in the area of Qingcheng County. Between 2003 and 2006, the centroid was shifted towards southwest direction with relatively large level of displacement. This is mainly due to relatively high levels of sustainability degree in Xifeng, Qingcheng and Zhengyuan counties, all of which are located in the southwestern part of Qingyang. Between 2006 and 2009, the centroid was shifted slightly towards the north direction, indicating that the rate of increase regarding the sustainability degree is slightly higher in the northern area than that of the southern area, with the latter displaying smaller magnitudes. Between 2009 and 2012, the centroid was shifted towards the southeast direction to some extent. During this period, the sustainability degree of every country and district in Qingyang showed certain level of decrease. However, since the level of decreasing amplitude in Ning and Zhengning in the southeastern parts was relatively small, the overall sustainability degree remained at a relatively large level, and thus the centroid was shifted towards the southeast direction.

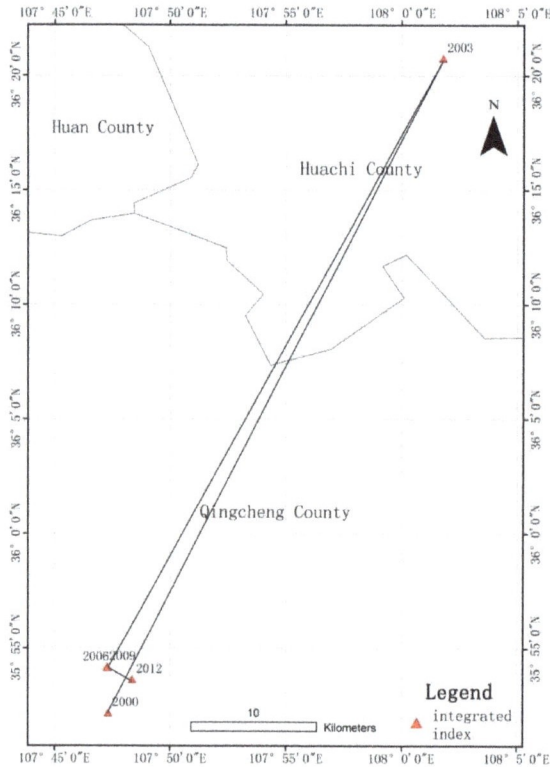

Figure 2. The centroid distribution of integrated index of sustainable development level.

The centroid distribution range of sustainable development coordination degree in Qingyang region is relatively narrow with small magnitude of fluctuations over time. All the centroids of coordination degree are distributed in the area of Qingcheng County. Obviously, the spatial difference regarding the change of coordination degree is not very large (Figure 3). Between 2000 and 2003, the centroid was slightly shifted towards the southwest direction. Between 2003 and 2006, or between 2006 and 2009, the centroid was shifted towards the northeast direction with relatively small magnitude of displacement. Between 2009 and 2012, the centroid was shifted again towards the southwest direction. Overall, in the recent decade, the spatial distribution pattern regarding the coordination degree in every county and district of Qingyang did not change significantly, and the subtle differences could only be detected along the southwest-northeast direction.

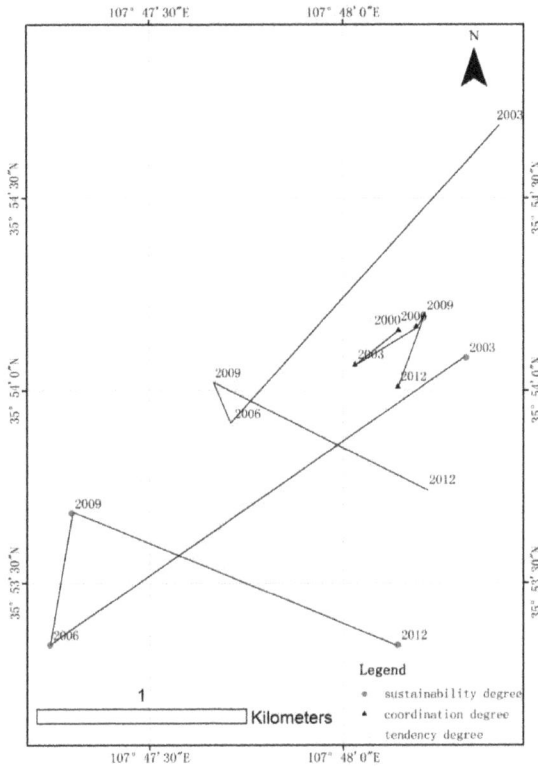

Figure 3. The centroid distribution of sustainability degree, coordination degree, and tendency degree.

The centroid of the tendency degree regarding sustainable development in Qingyang displays certain levels of fluctuations with relatively small magnitudes (Figure 3). All the centroids of tendency degree are distributed in the area of Qingcheng County. Between 2003 and 2006, the centroid was shifted towards the southwest direction with relatively large magnitude of displacement. This is due to the rapid growth rate of tendency degree in Xifeng, Qingcheng and Zhengyuan, all of which are located in the southwest area of Qingyang and with high levels of tendency degree. Between 2006 and 2009, the centroid was slightly shifted towards the north direction, indicating that the rate of increase regarding tendency degree is slightly higher in the northern area than that of the southern area, with the latter displaying smaller magnitudes. Between 2009 and 2012, the centroid was shifted towards the southeast direction. This is due to the rapid growth rate of tendency degree in Ning and Zhengning located in the southeast area of Qingyang, and both shared relatively large magnitude of increase.

4. Conclusions

Since 2000, the development pattern of Qingyang has fitted the mode of unsustainable development in the presence of certain levels of spatial differences. Under current development scenarios with rapid economic and social development, the natural resources and environment are gradually degrading. Because such development is at the expense of natural resources and environment, the coordination degree among population, economy, society, resources and environment is damaged, and the current state of development is basically uncoordinated. Currently, although the level of regional sustainable development and associated tendency degree remains arising, such a trend is solely temporary with severe hidden crises, and the tendency degree begins to show a declining trend. In the long run, such pattern would lead to the overall decrease of all indexes related to regional sustainable development, and Qingyang would not be able to achieve its healthy, long-term sustainable development. In terms of spatial structure, the spatial differences of integrated indexes in terms of regional sustainable development represent the largest ones, followed by the sustainability and tendency degree, and the spatial differences of the coordination degree represent the smallest ones. Also, the spatial differences along the northeast-southwest direction are predominant.

The challenges that Qingyang faces during the critical period of its development are typically found in ecologically fragile area of China's Loess Plateau, known for its non-optimistic sustainable development reality. In such area, water scarcity is a big problem, the ecological environment is fragile, and the poorest or most backward districts could be easily found. To get rid of poverty, local government often emphasizes the rapid growth rate of economy, and is desperate to promote extensive economic growth at the expense of natural resources and environment, causing uncoordinated status among subsystems of regional sustainable development, which would further restrict the extent of regional sustainable development. Meanwhile, with the accelerated pace of industrialization process, the economy has been spurred in such area, which might counterbalance the restraining effect imposed by environmental degradation on regional sustainable development, and enable the level of regional sustainable development to keep increasing. However, such enhancement is abnormal. With the rapid deterioration of the environment and natural resources, their inhibitory effect on economy and society would be expanded, eventually leading to the slow development rate, or the recession, of the entire system.

For the ecologically fragile area of China's Loess Plateau, the only solution is to change the traditional mode of economic development, to follow the guide of ecological economic conception for regional development, to walk the road of circular economy in order to make the population, economy, society, resources and environment systems perfectly integrated, to realize the optimal allocation of regional sustainable development, and, ultimately, to achieve regional sustainable development strategies. Meanwhile, spatial differences deserve more attention, and the characteristics of different regions should be taken into account in order to achieve optimal spatial structure.

Acknowledgments

We are grateful for financial supports of Natural Science Foundation of China (NO. 41261112; NO. 41301652; NO. 41471116), the Science and Technology Support Program of Gansu Province

(NO. 1304FKCA067) and the Program of Promoting the Scientific Research Capability of Young Teachers in Northwest Normal University (NO. NWNU-LKQN-10-20).

Author Contributions

Chenyu Lu and Yongjin Li designed the study and wrote the paper. Chengpeng Lu and Hengji Li analyzed the data. Chunjuan Wang and Weili Zhu contributed to data collection and processing. All authors have read and approved the final manuscript.

Conflicts of Interest

The authors declare no conflict of interest.

References

1. Lu, C.Y.; Zhang, L.; Zhong, Y.G.; Ren, W.X.; Tobias, M.; Mu, Z.L.; Ma, Z.X.; Geng, Y.; Xue, B. An overview of e-waste management in China. *J. Mater. Cycles Waste Manag.* **2015**, *17*, 1–12.
2. Xue, B.; Mitchell, B.; Geng, Y.; Ren, W.X.; Müller, K.; Ma, Z.X.; Puppim de Oliveira, J.A.; Fujita, T.; Tobias, M. A review on China's pollutant emissions reduction assessment. *Ecol. Indic.* **2014**, *38*, 272–278.
3. Bruyn, S.M.; Opschoor, J.B. Developments in throughput-income relationship: Theoretical and empirical observations. *Ecol. Econ.* **1997**, *20*, 255–268.
4. Xue, B.; Chen, X.P.; Geng, Y.; Guo, X.J.; Lu, C.P.; Zhang, Z.L.; Lu, C.Y. Survey of officials' awareness on circular economy development in China: Based on municipal and county level. *Resour. Conserv. Recycl.* **2010**, *54*, 1296–1302.
5. Mao, H.Y. *A Study on Regional Human and Sustainable Development of Region*, 1st ed.; China Science and Technology Press: Beijing, China, 1995; pp. 46–58. (In Chinese)
6. Chaharbaghi, K.; Willis, R. Study and practice of sustainable development. *Eng. Manag. J.* **1999**, *9*, 41–48.
7. Lambin, E. Conditions for sustainability of human–environment systems: Information, motivation, and capacity. *Glob. Environ. Chang.* **2005**, *15*, 177–180.
8. Graymore, M.L.M.; Sipe, N.G.; Rickson, R.E. Regional sustainability: How useful are current tools of sustainability assessment at the regional scale. *Ecol. Econ.* **2008**, *67*, 362–372.
9. Sagar, A.D.; Najam, A. The human development index: A critical review. *Ecol. Econ.* **1998**, *25*, 249–264.
10. Liu, J.G.; Dietz, T.; Carpenter, S.R.; Alberti, M.; Folke, C.; Moran, E.; Pell, A.N.; Deadman, P.; Kratz, T.; Lubchenco, J.; *et al.* Complexity of coupled human and natural systems. *Science* **2007**, *317*, 1513–1516.
11. Xue, B.; Chen, X.P.; Geng, Y.; Yang, M.; Yang, F.X.; Hu, X.F. Emergy-based study on eco-economic system of arid and semi-arid region—Case of Gansu province, China. *J. Arid Land* **2010**, *2*, 207–213.

12. Leng, S.Y.; Song, C.Q. Sustainable development research sponsored by geography division, national natural science foundation of China. *Bull. Natl. Nat. Sci. Found. China* **2002**, *3*, 158–160. (In Chinese)

13. Ye, Z.B. *Assess the Theory and Practice of Sustainable Development*, 1st ed.; China Environmental Science Press: Beijing, China, 2002; pp. 37–50. (In Chinese)

14. Lu, D.D. A tribute to the geographers for their contributions to China and mankind: Centennial celebration on the geographical society of China. *Acta Geogr. Sinica* **2009**, *64*, 1155–1163. (In Chinese)

15. Niu, W.Y. *System Analytical Theory of Sustainable Development*, 1st ed.; Hubei Science and Technology Press: Wuhan, China, 1998; pp. 89–101. (In Chinese)

16. Guan, W. The data analysis of regional water resources and economic society coupling system sustainable development. *Geogr. Res.* **2007**, *26*, 685–692. (In Chinese)

17. Gong, S.S. The three main relationships of regional sustainable development system. *J. Central China Norm. Univ. (Nat. Sci.)* **1999**, *33*, 596–604. (In Chinese)

18. Ye, D.F. The interactive mechanism of man-earth areal system and the sustainable development. *Geogr. Res.* **2001**, *20*, 307–314. (In Chinese)

19. Xu, Z.M.; Zhang, Z.Q.; Chen, G.D. Review indicators of measuring sustainable development. *China Popul. Resour. Environ.* **2000**, *10*, 60–64. (In Chinese)

20. Zhang, J.G.; Li, Z.W. A review of research method about the quantitative analysis for sustainable development. *J. China Univ. Geosci. (Soc. Sci. Ed.)* **2003**, *3*, 32–35. (In Chinese)

21. Song, C.Q.; Leng, S.Y. Characteristics and trend of modern geography and progresses of geographical research in China. *Adv. Earth Sci.* **2005**, *20*, 595–599. (In Chinese)

22. Chinese Academy Sustainable Development Research Group. *China Sustainable Development Strategy Report in 2009*, 1st ed.; Science Press: Beijing, China, 2009; pp. 8–13. (In Chinese)

23. Li, S.F. *China Sustainable Development Experimental Zone Research*, 1st ed.; Qilu Press: Shandong, China, 2004; pp. 22–25. (In Chinese)

24. Lin, T.; Guo, X.H.; Zhao, Y. A study of residents' environmental awareness among communities in a peri-urban area of Xiamen. *Int. J. Sustain. Dev. World Ecol.* **2010**, *17*, 285–291.

25. Bao, H.J.; Li, Z.S.; Wang, T. Conceptual model of sustainable development in sand-desertification region, China. *Sci. Geogr. Sinica* **2007**, *27*, 173–176. (In Chinese)

26. Li, H.; Li, N. The analysis on the sustainable development level in Xinjiang area. *J. Xinjiang Agric. Univ.* **2009**, *32*, 83–89. (In Chinese)

27. Yang, H.Z.; Li, A.M.; Ye, T. Regional sustainability evaluation based on modified ecological footprint. *J. Tongji Univ. (Nat. Sci.)* **2010**, *38*, 1188–1193. (In Chinese)

28. Mao, H.Y. Study on comprensive regulation of sustainable development when Shandong Province step into next century. *Acta Geogr. Sinica* **1998**, *53*, 413–427. (In Chinese)

29. Zhao, Y.J.; Cao, K.L.; Sun, X.G.; Wang, P. The analysis of the PRED system in Guangdong Province. *Geogr. Territ. Res.* **2000**, *16*, 41–45. (In Chinese)

30. Zhang, J.P.; Qin, Y.C.; Zhang, E.X. Quantitative study methods of regional sustainable development in China: A review. *Acta Ecol. Sinica* **2009**, *29*, 6702–6711. (In Chinese)

31. Li, L.F.; Zheng, D. An assessment of regional sustainable development: Progress and perspectives. *Progr. Geogr.* **2002**, *21*, 237–248. (In Chinese)
32. Shen, Y.M.; Mao, H.Y. The systematic study of theoretical problems on regional sustainable development. *Progr. Geogr.* **1999**, *18*, 287–293. (In Chinese)
33. Ye, M.Q.; Zhang, S.Y. Regional sustainable development system and target implementation process. *Sci. Technol. Progr. Policy* **2001**, *18*, 27–29. (In Chinese)
34. Zhang, Z.L.; Chen, X.P.; Heck, P. Emergy-Based Regional Socio-Economic Metabolism Analysis: An Application of Data Envelopment Analysis and Decomposition Analysis. *Sustainability* **2014**, *6*, 8618–8638.
35. Lu, C.Y.; Chen, X.P.; Xue, B. Synthetic assessment of regional sustainable development in Qingyang based on the theory of meta index. *J. Arid Land Resour. Environ.* **2011**, *25*, 18–23. (In Chinese)
36. Zhang, Z.L.; Chen, X.P.; Jiao, W.T.; Lu, C.P. Dynamic evolution of the coupled environmental economic system of Qingyang, Gansu Province based on emergy theory and econometric methods. *Acta Sci. Circumst.* **2010**, *30*, 2125–2135. (In Chinese)
37. Lu, C.Y. A Study on the Regional Sustainable Development Based on GIS&SD—Case of Qingyang. Ph.D. Thesis, Lanzhou University, Lanzhou, China, 2009. (In Chinese)
38. Zhou, X.; Teng, F.X. A study on the model of sustainable development and evaluation system. *Econ. Rev.* **2007**, *9*, 48–50. (In Chinese)
39. Cao, B.; Lin, J.Y.; Cui, S.H. Review on assessment index of sustainable development. *Environ. Sci. Technol.* **2010**, *33*, 99–105. (In Chinese)
40. Zeng, Z.X.; Gu, P.L. *Systems Analysis and Evaluation of Sustainable Development*, 1st ed.; Science Press: Beijing, China, 2002; pp. 78–92. (In Chinese)
41. Li, J.; Li, X.M.; Liu, Z.Q. Assessment and analysis on urban ecological environment based on degree of urbanization development: Case study of Dalian, Liaoning Province. *China Popul. Resour. Environ.* **2009**, *19*, 156–161. (In Chinese)
42. Zhu, Y.N. A study on evaluation of the level of sustainable development in Ordos of Inner Mongolia. *J. Inner Mong. Agric. Univ. (Soc. Sci. Ed.)* **2010**, *49*, 43–46. (In Chinese)
43. Ren, X.H.; Cui, L.F.; Li, Y.; Gao, X. Quantitative study to the urban sustainable development based on the view of system coordination. In Proceedings of the International Conference on Internet Technology and Applications, Wuhan, China, 21–23 August 2010.
44. Huang, J.S.; Feng, Z.X. The variation track and contrastive analysis on the social economic gravity center and environmental pollution gravity center in Shanxi Province. *Hum. Geogr.* **2006**, *18*, 117–122. (In Chinese)
45. Qiao, J.J.; Li, X.J. The shift route of Chinese economic gravity center in recent 50 years. *Areal Res. Dev.* **2005**, *24*, 12–16. (In Chinese)

Sustainability in China: Bridging Global Knowledge with Local Action

Bing Xue and Mario Tobias

Abstract: As the biggest emerging and developing country, and the second largest economy on the planet, China's road to sustainability has attracted global attention; therefore, we need to have a deeper understanding to address this issue at very different levels. This editorial mainly reviews the contributions of the published papers in the Special Issue of *"Sustainability in China: Bridging Global Knowledge with Local Action"*, the main findings in this special edition suggest that the concept of sustainability is more comprehensive and complex, and the transformation process from scientific knowledge to local action still has a long way to go, not only in China, but also in many developing countries. More research on the fundamental and innovative processes of sustainable transformations should be conducted. China needs to make more efforts to strengthen its road to sustainability, by merging all relevant types of knowledge, both within and outside science, as well as locally and globally.

Reprinted from *Sustainability*. Cite as: Xue, B.; Tobias, M. Sustainability in China: Bridging Global Knowledge with Local Action. *Sustainability* **2015**, *7*, 3714-3720.

1. Introduction

As the biggest emerging and developing country on the planet, China's rapid pace of both urbanization and industrialization over the past few decades has attracted global attention, while a heavy environmental price has been paid for being the world's second-largest economy [1]. By announcing future reforms toward long-run sustainable development, China's leaders have offered green hope to the public [2], aiming to build a modern ecological-civilization society in the coming decades. However, just as Liu concluded that "any individual force can cause positive and negative impacts on sustainability directly or in-directly" [3], therefore, we need to have a deeper understanding into China's sustainability at very different levels. Both spatially and structurally, concerns range from generating sustainable household livelihoods to global climate change, to developing technological applications to generate institutional changes. Moreover, China needs both local, case-based empirical studies, as well as global, experience-based learning to inform itself of the best route towards sustainability. This Special Issue, *"Sustainability in China: Bridging Global Knowledge with Local Action"* of *Sustainability* aims to investigate the intended and spontaneous issues concerning China's road to sustainability in a top-down or bottom-up manner. Thus, based on the peer-reviewing results, this Special Issue selected papers trying to address some key aspects related to China's sustainability. More specifically, the reader can expect more contributions, as presented in Section 2.

2. Contributions

During the past thirty years, China's rapid economic development has been mainly driven by an abundant supply of cheap labor (the so called "demographic dividend"), however, the arrival of an aged society, with a still quite undeveloped economy, will make it more difficult for China to establish a comprehensive social pension system. Wang and Béland's paper on *Assessing the Financial Sustainability of China's Rural Pension System*, finds that the funding gap of China's rural pension system would rise from 97.80 billion Yuan in 2014 to 3062.31 billion Yuan in 2049, which implies that the rural pension system in China will not be financially sustainable [4]. Even in this article, the authors explained how this "gap" could be fixed through policy intervention by referring to recent international experiences; however, further risks are still hidden behind the realities.

As a major force behind anthropogenic carbon emissions, China accounted for 29% of global carbon dioxide emissions in 2012 and 80% of the world's increase in CO_2 emissions since 2008 [5], and carbon emission has been one of the biggest challenges on China's road to long-term sustainability [6], however, the uncertainty of China's CO_2 emission are always being discussed [7], therefore, more efforts should be made in uncovering China's carbon emission in various sectors. Liu and his colleagues' paper [8] examines the greenhouse gas (GHG) emission of the industrial process by taking Shenyang, a typical Chinese industrial city, located in the Liaoning province, as a study case. One of the key findings is that the cement, iron, and steel industries will be the largest emission sources, and the total carbon emissions under the business as usual (BAU) scenario will be doubled in 2020 compared with those of 2009, however, when counter measures are taken, the GHG emissions would be reduced significantly.

Water management is another of the key challenges in China, due to the increasing water demand, both from industrial and agricultural sectors, along with rapid urbanization and industrialization [9]. Yuan and his colleagues find that, under the BAU scenario, water consumption for coal power generation in Western China provinces would increase from 1130 million m^3 in 2012 to 2085 million m^3 in 2020, therefore, an integrated energy-water resource plan with regionalized environmental carrying capacity as constraints should be developed [10]. The paper "*Water Quality Changes during Rapid Urbanization in the Shenzhen River Catchment: an Integrated View of Socio-Economic and Infrastructure Development*", submitted by Qin *et al.*, investigates the causes of water quality change over the rapid urbanization period of 1985–2009 in the Shenzhen River catchment of China, and examines the correlation with infrastructure development and socio-economic policies. They conclude that water quality in urbanization could be significantly improved by implementing integrated methods [11].

Concerning the energy sector, reflecting its rapid industrialization and economic growth, China has become a voracious consumer of energy. For example, the residential energy consumption of China in 2012 was about 400 million tons of coal-equivalent, which approximately equals the total amount of energy consumption of Brazil in the same year [12]. Great efforts have been made in China to reduce energy consumption [1], however, just as Lin *et al.* stated in their paper, "China is a fast developing country with a vast size, and there are great differences in both the amount and structure of residential energy consumption at the provincial level" [12], therefore, they conducted a factor analysis and found that population, economic development level, energy resource endowment, and climatic conditions are the main factors driving residential energy consumption [12], their efforts could contribute to a deeper understanding on Chinese residents' energy consumption demands in the future. Another paper submitted by Yan and Tao [13] focuses on evaluating the performance of the biomass power plants in China in 2012, by developing and employing two new DEA (data envelopment analysis) models they found that there is a great technology gap between the biomass power plants in the northern part of China and those in the southern part of China. The results from these two papers show that, regardless of energy consumption or energy technologies [12,13], regional difference should be considered as a basic factor in sustainability policy-making.

Regarding the governance system in pushing forth China's sustainability, such as environmental legislation, performance evaluation, policy implementation, and hidden barriers, there are four interesting papers in this Special Issue that are targeting this question. Mu and his colleagues investigate the achievements, challenges, and trends in China's environmental legislation, and they conclude that China's environmental legislation still faces a series of challenges, such as the imbalance between rights and obligations and less effort in engaging public participation, therefore, more effort should be made in revising environmental law [14], fortunately, we are glad to say that on 1 January 2015, a new environmental protection law (EPL) took effect in China, which is the nation's first attempt to harmonize economic and social development with environmental protection. However, some gaps still exist in the new law, such as implementation and accountability, especially at the local level [15].

By focusing on the policy of cleaner production, Guan, Grunow, and Yu conducted comparative research to examine local cleaner production policy implementation in China and they find that the location-based incentives of local governments strongly influence the implementation strategies, and that the choices of different strategies can bring out various policy results. They suggest that multi-level approaches should be employed for addressing successful policy implementation [16]. Liu et al. focus on the environmental impact assessment (EIA) policy, in their paper "*Environmental Justice and Sustainability Impact Assessment: In Search of Solutions to Ethnic Conflicts Caused by Coal Mining in Inner Mongolia, China*" [17], they note that existing environmental assessment tools are inadequate to address sustainability, which is concerned with environmental protection, social justice, and economic equity, therefore, it is necessary to develop a sustainability impact assessment (SIA) to fill in the gap. Guo et al. focus on investigating the impacts of air pollutant emission policies on thermal coal supply chain enterprises in China [18], the policy-simulated results imply that the energy conservation and emission reduction policies, and sustainable energy policies, can work more efficiently, which provide evidence that a co-benefits approach should be suggested in policy integration when facing more challenges with limited capacity and resources [1,19].

Synthetic measurement of regional sustainable development has been one of the key issues in the research field of sustainability [20]. Zhang, Chen, and Peter conducted a socio-economic metabolism analysis by means of the Emergy Accounting method, coupled with a DEA model and decomposition analysis techniques to assess sustainability assessment at the city level. They conclude that the integrated approach is suggested as a tool to design future scenarios of resource-use and ecological efficiency, and that the result of socio-economic metabolism analysis implies that more efforts should be made to promote the efficiency of resource utilization and to optimize natural resource use [21]. In contrast, Ma, Eneji, and Liu explored the agro-ecosystem, according to the results based on Emergy Analysis, they conclude that the agro-ecosystem maintained provisioning and regulating services but with an increasing volatility under continued growth in production inputs and disservice outputs [22]. Lu *et al.* developed an integrated model by combining an assessment index system, assessment model, and GIS approach, and successfully applied this model to investigate the temporal-spatial characters and the trends of the sustainable development degree in Loess Plateau Ecologically Fragile of China [23]. By focusing on industrial solid waste and municipal solid waste, Chen *et al.* investigated the sustainability performance of solid waste management by applying a decoupling analysis and further identified the main drivers of solid waste change in China by adopting the Logarithmic Mean Divisia Index model [24]. The paper by Feng *et al.* [25], "*An Entropy-Perspective Study on the Sustainable Development Potential of Tourism Destination Ecosystem in Dunhuang, China*", addresses the sustainability issues in developing tourism from the perspective of Entropy analysis. They propose an evaluation index system, based on dissipative structure and entropy change for the tourism destination ecosystem, and then build up the evaluation model, based on the methodology of Information Entropy, and, finally, applied this model to investigate the sustainability degree in local tourism development. Integrated analysis on the human-nature system could provide a scientific basis for understanding and optimizing regional sustainability [26], therefore, more tools on sustainability-evaluation should be encouraged to be developed and applied for meeting regional sustainable development.

Li, López-Carr, and Chen conduct research on ecological migration, based on case study in the arid northwest of China [27]. The history of China's eco-migration can be traced back as early as the 1980s, and those eco-migrations were usually performed as the resettlements of million individuals or families moved from poor areas with harsh environment and fragile ecology to environmental livable areas. In recent years, in order to protect the local culture and maintain the social stability of immigrant communities, local governments started to implement the new policy of "resettlement of entire village" instead of the previous individual or family-based resettlement. However, most residents actually do not intend to migrate, despite rigid eco-environmental conditions and governance polices threatening livelihood-sustainability [27], therefore, both horizontal and vertical interactions, as well as the dynamics between migrants and their resettlements should be illustrated, based on disciplinary approaches by taking China as a study case. We may draw lessons from China's practices regarding ecological-migration and then adapt these experiences to develop a better management approach on international climate-induced resettlement [28].

3. Outlook

The main findings of the papers in the special edition suggest that the concept of sustainability is more comprehensive and complex, and that the transformation process from scientific knowledge to local action is still a long way away, not only in China, but also in many developing countries. Considering that a new set of UN Sustainable Development Goals, which build upon the Millennium Development Goals and converge with the post-2015 development agenda will be effected from

2016 [29], more investigations on the fundamental and innovative processes of sustainable transformations should be conducted to secure effective, equitable, and durable solutions to some of the most urgent problems of global change and local sustainability, including climate change, water security, energy consumption, *etc.*, [30]. The Potsdam Nobel Laureates Symposium, "*Global Sustainability—A Nobel Cause*", has identified the need for a new "global contract" to bring together relevant knowledge from inside and outside the scientific community in order to meet the challenges of increasing sustainability in the age of the Anthropogenic [31]. As the biggest developing country, the second largest economy, and the largest energy consumer and carbon emitter, China needs to make more efforts to strengthen its road to sustainability, by merging all relevant types of knowledge, both within and outside science, as well as both locally and globally.

Acknowledgments

The initiative and organization of this Special Issue was jointly supported by the Fellowship of the Institute for Advanced Sustainability Studies Potsdam, the fellowship of the Alexander von Humboldt Foundation, the International Postdoctoral Exchange Fellowship Program under China Postdoctoral Council (20140050), and the Natural Science Foundation of China (41471116, 41101126, 71303230). We also thank the support from the Science and Technology Department of Shenyang (F12-182-9-00, F13-172-9-00) and the Science and Technology Department of Liaoning Province (2014416025).

Conflicts of Interest

The authors declare no conflict of interest.

References

1. Xue, B.; Mitchell, B.; Geng, Y.; Ren, W.; Müller, K.; Ma, Z.; de Oliveira, J.A.P.; Fujita, T.; Tobias, M. A Review on China's Pollutant Emissions Reduction Assessment. *Ecol. Indicat.* **2014**, *38*, 272–278.
2. Yang, H.; Flower, R.J.; Thompson, J.R. Pollution: China's New Leaders Offer Green Hope. *Nature* **2013**, *493*, doi:10.1038/493163d.
3. Liu, J.G. China's Road to Sustainability. *Science* **2010**, *328*, 50.
4. Wang, L.; Béland, D. Assessing the Financial Sustainability of China's Rural Pension System. *Sustainability* **2014**, *6*, 3271–3290.
5. Liu, Z.; Guan, D.; Crawford-Brown, D.; Zhang, Q.; He, K.; Liu, J. A Low-carbon Road Map for China. *Nature* **2013**, *500*, 143–145.
6. Geng, Y.; Sarkis, J. Achieving National Emission Reduction Target—China's New Challenge and Opportunity. *Environ. Sci. Technol.* **2012**, *46*, 107–108.
7. Xue, B.; Ren, W. China's Uncertain CO_2 emissions. *Nat. Clim. Change* **2012**, *2*, doi:10.1038/nclimate1715.
8. Jiao, L. Water shortages loom as Northern China's aquifers are sucked dry. *Science* **2010**, *328*, 1462–1463.

9. Yang, H.; Wright, J.A.; Gundry, S.W. Water accessibility: Boost water safety in rural China. *Nature* **2012**, *484*, doi:10.1038/484318b.

10. Yuan, J.; Lei, Q.; Xiong, M.; Guo, J.; Zhao, C. Scenario-Based Analysis on Water Resources Implication of Coal Power in Western China. *Sustainability* **2014**, *6*, 7155–7180.

11. Qin, H.-P.; Su, Q.; Khu, S.-T.; Tang, N. Water Quality Changes during Rapid Urbanization in the Shenzhen River Catchment: An Integrated View of Socio-Economic and Infrastructure Development. *Sustainability* **2014**, *6*, 7433–7451.

12. Lin, W.; Chen, B.; Luo, S.; Liang, L. Factor Analysis of Residential Energy Consumption at the Provincial Level in China. *Sustainability* **2014**, *6*, 7710–7724.

13. Yan, Q.; Tao, J. Biomass Power Generation Industry Efficiency Evaluation in China. *Sustainability* **2014**, *6*, 8720–8735.

14. Mu, Z.; Bu, S.; Xue, B. Environmental Legislation in China: Achievements, Challenges and Trends. *Sustainability* **2014**, *6*, 8967–8979.

15. Zhang, B.; Cao, C. Policy: Four Gaps in China's New Environmental Law. *Nature* **2015**, *517*, 433–434.

16. Guan, T.; Grunow, D.; Yu, J. Improving China's Environmental Performance through Adaptive Implementation—A Comparative Case Study of Cleaner Production in Hangzhou and Guiyang. *Sustainability* **2014**, *6*, 8889–8908.

17. Liu, L.; Liu, J.; Zhang, Z. Environmental Justice and Sustainability Impact Assessment: In Search of Solutions to Ethnic Conflicts Caused by Coal Mining in Inner Mongolia, China. *Sustainability* **2014**, *6*, 8756–8774.

18. Guo, X.; Guo, X.; Yuan, J. Impact Analysis of Air Pollutant Emission Policies on Thermal Coal Supply Chain Enterprises in China. *Sustainability* **2015**, *7*, 75–95.

19. Xue. B.; Ma, Z.; Geng, Y.; Heck, P.; Ren, W.; Tobias, M.; Maas, A.; Jiang, P.; de Oliveira, J.A.P.; Fujita, T. A life cycle co-benefits assessment of wind power in China. *Renew. Sustain. Energy Rev.* **2015**, *41*, 338–346.

20. Zhang, L.; Xue, B.; Geng, Y.; Ren, W.; Lu, C. Emergy-based city's sustainability assessment: Indicators, features and findings. *Sustainability* **2014**, *6*, 952–966.

21. Zhang, Z.; Chen, X.; Heck, P. Emergy-Based Regional Socio-Economic Metabolism Analysis: An Application of Data Envelopment Analysis and Decomposition Analysis. *Sustainability* **2014**, *6*, 8618–8638.

22. Ma, F.; Eneji, A.E.; Liu, J. Understanding Relationships among Agro-Ecosystem Services Based on Emergy Analysis in Luancheng County, North China. *Sustainability* **2014**, *6*, 8700–8719.

23. Lu, C.; Wang, C.; Zhu, W.; Li, H.; Li, Y.; Lu, C. GIS-Based Synthetic Measurement of Sustainable Development in Loess Plateau Ecologically Fragile Area—Case of Qingyang, China. *Sustainability* **2015**, *7*, 1576–1594.

24. Chen, X.; Pang, J.; Zhang, Z.; Li, H. Sustainability Assessment of Solid Waste Management in China: A Decoupling and Decomposition Analysis. *Sustainability* **2014**, *6*, 9268–9281.

25. Feng, H.; Chen, X.; Heck, P.; Miao, H. An Entropy-Perspective Study on the Sustainable Development Potential of Tourism Destination Ecosystem in Dunhuang, China. *Sustainability* **2014**, *6*, 8980–9006.

26. Liu, J.-G.; Mooney, H.; Hull1, V.; Davis, S.J.; Gaskell, J.; Hertel, T.; Lubchenco, J.; Seto, K.C.; Gleick, P.; Kremen, C.; *et al.* Systems integration for global sustainability. *Science* **2015**, *347*, doi:10.1126/science.1258832.

27. Li, Y.; López-Carr, D.; Chen, W. Factors Affecting Migration Intentions in Ecological Restoration Areas and Their Implications for the Sustainability of Ecological Migration Policy in Arid Northwest China. *Sustainability* **2014**, *6*, 8639–8660.

28. López-Carr, D.; Marter-Kenyon, J. Human adaptation: Manage Climate-induced Resettlement. *Nature* **2015**, *517*, 265–267.

29. United Nations (UN). Sustainable Development Goals. Available online: https://sustainabledevelopment.un.org/topics/sustainabledevelopmentgoals (accessed on 15 March 2015).

30. The International Social Science Council (ISSC). Transformations to Sustainability. Available online: http://www.worldsocialscience.org/ (accessed on 15 March 2015).

31. Institute for Advanced Sustainability Studies (IASS). Articles of Association of the IASS. Available online: http://www.iass-potsdam.de/en/institute (accessed on 15 March 2015).

MDPI AG
Klybeckstrasse 64
4057 Basel, Switzerland
Tel. +41 61 683 77 34
Fax +41 61 302 89 18
http://www.mdpi.com/

Sustainability Editorial Office
E-mail: sustainability@mdpi.com
http://www.mdpi.com/journal/sustainability

www.ingramcontent.com/pod-product-compliance
Lightning Source LLC
Chambersburg PA
CBHW050345230326
41458CB00102B/6397